T0338189

HETEROGENEOUS CELLULAR NETWORKS

HETEROGENEOUS CELLULAR NETWORKS

Editors

Rose Qingyang Hu
Utah State University, USA

Yi Qian
University of Nebraska-Lincoln, USA

A John Wiley & Sons, Ltd., Publication

This edition first published 2013

Registered office
John Wiley & Sons Ltd, The Atrium, Southern Gate, Chichester, West Sussex, PO19 8SQ, United Kingdom

For details of our global editorial offices, for customer services and for information about how to apply for permission to reuse the copyright material in this book please see our website at www.wiley.com.

Library of Congress Cataloging-in-Publication Data

Heterogeneous cellular networks / editors, Rose Qingyang Hu, Yi Qian.
pages cm
Includes bibliographical references and index.
ISBN 978-1-119-99912-6 (cloth)
1. Cell phone systems. 2. Internetworking (Telecommunication) I. Hu, Rose Qingyang, editor of compilation. II. Qian, Yi, 1969– editor of compilation.
TK5103.2.H48 2013
621.3845′6–dc23

2012043611

A catalogue record for this book is available from the British Library.

ISBN: 9781119999126

Typeset in 10/12pt Times by Aptara Inc., New Delhi, India
Printed and bound in Singapore by Markono Print Media Pte Ltd

To our respective families.

Contents

Part III Deployment, Standardization and Field Trials

Contributors

Stefano Busanelli, Guglielmo Srl, Italy

Hsiao-Hwa Chen, National Cheng Kung University, Taiwan

Kwang-Cheng Chen, National Taiwan University, Taiwan

Wei-Peng Chen, Fujitsu Laboratories of America, USA

Yanjiao Chen, Hong Kong University of Science and Technology, China

Shin-Ming Cheng, National Taiwan University of Science and Technology, Taiwan

Andreas Czylwik, University of Duisburg-Essen, Germany

Jaime Ferragut, Centre Tecnològic de Telecomunicacions de Catalunya, Spain

Gianluigi Ferrari, University of Parma, Italy

Bo Hagerman, Ericsson Research, Sweden

Jinkyu Han, Samsung Telecommunications America, USA

Nageen Himayat, Intel Corporation, USA

Honglin Hu, Shanghai Research Center for Wireless Communications, China

Rose Qingyang Hu, Utah State University, USA

Nicola Iotti, Guglielmo Srl, Italy

Zichao Ji, New Postcom Equipment Co. Ltd., China

Ming Jiang, New Postcom Equipment Co. Ltd., China

Kerstin Johnsson, Intel Corporation, USA

Zander Zhongding Lei, Institute for Infocomm Research, Singapore

Qian (Clara) Li, Intel Corporation, USA

Ying Li, Samsung Telecommunications America, USA

Peng Lin, Hong Kong University of Science and Technology, China

Lingjia Liu, University of Kansas, USA

Josep Mangues-Bafalluy, Centre Tecnològic de Telecomunicacions de Catalunya, Spain

Marco Martalò, University of Parma, Italy

Pantelis Monogioudis, Alcatel-Lucent, USA

Young-Han Nam, Samsung Telecommunications America, USA

Boon Loong Ng, Samsung Telecommunications America, USA

Zhouyue Pi, Samsung Telecommunications America, USA

Yi Qian, University of Nebraska-Lincoln, USA

Manuel Requena-Esteso, Centre Tecnològic de Telecomunicacions de Catalunya, Spain

Meryem Simsek, University of Duisburg-Essen, Germany

Giovanni Spigoni, University of Parma, Italy

Shilpa Talwar, Intel Corporation, USA

Carl Weaver, Alcatel-Lucent, USA

Zhenzhen Wei, State Grid Electric Power Research Institute, China

Wenkun Wen, New Postcom Equipment Co. Ltd., China

Karl Werner, Ericsson Research, Sweden

Sai Ho Wong, Institute for Infocomm Research, Singapore

Geng Wu, Intel Corporation, USA

Jin Yang, Verizon Communications Inc., USA

Yang Yang, Shanghai Research Center for Wireless Communications, China

Shu-ping Yeh, Intel Corporation, USA

Jiantao Yu, Shanghai Research Center for Wireless Communications, China

Jianzhong (Charlie) Zhang, Samsung Telecommunications America, USA

Jin Zhang, Hong Kong University of Science and Technology, China

Qian Zhang, Hong Kong University of Science and Technology, China

Xiaoying Zheng, Shanghai Research Center for Wireless Communications, China

Chenxi Zhu, Fujitsu Laboratories of America, USA

Preface

Wireless data traffic has been increasing exponentially in recent years. Driven by a new generation of devices (smart phone, netbooks, etc.) and highly bandwidth-demanding applications such as video, capacity demand increases much faster than spectral efficiency improvement, in particular at hot spots/area. Also as service migrates from voice centric to data centric, more users operate from indoor, which requires increased link budget and coverage extension to provide uniform user experience. Traditional networks optimized for homogeneous traffic face unprecedented challenges to meet the demand cost-effectively. More recently, 3GPP LTE-advanced has started a new study item to investigate heterogeneous cellular network deployments as an efficient way to improve system capacity as well as effectively enhance network coverage. Unlike the traditional heterogeneous networks that deal with the interworking of wireless local area networks and cellular networks, in which the research community has already been studied for more than a decade, in this new paradigm in cellular network domain, a heterogeneous network is a network containing network nodes with different characteristics such as transmission power and RF coverage area. The low power micro nodes and high power macro nodes can be maintained under the management of the same operator. They can share the same frequency carrier that the operator provides. In this case, joint radio resource/interference management needs to be provided to ensure the coverage of low power nodes. In some other cases, the low power and high power nodes can be coordinated to use more than one carrier, e.g., through carrier aggregation, so that strong interference to each other can be mitigated, especially on the control channel. The macro network nodes with a large RF coverage area are deployed in a planned way for blanket coverage of urban, suburban, or rural areas. The local nodes with small RF coverage areas aim to complement the macro network nodes for coverage extension and/or capacity enhancement. In addition to this, global coverage can be further provided by satellites (macro-cells), according to an integrated system concept.

There is an urgent need in both industry and academia to better understand the technical details and performance gains that are made possible by heterogeneous cellular networks. To address that need, this edited book covers the comprehensive research topics in heterogeneous cellular networks. This book focuses on recent advances and progresses in heterogeneous cellular networks. This book can serve as a useful reference for researchers, engineers, and students to understand heterogeneous cellular networks in order to design, build and deploy highly efficient wireless networks.

The scope of topics covered in this book is timely and is a growing area of high interest. The book contains 15 referred chapters from researchers working in this area around the world.

It is organized along three parts, together with the preface, with each focusing on a different research topic for heterogeneous cellular networks.

In Chapter 1, Wu et al. give acomprehensive overview of the current activities and future trends in heterogeneous cellular networks. More specifically, the chapter provides a technology and business overview of the heterogeneous networks, the state of the art in technology development, the main challenges and tradeoffs, and the future research and development directions.

Part I: Radio Resource and Interference Management

Heterogeneous networks are usually operated in the interference limited regime due to the overlaid coverage areas of various base stations. Accordingly, radio resource management and interference management are of critical importance to the success of heterogeneous networks. In part I of this book, the recent developments on radio resource and interference management are presented.

In Chapter 2, Liu et al. discuss various deployment scenarios and corresponding interference management categories for heterogeneous networks. For multi-carrier scenario, carrier partitioning, power control, and carrier aggregation based approaches are introduced. For co-channel scenario, time-domain solution and the power setting solutions are discussed.

In Chapter 3, Talwar et al. describe a heterogeneous network architecture, which is composed of a hierarchy of multiple types of infrastructure elements, and one or more radio access technologies. They focus on several use cases, outline the challenges and present a number of promising interference mitigation solutions. This chapter also describes industry trends, standardization activities and future research directions for this rich area of investigation.

In Chapter 4, Wen et al. discuss the interference management issues in the context of LTE-Advanced HetNet scenarios. They study the ICI (Inter-Cell Interference) management techniques for LTE-Advanced HetNet deployments. It is concluded that the existence of cross-tier interference invalids the effectiveness of conventional frequency-domain inter-cell interference coordination (ICIC) methods such as FFR. Therefore, the time-domain based ICIC solution, enhanced inter-cell interference coordination (eICIC), have been proposed and standardized in LTE-Advanced for tackling the ICI issue in HetNet scenarios.

In Chapter 5, Wong and Lei summarize enhanced ICIC techniques that are suitable for handling interference in heterogeneous networks deployment. These techniques are divided into frequency, time, power and spatial domains, and they can be combined when necessary. Information exchange among different cells is performed over the backhaul, and when its latency is very small, dynamic enhanced ICIC is possible.

In Chapter 6, Cheng and Chen provide an overview of the possible cognitive radio-enabled interference mitigation approaches to control cross-tier and intra-tier interference in OFDMA femtocell heterogeneous network. Various approaches have been investigated, including orthogonal radio resource assignment in time-frequency and antenna spatial domains, as well as interference cancellation via novel decoding techniques.

In Chapter 7, Zhu and Chen present a distributed bandwidth allocation scheme based on non-cooperative game theory for OFDMA-based femtocell networks such as LTE or WiMAX. The bandwidth allocation scheme can be implemented using different network control architectures, including fully distributed, hybrid, and centralized architectures.

Part II: Mobility and Handover Management

Mobility management is a key component of the next generation cellular networks, which are expected to support high mobility and high data rates, and are becoming more heterogeneous. Poor mobility management will result in unnecessary handovers, handover failures, radio link failures, and the unbalanced load among cells, where system resources are wasted and user experiences are deteriorated. In Part II of this book, the recent developments on mobility and handover management in the heterogeneous cellular networks are presented.

In Chapter 8, Zheng et al. investigate algorithms and technologies to address the mobility robustness optimization and the mobility load balance optimization, respectively. The two mobility functionalitics are coupled since they both need to adjust the handover settings. When they operate together, there might be some conflicts between their respective decisions. The authors propose a coordinated solution to avoid these conflicts and make them collaborated.

In Chapter 9, Weaver and Monogioudis consider networks with Open Subscriber Group (OSG) heterogeneous nodes of two types: macro and metro eNBs, and study connected-mode mobility in LTE heterogeneous networks.

In Chapter 10, Simsek and Czylwik give an overview on various cell selection methods for femto and macro mobile stations together with discussions on their benefits and drawbacks. The chapter provides a study on the impacts of deploying a large number of femtocells into a macro-cellular system.

In Chapter 11, Mangues-Bafalluy et al. extend the concept of heterogeneous cellular networks by introducing additional degrees of heterogeneity, i.e., 3GPP and non-3GPP technologies, as well as the combination of data networking and 3GPP architectures. They conclude that innovative HetNet deployments are feasible if the traditional HetNet vision is generalized.

In Chapter 12, Spigoni et al. study the role of vertical handover (VHO) in future HetNets. In particular, on the basis of internet working experimental results obtained with low-complexity novel VHO algorithms (relying on RSSI and goodput measurements), they draw some conclusions on the potential and limitations of VHO in HetNets.

Part III: Standardization and Field Trials

Standardization, deployments and field trials are the key steps for the success of the large commercial deployments of heterogeneous cellular networks. In Part III of this book, overviews on recent standardization activities, deployment approaches and field trials are given on heterogeneous cellular networks.

In Chapter 13, Nam et al. provide an overview on the evolution of HetNet technologies in LTE-Advanced Standards, in particular, the enhancement on the ICIC techniques introduced in LTE Release 10. They describe the HetNet deployment scenarios and provide detailed descriptions on two newly introduced techniques, namely Carrier Aggregation (CA) based ICIC and Time-domain ICIC. They also provide a glimpse of further evolution of ICIC for the future release of the LTE-Advanced currently being standardised.

In Chapter 14, Lin et al. propose three frameworks for heterogeneous network deployment and management according to the deployment types of femtocells, which are joint-deployment, WSP-deployment and user-deployment frameworks. The unique characteristics, corresponding challenges and potential solutions of these frameworks are further investigated to provide a deeper insight systematically.

In Chapter 15, Hagerman et al. present a field trial in a pre-commercial LTE network with the purpose of investigating how well MIMO works with realistically designed handhelds in 750 MHz band. The trial comprises test drives in urban and suburban areas with different network load levels. The effects of hands holding the devices and the effect of using the device inside a test vehicle are also investigated. The trial has proven that MIMO works very well and gives a substantial performance improvement at the 750 MHz carrier frequency.

This book has been made possible by the great efforts and contributions of many people. First of all, we would like to thank all the contributors for their excellent chapter contributions. Second, we would like thank all the reviewers for their dedicated time in reviewing the book, and for their valuable comments and suggestions for improving the quality of this book. Finally, we appreciate the advice and support of the staff members from Wiley, for putting this book together.

Rose Qingyang Hu
Logan, Utah, USA

Yi Qian
Omaha, Nebraska, USA

1

Overview of Heterogeneous Networks

Geng Wu,[1] Qian (Clara) Li,[1] Rose Qingyang Hu,[2] and Yi Qian[3]
[1] *Intel Corporation, USA*
[2] *Utah State University, USA*
[3] *University of Nebraska – Lincoln, USA*

We are living in a rapidly changing world. Every two days now we create as much information as we did from the dawn of civilization up until 2003 [1]. Users want to communicate with each other at any time, anywhere and through any media, including instant messages, email, voice and video. Users want to share their personal life experience, ideas and news with friends through social networking, and use their intelligent mobile devices to produce and to consume content generated by users or by commercial media. In the meantime, mobile internet is rapidly evolving towards embedded internet, expanding its reach from people to machines [2]. In fact, the wireless industry now expects 50 billion machine-type devices connected to the global network by 2020 [3], truly forming an internet of everything.

The advancement of a number of fundamental technologies powers the rapid market growth. Moore's Law continues to provide more transistors and power budget, enabling the semiconductor industry to deliver more powerful signal processing capabilities at lower power consumption and lower cost. Application developers continue to innovate and maximize the benefits of the signal processing technology, with user interface evolving from keypad to touch to gesture, and applications from voice to video to augmented reality. As our society enters the age of 'Big Data' [4], our communication infrastructure also needs to evolve to meet the overwhelming demands for capacity and bandwidth. The migration from homogenous to heterogeneous network architecture is therefore essential to support a broad range of connectivity and to deliver unprecedented user experience. The future is coming today.

As one of the main pillars and the future trends of mobile communication technology, heterogeneous networks have received a lot of attention in the wireless industry and in the

Heterogeneous Cellular Networks, First Edition. Edited by Rose Qingyang Hu and Yi Qian.

academic research communities. This chapter is intended to provide a technology and business overview of heterogeneous networks, the state of the art in technology development, the main challenges and tradeoffs, and the future research and development directions. However, we are still at an early stage of development of heterogeneous network technology. As you will find throughout this chapter, there are many more questions than answers at this time, and many questions may have more than one valid answer, depending on the market, the target applications and the exact deployment scenarios and competitive environment. We expect that heterogeneous network technology will continue to evolve along with the convergence of information technology and telecommunication, and increasingly intelligent mobile devices.

1.1 Motivations for Heterogeneous Networks

There are significant economic and technological reasons for the rapid development of heterogeneous networks. The outcomes of this technological development are expected to have profound impacts on the future of telecommunications.

1.1.1 Explosive Growth of Data Capacity Demands

In recent years, mobile internet has witnessed an explosive growth in demand for data capacity [5]. This is largely fuelled by the proliferation of more intelligent mobile devices. Market studies have shown that the data traffic volume is a direct function of the device's screen size, the user-friendliness of its operating system and the responsiveness of wireless network that the device is connected to. For example, a 3G smartphone on average consumes about 30 times the system capacity of a 2G voice phone, and a tablet consumes five times the system capacity of a smartphone. As the mobile devices continue to increase in screen size, image resolution and battery life, and as the network infrastructures continue to improve in peak data rate and network latency, the growth in data capacity demand will continue.

In addition to this organic growth in capacity, demand from the improved mobile devices and communication infrastructure, user-generated content and social networking add significant additional burden to the network. In fact, mobile devices are an ideal platform for social networking applications such as Facebook since they offer ubiquitous coverage with its always-on and always-connected connectivity. Social networking and other similar applications usually produce small but frequent data transmissions. A network may have to frequently set up and tear down the radio links to conserve precious radio resources in order to accommodate a large number of users. This often results in an excessive amount of control messages over the control plane. On the other hand, as watching YouTube videos on mobile devices gains popularity, the capacity demand on the data plane is also growing rapidly, and often in an asymmetric fashion between the uplink and the downlink. Finally, depending on how cloud and client partition the signal processing load, cloud-based services may further accelerate demand, as information is shipped between the mobile devices to the cloud for cloud computing and network storage. One such example is Apple's Siri voice reorganization application software. Since the popularity of mobile applications is often difficult to predict, we start to see drastically different capacity demands between the control plane and the data plane, between the uplink and the downlink. We also start to see network congestion expanding from the access network (the traditional capacity bottleneck) to the core network and even to the backbone network and connections.

Machine-type communications add yet another complexity to the future generations of wireless networks. With mobile internet evolving towards embedded internet, future networks need to scale up in size and complexity in order to accommodate an unprecedented number of connected devices with vastly different traffic characteristics, usage models and security requirements. The capacity demands from these machine-type devices range from very low traffic volume monthly meter reading to high speed real-time video surveillance. In addition, securely managing billions of such connected devices across many different types of networks and operating environments adds to the complexity of capacity planning.

The combined capacity demands from organic traffic growth, user-generated contents, social networking and machine-type connected devices require orders of magnitude capacity increase in future wireless networks. This heterogeneous data traffic growth also mandates a paradigm shift in network architecture design and provisioning.

1.1.2 From Spectral Efficiency to Network Efficiency

The wireless industry has several options for meeting the explosive data traffic growth. After decades of relentless air interface innovations, today we are practically reaching the theoretical limit of radio channel capacity, commonly known as the Shannon limit. Although air interface improvement will continue to maximize the benefits of advanced wireless communication research and take full advantage of advanced signal processing technologies for an even higher spectral efficiency, we need several orders of magnitude greater system capacity than what the air interface spectral efficiency improvement can offer. The future capacity increases therefore need to come from a combination of technology solutions, including, in particular, maximizing the overall network efficiency instead of solely relying on the spectral efficiency improvement at the radio link level (Figure 1.1). Heterogeneous networks are a fundamental technology behind most of these solutions.

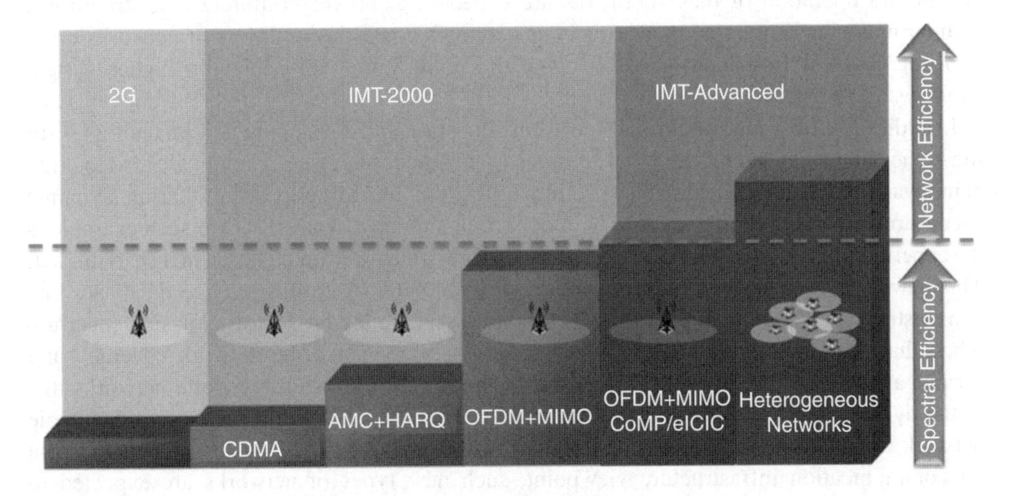

Figure 1.1 Wireless technology evolution.

In the near term, mobile network operators are looking at limiting the monthly data usage of each subscriber over the wireless wide areas networks (WWAN), and throttling the data rate of heavy usage users when necessary. However, limiting usage or throttling capacity demand is in general only a temporary fix to the immediate network overloading problems. We need more proactive solutions to encourage and enable future sustained data traffic growth, and to provide mobile broadband access to all users, and to enrich every person's life on earth.

One such solution that mobile network operators are looking at is the data offloading strategy. This includes (but is not limited to) facilitating and encouraging subscribers to offload their traffic from macro base stations to the alternative small-cell networks, essentially forming a basic heterogeneous network. Since the capacity bottleneck varies from market to market and from network to network, there are many flavours and technical options for offloading strategy, including macrocell network and small-cell network of the same air interface technology, between networks of different air interface technologies, or between mobile operator core network and public internet. There is no single answer to the mobile data offloading question. These options are complementary, and all of them will continue to develop to meet the ever-increasing capacity demands.

Another obvious answer to the growing demand in data capacity is to add more spectrum. The wireless industry and regulators are working together to investigate the possibility of adding more frequency bands, both licensed and unlicensed, for mobile internet applications. However, since there is a limited supply of spectrum, and there is the strong desire for globally harmonized frequency allocation to maximize the economy of scale, the progress in new frequency allocation has been slow. As many densely populated markets are already on the verge of running out of spectrum, we see increased pressure to re-farm the existing frequency bands and for the rapid deployment of small cells for high spatial frequency reuse. In addition, the wireless industry has also started to look at high frequency bands such as millimeter wave for mobile internet applications. Since these bands have very different radio propagation characteristics from the traditional lower frequency bands (usually below 3 GHz) used for high mobility cellular networks, the technology, design and operation of these networks are expected to be very different from traditional cellular networks. Therefore, heterogeneous networks consisting of layers of networks operating at different frequency bands become the main venue for achieving higher system capacity.

In addition to obtaining additional spectrum allocation and developing new technologies for the higher frequency bands, the wireless industry and the research community are also looking at innovative ways for more flexible spectrum utilization, including spectrum sharing, dynamic spectrum access and cognitive radio with opportunistic network access. One such example is the experimental use of TV white space spectrum for wireless communication in the US market. This new type of spectrum access requires additional network entities such as databases that administrate the alternative radio transmitters to operate in the broadcast television spectrum when that spectrum is not used by the licensed service. Since the network coverage and service availability are different from those of the traditional wireless mobile networks due to the dynamic nature of the spectrum availability, the industry is still investigating suitable network architecture and business models to achieve viable return on investment. From a telecommunication infrastructure viewpoint, such new types of networks are expected to become part of the global heterogeneous networks.

1.1.3 Challenges in Service Revenue and Capacity Investment

In recent years, mobile service revenue growth has shifted from circuit-switched voice and short message service (SMS) to data services. This shift adds significant pressure to mobile network operators' profitability for three main reasons. First, mobile data in general yields a lower revenue per bit compared to the traditional voice services and SMS. Secondly, the highly profitable operator walled-garden mobile applications are facing stiff competition from over-the-top mobile applications. Finally, as mobile data traffic explodes, operators need extensive capital investment in new network capacity to meet the demand. Since mobile network operators are instrumental in investing, operating and maintaining global mobile internet infrastructure, it is crucial for the wireless industry and the academic research communities to develop new networking technologies that allow operators to remain profitable and competitive so that they can continue to invest in capacity and new services. Heterogeneous networking is considered one of the most important technologies that not only deliver tens- to thousands-fold system capacity increase but also enable new generations of services to replace the revenue from traditional but diminishing voice-centric telecom services.

To summarize, while the demand for data capacity is exploding and the improvement in spectral efficiency in homogeneous networks is slowing down due to the approaching Shannon limit, it becomes essential that the future focus of wireless technology shifts from further increasing the spectral efficiency of the radio link to improving the overall network efficiency through heterogeneous network architecture and related signal processing technologies. We need heterogeneous networks to deliver a higher system capacity to meet the higher traffic density. We want to leverage heterogeneous network architectures to expand network coverage, to improve service quality and fairness throughout the network coverage areas, in particular at the cell edge. We also want to use heterogeneous networks as a platform for future technological innovations, including the integration of new types of networks, new types of connectivity and new types of connected devices and applications.

1.2 Definitions of Heterogeneous Networks

Heterogeneous networking is one of the most widely used but most loosely defined terms in today's wireless communications industry. Some people consider the overlay of macro base station network and small cell network (e.g., micro, pico and femtocells) of the same air interface technology as heterogeneous networks. Others consider cellular network plus WiFi network as a main use case. There are also those who consider the inclusion of new network topologies and connectivity as part of the heterogeneous networks vision, such as personal hotspot, relay, peer-to-peer, device-to-device, near field communication (NFC) and traffic aggregation for machine type devices. In fact, as flexible sharing and dynamic access of spectrum become part of the network infrastructure, we can expect heterogeneous networks to also include cognitive radios.

Despite these diverse definitions and understandings, heterogeneous network research and deployment have made significant progress in the past several years, in particular in the area of data offloading through small cells (including WiFi access points). In practice, heterogeneous deployments are defined as mixed deployments consisting of macro, pico, femto and relay nodes. To the authors, heterogeneous networking is about a set of essential technologies and capabilities that deliver unprecedented large system capacity through the integration of heterogeneous architectures from WAN to LAN to PAN, provide always-on and always-best-connected connectivity for compute continuum, and offer innovative services and significantly better user experiences through the introduction of improved network efficiency.

In general, a heterogeneous network consists of multiple tiers (or layers) of networks of different cell sizes/footprints and/or of multiple radio access technologies [6]. An LTE macro base station network overlaying an LTE pico base station network is a good example of a multi-tier heterogeneous network. In this case since the same LTE air interface technology is used across different layers/tiers of networks, 3GPP (the standards body that created LTE) has developed solutions and design provisions to facilitate the interaction and the integration of such a heterogeneous network, including an extensive performance evaluation methodology that models a variety of deployment scenarios. On the other hand, a heterogeneous network consisting of a macrocellular LTE network and a WiFi network is a good example of multi-tier and multi-air-interface networks. Since the air interfaces were developed by different standards bodies (in this case 3GPP for LTE and IEEE802.11/Wi-Fi Alliance for WiFi), collaboration between standards development bodies is necessary to make the heterogeneous network work. The Hotspot 2.0 specification developed by Wi-Fi Alliance is one such example.

In addition to small cells, future heterogeneous networks may also include super big base stations. Cloud-RAN is one example [7]. Through high-speed optical fibre connections, a cloud-RAN base station relocates all or most of the baseband signals from tens to hundreds of traditional stations to a centralized server platform for massive signal processing. This architecture may significantly reduce the network's energy consumption since air conditioning at each cell site may no longer be required. Furthermore, due to its large size, a super base station can dynamically allocate its signal processing resource to adapt to the varying traffic loading within its geographical coverage during a day, a phenomenon often referred to as the 'tidal effect'. This further reduces hardware requirements and energy consumption. These super base stations may facilitate tighter couplings between different types of base stations including macro and small cells. They may also serve as a platform for cost-effective implementation of advanced air interface and network features to significantly increase network efficiency, which we will discuss in more details in future sections.

Although the sizes of base stations in each tier or layer of a hybrid network may differ significantly, ranging from femto station to picocell, microcell, macrocell and cloud-RAN, as shown in Figure 1.2, and although the radio access technology used in each tier may be the same or different, there is a common set of challenges and techniques to integrate them together to form a high performance heterogeneous network. This chapter will discuss a number of fundamental issues and solutions. It should be noted that due to the broad technical scope and highly complex economic tradeoffs, the designs of a heterogeneous network may have different flavours and focuses, depending on the existing installed network equipment, the choices of network transport, the availability of the required multi-mode devices, and the main set of applications expected to be supported by a particular heterogeneous network. As the requirements from users and mobile network operators continue to evolve, the definitions and the technical focuses of heterogeneous networks are expected to also evolve with time.

1.3 Economics of Heterogeneous Networks

There are various aspects to the economics of heterogeneous networks: the total cost of ownership to mobile network operators, the performance and cost benefits to end users and the

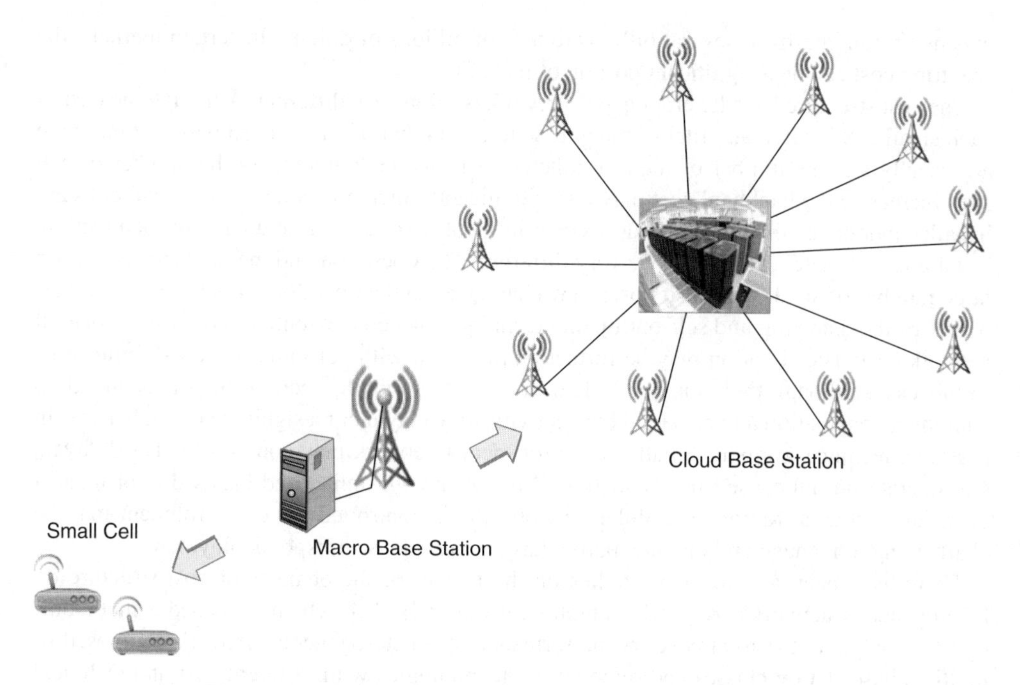

Figure 1.2 Base stations are becoming both bigger and smaller.

increase of the size of both total and addressable markets for telecommunication equipment manufacturers.

1.3.1 Total Cost of Ownership

The most important elements in the total cost of ownership (TCO) to a mobile wireless network operator are the capital expenditure (CAPEX) for network construction and the operating expenditure (OPEX) for network operation.

The cost structure for a traditional wireless network is relatively well understood [8]. For a typical cellular network, the CAPEX usually includes the cost of the radio access network (the base stations and the radio network controllers), the mobile core network (the gateways and the IMS platforms), the backhaul infrastructure and site acquisition, construction, engineering and integration. The radio access network represents about 60% of CAPEX, followed by the core network at about 15%, backhaul at about 5% and site acquisition, construction and engineering at 10–20%. There are of course large variations from deployment to deployment; for example, the radio access network cost can be reduced when an operator can upgrade its existing base station equipment to support a new air interface technology, or the site acquisition cost can be avoided if an operator can overlay a new network on its existing cell sites. The OPEX is mainly associated with network operation and management, including the expense of site rental, backhaul transmission, operation and maintenance of the network, and electric power. Given a 7-year depreciation period of base station equipment, OPEX may account for up to 60% of the TCO [7]. The cost of acquiring wireless licences is normally excluded from CAPEX or OPEX. This is because such acquisitions are infrequent and sometimes extremely expensive, ranging from several billions to tens of billions of dollars. In certain markets, the spectrum cost can be a significant portion of the TCO.

The cost structure for a heterogeneous network is often very different. Although the cost of each small cell is often an order of magnitude less than that of a macrocell base station, there are usually a large number of them in a heterogeneous deployment. As the number of cell sites increases, the backhaul cost may also significantly increase. Since many small cells are installed indoors on the wall or outdoors on utility poles, the cost structure for site acquisition, installation and site leasing may be very different. The operation and maintenance cost for a large number of small cells also poses new challenges to the operators of heterogeneous networks. Self-organizing and self-optimizing techniques become essential to reduce the overall network cost. The situation may be further complicated with consumer-installed femtocells. In this case, although there may be little or no cost to a mobile operator for site acquisition, equipment installation and backhaul connectivity, they may incur a significantly higher cost in customer technical support. Finally, a heterogeneous network may consist of network layers that operate on unlicensed bands such as WiFi. Although unlicensed bands do not incur a licensing cost as a traditional cellular network, the uncontrolled radio environment may be challenging to manage and operate, particularly in very dense hotspot deployments.

There have been a number of studies on the impact of the changes of cost structure on heterogeneous networks. A good example is given in [9, 10], which proposed a methodology for analyzing the total cost and performance of a heterogeneous network composed of multiple base station classes and radio access technologies with different cost and technical characteristics. It also demonstrated the impacts of the shift in cost structure and business model. Another good example is given in [7], which highlighted the fact that electricity cost at macrocell sites is about 41% of the OPEX per year, of which 48% is consumed by the air conditioning. As the industry continues the push for higher energy efficiency, there is a strong desire to significantly reduce the energy consumption through advanced base station design and network architectures, including fan-less cooling small cells and centralized baseband processing of large-scale cloud base stations.

It should be noted that the evolution towards heterogeneous networks presents both opportunities and challenges in terms of TCO for the mobile wireless industry. This is partly due to the significant shift of cost structure on base stations, backhaul, network installation and maintenance, energy cost and radio spectrum cost, and partly due to the significant shift of usage model from outdoor to indoor/hotspot, from high system capacity to both high system capacity and high data rate.

1.3.2 Heterogeneous Networks Use Scenarios

Heterogeneous networks have many architectural flavours and implementation variations to meet different market requirements and cost considerations [11]. However, the goals are similar. For consumers, heterogeneous networks need to provide ubiquitous coverage, secure, high data rate, high capacity, always-on, and always-connected-to-the-best-network user experience. For mobile operators, heterogeneous networks need to provide fast time-to-market, optimal network utilization, and operator control and network manageability.

The most classical heterogeneous network deployment is home femtocells. They use mobile operator licensed spectrum and are primarily deployed by end users for network coverage extension in an indoor environment or remote rural areas where outdoor macrocells have difficulty in providing coverage. As a cellular coverage extension, they are primarily used for voice services. These home femtocells are typically connected to the mobile core network through the consumer's own internet service such as DSL, and therefore there is usually no backhaul transmission cost to the mobile operator. A home femtocell often limits its access to a 'closed subscriber group', which makes economic sense to an end user since he/she pays for the backhaul transmission cost, and there is little reason to share it with others. One unique characteristic of home femtocells is that they are mostly installed by the consumers who do not necessarily have adequate knowledge about radio technologies. As a result, the RF interference from/to a home femtocell may be difficult to manage. A popular solution is to assign home femtocells to a different carrier frequency than the macrocellular network, provided that the mobile operator has sufficient spectrum for a standalone RF carrier. Since home femtocells are considered consumer products, the average selling price is in the order of hundreds of dollars.

Another class of small cells is the so-called picocell, sometimes also referred to as an enterprise femtocell or metro femtocell. They use the same air interface technology over mobile operator's licensed spectrum, and serve as an extension of the macrocellular network. This class of small cells usually has a larger subscriber capacity compared to home femtocells and provides voice and data services in office environments, indoor coverage in places like shopping centres or outdoor hotspot coverage such as a busy shopping street or sports stadium. They are often environmentally hardened in particular for outdoor deployment, professionally installed with more advanced antennas, with service open to all qualified subscribers instead of only to the members of a closed subscriber group. It should be noted that a home femtocell and a picocell may not be very different in terms of subscriber capacity and transmission power. Table 1.1 shows a typical example. The main differences are actually related to how they are connected to mobile operator's core network, an important issue that we will discuss in a later section. The average cost of a picocell is in the order of one to several thousand dollars since it often needs carrier-grade equipment.

In the past few years, data traffic offloaded from cellular networks to WiFi has gained significant momentum. This type of heterogeneous network operates in both licensed (for

Table 1.1 A comparison of home femto and public picocell key features

	Femtocell	Picocell
TX power	low power, <250 mW	higher power, 250 mW to 2 W
capacity	low capacity, <8 users	higher capacity, 16–32 users
backhaul	consumer grade, paid for by user	carrier grade, paid for by operator
equipment	owned by consumer	owned by operator
cell 'site'	consumer installation	operator professional installation
deployment	unplanned, consumer deployment	planned, deployment by operator
user access	closed, restricted access	open to all qualified subscribers
handover	loose coupling at network layer	tight coupling, intra-network handover
security	not trusted by operator network	trusted by operator network
enclosure	consumer electronics	often environmentally hardened

2G/3G/4G cellular) and unlicensed band (for WiFi). Since cellular networks and WiFi are very different in user/device credentials, authentication, air interface characteristics, network architectures, interference environments and subscriber management and billing systems, the integration of WiFi and cellular network requires careful economic consideration and therefore varies from market to market. Some mobile operators choose to deploy their own WiFi networks, while others offer services through third-party WiFi service providers either under mobile operator's own brand or under a third-party brand. The size of the WiFi networks also varies, from several thousand access points (APs) for selected hotspot coverage at airports to more than 2 million APs for major city coverage across a country. Due to such wide variations in network equipment ownership, in backhaul transmission provider and in cell site real estate arrangement, the cost structure tends to be complex. The business model is further complicated by the exact data offloading strategy, since some operators choose to offload the radio link only and bring the traffic over WiFi to mobile operator's core network, while others choose to offload both the radio access and the core networks, with the WiFi traffic directly going into the public internet.

1.3.3 General Tends in Heterogeneous Networks Development

There are several general trends in heterogeneous network development, including the integration of WiFi and cellular air interfaces in the same small-cell platform, the techniques for dense deployment of small cells in extremely heavy loading conditions such as a sports stadium, the convergence of Hotspot 2.0 and Access Network Discovery and Selection Function (ANDSF) for network interoperability, the evolution from providing coverage or capacity to offering value-added services by integrating location and proximity services into the integrated small cells, and the possibility of developing super control nodes that coordinate the radio resources across different layers of networks. The debate on femto station versus WiFi access point is practically settled. It is now widely recognized that they are complementary.

Ultimately, a practical heterogeneous network must deliver three things to realize its economic potential: technically, it needs to fill the capacity gap created by the data tsunami; business-wise it must help mobile operators with in-service differentiation and new revenue opportunities, while minimizing CAPEX and OPEX; and to end users, it must offer superior user experience through always-on and always-best-connected. To achieve these goals, a heterogeneous network solution needs to leverage existing technologies and deployment, to enrich and to expand existing applications, and to enable future service innovations [12].

1.4 Aspects of Heterogeneous Network Technology

In this section we discuss different technical aspects of heterogeneous network technology, the challenges, the tradeoffs and the future technology directions.

1.4.1 RF Interference

When deploying 3G femtocells within the coverage area of a 3G macrocell network, these two layers of network may share the same carrier frequency, or they can be deployed on two separate carrier frequencies.

Sharing the same frequency has the advantage of minimum spectrum usage, which is particularly important to operators with a limited spectrum supply. However, the interference could be severe between these two networks if they are not properly engineered. For example, a UE connected to a macro base station may produce excessive amount of interference to nearby femtocells, in particular when the UE is at the cell edge transmitting at close to its maximum power level. One practical solution to address this problem is to desensitize the femto station receiver to avoid RF saturation. In general, it is always desirable to place the femto stations in locations that provide certain natural RF signal isolation with the macro network, and to carefully engineer the handover triggers for reliable UE mobility between these two networks.

When macro stations and femto stations are deployed on separate carrier frequencies, there are still challenges for the UEs to associate with the most appropriate network. Traditional cellular networks were designed for homogeneous network deployment, where the pilot strength measured at a UE is a good indication of its distance from the base station. However, this is generally no longer true for a heterogeneous deployment, where the pilot signal from the macro station can overwhelm the much weaker pilot signal from a femto station, even when the UE is already well within the coverage area of the femto station. A much better measurement in this case should be the actual path loss between the UE and the base station instead of the pilot strength. However, since such change would require some fundamental modifications to the existing standards and the deployed equipment, the industry adopted the 'range extension' technique which essentially adds a bias value to compensate for the weaker femto pilot, and therefore 'extend' the coverage area of a femto station [13].

In actual consumer femtocell deployment, the closed-subscriber-group (CSG) poses yet another challenge to RF interference management [14]. Since consumers usually use their own internet subscription for femtocell backhaul connection, the access to the femto station is restricted to a predefined group of users, usually the family members or house guests. A UE will not be able to access a consumer-deployed femto station even in close proximity unless it is part of the CSG. This can produce an excessive amount of interference to the networks, in particular in a dense femtocell deployment environment such as within an apartment building. Fortunately this is less of a concern for enterprise environments, since all users usually have access to all femto stations in the office space. There have been discussions on open subscriber group for femtocells, but it is more of a business model issue than a technical solution.

The situation with WiFi networks is very different. On the one hand, there is no 'macro' WiFi station to worry about, but on the other hand, the interference among WiFi access points and devices is more unpredictable due to the unlicensed-band nature of the network. As capacity demand continues to grow, more and more hot spot areas are served by multiple WiFi networks, which can cause excessive interference to each other. The wireless industry is looking at network sharing (e.g., several service providers sharing one access point by broadcasting several SSIDs over the air), local WiFi channel coordination and new deployment in the higher unlicensed spectrum including the 5 GHz band.

Future heterogeneous networks are also expected to support new types of connectivity and network architectures. These include layer 2 and layer 3 relay (including in-band backhaul application similar to a mess network), mobile relay (e.g., a relay station installed in a high speed train to provide relay service between the passengers in the train and the base station network along the trackside [15]), D2D (device-to-device direct communication [16]), traffic aggregation point for machine-type communications (a small radio station that provides short-range connectivity such as ZigBee to local sensors and relays the aggregated traffic to the network through a long-range cellular connection) and cognitive radio for spectrum sharing (e.g., in TV white space bands). Although the RF interference issues and challenges are very different in each case, there is a common set of requirements that a heterogeneous network RF engineer needs to consider, such as frequency planning, co-existence, transmission power level and receiver sensitivity within a heterogeneous network and within a neighbouring homogeneous or heterogeneous network operated by a different mobile network operator. At this moment, most heterogeneous network studies are focused on developing solutions for a specific combination of network layers. There is a need for longer-term vision and framework to scale up the heterogeneous network design and deployment through an optimal distribution of functions among devices, access points and cloud.

1.4.2 Radio System Configuration

Traditional cellular networks follow strict managed-network design principles. The cell sites are individually selected and base stations are manually engineered with sophisticated RF planning tools that model the actual deployment environment. Mobile devices provide real-time channel quality, interference level and resource request as inputs, but the network makes final decisions on radio resource management. This network-centric design has been effective in achieving optimal balance between coverage and capacity in homogeneous networks. The homogenous nature also simplifies mobile reporting, since the pilot to interference ratio can be directly used to estimate the electronic distance between the base station and a mobile device.

This traditional approach faces many challenges in a heterogeneous network environment. The first is site selection. Small cells are deployed in more versatile environments, ranging from residential, office, shopping centres and sports stadiums, both indoors and outdoors. The number of sites and environments are too costly to apply traditional deployment methodology. The second is carrier frequency coordination. Cellular operators have control over small-cell deployment due to its ownership of the licensed band. For unlicensed systems such as WiFi, the frequency planning needs to not only coordinate the channel allocation among access points, but also to adapt to the local interference environment. This is further complicated by consumer self-installation or professional installation without any network optimization.

The self-organizing network (SON) technology is therefore widely used in small cells [17]. SON requires interference sensing capability at the small-cell stations. For 3G/4G small cells operating in the FDD bands, this means the addition of an interference sensor to monitor its transmit frequency. For WiFi access points, this capability is inherited from the time division duplex(TDD) nature of its air interface. Based on the sensed interference information, the network can self-configure its channel allocation, scheduling algorithm and even its transmit pattern. The latter is particularly useful in places such as an apartment building where two femto stations are unintentionally installed next to each other across a building wall [18]. SON can be implemented in a distributed or a more centralized manner. Centralized SON coordination usually yields better performance, in particular in the dense deployment environments.

SON technology will further evolve in the future with the proliferation of cloud-RAN and small cells. SON technology needs to extend its capability to optimize across different layers of networks, and also across different base stations in a geographic location. We expect mobile devices to take a bigger role, for two reasons. First, multi-mode devices have the visibility of the local RF environment across layers of networks. Secondly, mobile devices may add features to help to reduce SON complexity. We also expect SON to heavily leverage cloud-RAN technology for optimum performance in a heterogeneous network environment. One possible solution is to have the cloud provide policy and general information on the RF environment, based on which each network layer can configure its radio system, and move more decision-making to devices. Recent development in connection manager software at certain mobile devices can be considered an early experiment of this technology concept.

1.4.3 Network Coupling

Heterogeneous networking is about multiple layers of networks interconnecting and working together. The relationship between networks of different layers can be loosely described as a network coupling issue. Although there are no precise definitions, loose network coupling usually refers to two networks generally maintaining independence of each other when forming a heterogeneous network, while tight coupling usually means a greater level of integration between two networks.

To illustrate the concept of network coupling, let's use 3GPP-WiFi interworking as a concrete example. According to the 3GPP specification [19], an 'un-trusted' WiFi network can connect to a 3G cellular network through an 'enhanced packet data gateway' (ePDG), which provides backhaul security through IPSec tunnels, real-time packet processing, and subscribers, service and applications monitoring. We consider these two networks loosely coupled when the WiFi access network and 3GPP core network share only the AAA server for authentication. In this loose-coupling case, each network operates independently, and neither of the networks needs to change its architecture or protocol stack. However, since there is no support for service continuity during handovers, a user may experience longer handover latency and a certain amount of packet loss. In a tight-coupling case, a WiFi data flows through the 3GPP core network from ePDG to the 3GPP packet data network gateway (PDN-GW), in the same way as cellular data does. In addition to 3GPP authentication as in the loose-coupling case, WiFi users can also access 3GPP services with the potential support of guaranteed QoS and seamless mobility. Other 3GPP services may be enabled for WiFi users as well through tight coupling.

In a broader sense, network coupling is about the level of integration between two networks. Femto station is another good example of tight coupling. In addition to the tight coupling at the core network level as we just described, a heterogeneous network may also create a tight coupling at lower layers between two radio access networks, and allow one radio access network to have certain visibility and radio resource management control over the other network. A relevant example is the CDMA-LTE handover design through tight coupling at the radio link layer, which potentially offers the same level of fast handover performance as a homogeneous network [20]. In LTE-Advanced and beyond, this type of 'tighter coupling' can be further extended to air interface PHY and MAC layers to allow the implementation of joint radio resource management and traffic scheduling. There are already suggestions to add an X2-like interface between a macro base station and a femto station, and add the aggregation of carriers of different radio access technologies, which we will discuss later.

Network coupling is a complex technical issue, but in real life deployment, it is also very often a critical business decision driven by the tradeoffs of complexity, cost, security and performance. A wireless network operator may choose loose coupling over tight coupling despite its potential higher performance. For example, if a cellular operator uses a third-party

Table 1.2 Types of network coupling

Type of network coupling	Example design characteristics	Benefits and complexity
no coupling	• two networks operate independently • mobile device connection manager coordinates wireless connectivity.	• no change to existing networks, suitable for roaming • operator may perform offline billings consolidation for single bill
loose coupling	• two networks share user credential and AAA • data traffic goes through separate core networks	• common user/device credential for two networks; potentially eliminate user intervention for WiFi access • core network traffic offload to internet
tight coupling	• user traffic goes through mobile operator core network, (e.g., home femtocells connected through a femto gateway)	• opportunity for operator to offer value-added services • opportunity to offer service continuity or even seamless handover
very tight coupling	• two access networks are directly interconnected • real-time radio resource management across networks, (e.g., picocell connected through X2 interface)	• opportunity to implement carrier aggregation and interference coordination • requires significant design and implement efforts

WiFi network to form a heterogeneous network, loose coupling may offer the simplicity to both networks for both business relationship and network management. This may also be the case for operators who own both networks but do not want to make significant changes to the already installed equipment. The network coupling decision is also significantly driven by the expected use scenarios of an operator. Certain operators may choose to use the networks for different purposes, for example, one for voice calls and smartphone data access, and the other for data offloading to the internet to avoid core network congestion. In this case, the benefits of high performance service continuity of a tight couple may be secondary. Table 1.2 provides a summary of types of network coupling, the general design characteristics and the associated benefits and complexity.

From a standards development viewpoint, the industry tends towards tightly integrated solutions with secure data offloading and service continuity to offer operators a complete range of technology choices. These solutions will continue to evolve for different markets including home, enterprise and wireless carriers.

1.4.4 User and Device Credential

A key technology element in a heterogeneous network is the user or device credential used for authentication, authorization and accounting. The most common user/device credentials include SIM card in mobile phones, embedded SIM and soft SIM in certain machine type communication devices, certificate in laptop computers, and username/password for WiFi access. These fragmented user/device credential solutions were mainly due to the legacy of the telecom and the information technology industries, and they are now increasingly becoming a challenge to the implementation and deployment of heterogeneous networks.

Taking cellular and WiFi interworking as an example, while most mobile smartphones use SIM cards, most WiFi-only devices currently use username/password pair or certificate for access authentication. A unified user/device credential solution is highly desirable for seamless network access, improved user privacy and information protection, uniform service availability, consistent quality of experience and simplified device management. Unfortunately, it may take time to implement the required changes to the portable device platform design, and more importantly to operators; accounting and billing system. In tightly coupling heterogeneous networks, we see stronger motivations to unified user/device credential to facilitate fast network transition and seamless service continuity across networks. In loosely coupling cases, we anticipate slower convergence of solutions, or no convergence at all.

This situation is further complicated by the proliferation of machine-type communication devices. Compared to consumer electronics such as smartphones, some machine-type communication devices require a much longer life cycle (e.g., seven or more years) and need to be environmentally hardened and able to survive deployment with minimum physical protection for the devices. Therefore new forms of device credentials may be required.

Nevertheless, the wireless industry has taken steps to address the issues, and there are several initiatives currently at different stages of development [21]. The convergence of telecom and information technology may also help to accelerate the process.

1.4.5 Interworking

Interworking provides a degree of seamlessness to multi-mode devices capable of transiting between different networks. There are two forms of transiting. The first is between networks that are geographically separated, commonly known as roaming. The second is between networks that are co-located, such as handover between different layers of networks within a heterogeneous network. For heterogeneous networks and their associated multi-mode devices, roaming and handover support is one of the fundamental capabilities that enable end users to access basic services seamlessly.

Interworking between networks is often achieved through an interworking function node or the access gateway [19]. In a heterogeneous network, since different layers of the network may have different network architecture and protocol stack design, such a node provides protocol conversion or translation on the control plane and intermediate protocol termination on the data plane. This node may also serve as the security demarcation point between two networks, in particular when the ownership of two networks is different. The home NodeB gateway defined in 3GPP runs an IPsec tunnel to the home NodeB (commonly known as a femto station) over the public internet to provide information security and an interworking function for handover. Since some of the networks such as WiFi do not have core network architecture and the associated protocol stack, an interworking node (in this case the ePDG) provides WiFi devices the access to the AAA server in the 3GPP core network for common authentication and charging. Future development will also add unified QoS capability across different layers of the network to provide more seamless experience to end users.

The main challenge in interworking is that the design and implementation complexity goes up exponentially when we scale up the number of layers in a heterogeneous network. Since each access technology usually comes with its own control plane design, making all layers talk to each other requires extensive engineering efforts, in particular for tight coupling. We expect the challenge to further increase in the future due to the introduction of new types of connectivity (e.g., ZigBee for machine-type devices) and new network topologies (e.g., mobile relay, device-to-device communications).

The wireless industry is looking at two technology directions to address the interworking complexity issue. One is to put more intelligence in the mobile devices, therefore offloading some of the interworking complexities from the network infrastructure to the devices. One example is to design and deploy sophisticated connection manager software in the mobile devices. The other one is to improve data plane processing, moving some of the complexities from the control plane to the data plane. Since the device has detailed knowledge of the applications it is running, enhanced data plane processing is in particularly attractive for applications that require consistent QoS support throughout layers of a heterogeneous network.

1.4.6 Handover

Once networks can interwork with each other, the next technical requirement is the ability to handover a mobile device among base stations and/or networks. As in any homogenous network, the reliability and the performance of handover directly affect user experience in a heterogeneous network and are therefore the key performance metrics.

Before we discuss the details of handover techniques in heterogeneous networks, we first look at the differences. In a homogeneous network, handover is primarily triggered by a mobile device moving out of the coverage area of a serving base station or network. Occasionally a wireless operator may also choose to direct a device to handover to a specific base station for network maintenance or load balancing purposes. In a heterogeneous network, a high-power macro base station network usually provides ubiquitous network coverage and high-speed mobility support, while the low-power small cells offer data offload in hotspot areas at a lower cost.

For intra-system handovers within the same layer of a heterogeneous network, the handover techniques and procedures are similar to that of a homogeneous network. For inter-system handovers across different layers, however, the handover decision is much more involved. In addition to the traditional coverage-triggered handovers, many decisions in a heterogeneous network are also based on the service delivery cost to an operator, the data pricing to users, the QoS requirements of specific applications, and the non-ubiquitous nature of the hotspot network layer [22]. For example, for data offloading purposes it is highly desirable to handover a mobile device from the macro base station network to small cells as soon as it moves within the hotspot or home network coverage area. For real-time and conversational applications such as voice calls, it may be desirable to avoid inter-layer handover due to service continuity and quality of service requirements. The degree of network coupling also has a significant impact on the handover algorithm design. Due to its smaller interruption to services, a heterogeneous network with tight coupling between layers can afford more aggressive inter-layer handover decisions than a similar network with loose coupling. The handover algorithm design in a heterogeneous network environment is therefore significantly more complex and challenging than that of a homogeneous network.

Similar to the traditional designs widely used in homogeneous networks, a handover process in heterogeneous networks also involves five stages: (1) handover preparation, (2) trigger for handover, (3) handover decision, (4) handover execution and session transfer and (5) handover completion. Handover preparation includes the mobile device looking for possible handover targets according to certain pre-optimized mobile-network association rules [23]. To facilitate this searching process and potentially save the battery consumption of the mobile device, the serving network may offer certain information about the candidate networks, such as the type of air interface technology and the basic radio configuration information including carrier frequency. The serving network may also provide certain handover-related parameters to the mobile device to optimize performance. In addition, a mobile device may pre-register itself to the target networks to save time in authentication and service authorization when handover actually happens. However, all these come at a cost to a heterogeneous network, including different layers of the network having to have sufficient information about each other, and having to maintain consistency of registration from a mobile device. This complexity goes up rapidly as the number of layers increases or when the deployment environment is complex (e.g., a mix of indoor and outdoor networks, or different network topologies). Fortunately, the industry and university research teams are working on advanced network self-configuration techniques. The efforts in unifying user and device credentials may also reduce the complexity in managing device pre-registration.

The next step is to determine the need for handover when certain preset conditions are met, commonly known as handover triggers. Handover triggers are mostly generated at the mobile device, based on a set of rules and parameters that are pre-installed by the operator or downloaded from the network. A network may also trigger handovers for network maintenance or performance optimization purposes. It is probably one of the most important technologies in a wireless network. In a homogeneous network, wrong handover triggers result in radio link failure (e.g., call drops) and/or inefficient usage of radio resources. In a heterogeneous network, since small-cell coverage may not be ubiquitous, knowing where and when to switch from one network to another affects both network performance and user experience. The design is further complicated since in addition to radio link conditions the handover trigger algorithm for a heterogeneous network also needs to consider operator policy, subscriber usage cost and application quality of service requirements. These requirements may vary in time and/or may be location-dependent.

After handover is triggered, an entity in the system needs to make a handover decision. In traditional cellular system designs, the infrastructure network is responsible for making the final handover decision. As soon as the handover conditions are met, the mobile device is responsible for sending a handover request to the network, but the network can decide either to issue a handover command or to take no action. This network-centric handover decision process has shortcomings in heterogeneous network environments. For example, the serving layer of the heterogeneous network may not have the full information or capability to make a handover decision to other layers. This is particularly true when networks are loosely coupled. We expect mobile devices to take a more active role in handover decision-making in future evolved heterogeneous networks, and as a result, some of the handover message flows will change accordingly.

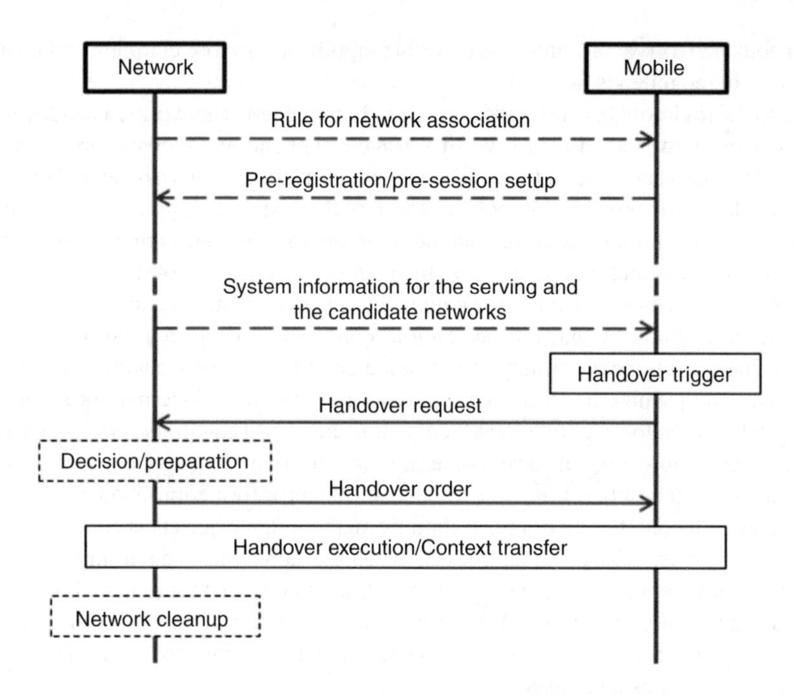

Figure 1.3 An example of a mobile-triggered network-decided generic handover process.

Heterogeneous networks can apply the same techniques used in homogeneous networks for fast handover execution and minimum service interruption. In a similar way to traditional handovers, the source network may prepare the target network(s) to pre-fetch the required context information before actual handover occurs. During handover execution, depending on the degree of network coupling, session information including the remaining user data in the transmit buffer may be transferred from the source to the target network to minimize data loss and to achieve session continuity [24].

Upon completion of the handover process, the heterogeneous network needs to perform certain clean-up to maintain its overall resource utilization efficiency and software robustness. An example of a mobile-triggered network-decided generic handover process is shown in Figure 1.3.

1.4.7 Data Routing

While handover is mostly a link layer issue, data packet routing is mostly a network layer function. There are two main aspects to data packet routing in a heterogeneous network: (1) the radio link selection and/or aggregation over the air and (2) IP packet routine in network. The actual design of data routing is based both on business and technical considerations.

Unlike a homogeneous network where a mobile device usually connects to the network through a single radio link, a heterogeneous network can allow multiple air interface technologies to operate concurrently, therefore potentially creating multiple radio links between a mobile device and the heterogeneous network, for example, a smartphone with both 3G and

WiFi connections turned on at the same time. One simple way of using this capability is to aggregate them, in order to achieve a higher data rate. The other way is to intelligently map user application flows to different radio links, according to their respective QoS requirements, operator policy and user preference. For example, the mobile device can use the cellular connection for voice calls and the WiFi link for high-speed best-effort internet access. This capability is supported in 3GPP IFOM (IP flow mobility) feature [25].

In addition to radio link selection and aggregation over the air, a heterogeneous network may also support different routing options within the infrastructure network to different layers. This flexibility becomes increasingly important as the industry is starting to experience more and more core network overloading issues as the air interface capacity rapidly improves due to the deployment of small cells and also due to more spectrally efficient air interfaces such as LTE and LTE-Advanced. 3GPP is working on a number of solutions including LIPA (Local IP Access, commonly known as local breakout for 3GPP femto stations) and SIPTO (Selective IP Traffic Offloading) to offload an operator's core network or to selectively route traffic to the public internet or to the core network according to operator business considerations [26].

The combination of multi-radio-link capability and the network routing flexibility is an area of intensive research and technology development in industry. It is one of the unique tools in a heterogeneous network to achieve an optimum balance between user experience and service revenue, and to implement an operator's data offloading strategy.

1.4.8 Quality of Service

Providing consistent quality of service to end users is extremely important to any wireless networks. Since it also offers opportunities for product differentiation and incremental revenue for network operators, it is one of the most mission-critical technology elements in future wireless networks.

Due to its end-to-end nature, delivering quality of service is complex and challenging, even in a traditional homogeneous network. First of all, the network needs to have the capability of establishing and managing service policy, and have the associated accounting and charging infrastructure. When a subscriber requests a service, the network authenticates the user and authorizes the service according to its subscription record. The network then configures each network node along the data path of this specific application. Once network nodes start performing the desired QoS treatment for this particular application IP flow, the network needs to have the capability of monitoring the actual service quality delivered to the end user. When the application session is ended, the network needs to collect all relevant accounting information and to perform system clean up.

This rather complex process faces several challenges in real life. First of all, quality of service is meaningless to a user unless it is delivered end to end, but a wireless network only has the span of control over the radio link and within the operator's own core network. This has not been a major issue, since traditionally the radio link was almost always the bottleneck for service quality, but recently the backbone networks have also started to experience congestion due to the proliferation of video and other media-rich contents. The second challenge is the lack of consistency in QoS implementation. Different segments of the network may have different interpretations of QoS marking, and may use different parameters in traffic scheduling algorithms. This is, in particular, an issue when traffic travels through networks of different operators. The third challenge is that QoS scheduling may not always produce as significant a result as users expect. This is partly due to the significant increase in air interface data rate and backhaul bandwidth in the past decade, and partly due to the fact that many popular contents and applications are the so-called 'over-the-top' traffic where operators have little real control of service quality (over-the-top traffic refers to contents or applications that are delivered directly from the provider to the customer using an open internet connection, independently of wireless operator network). Finally, end users see little reason to pay extra for QoS unless the service becomes practically unusable without it. As a result, current focus in the wireless industry is limited to VoLTE (VoIP over LTE), where operators own both the network and the service, and QoS support is essential for VoLTE service reliability to approach that of circuit switched voice [27].

QoS support in heterogeneous networks incurs additional complexity. There are multiple layers of networks involved, each of which may employ different air interface technology and network architecture, with different QoS mechanisms. Even for a heterogeneous network consisting of the same air interface technology – for example 3G macro stations with 3G femto stations – the backhaul transport over public internet for femto stations may affect the QoS management and delivery. Network coupling brings further variations, and some deployments may face limitations in delivering consistent QoS across loosely coupled networks. For heterogeneous networks with data routing capability among layers, the routing change itself may introduce additional latency in the data path. Finally, the RF interference condition in unlicensed band is often difficult to predict and changes with time, particularly in dense hotspots where QoS happens to be most needed.

As discussed earlier, due to the proliferation of mobile internet traffic, wireless operators have to invest heavily to meet the capacity demand while their revenues from traditional voice and SMS services are decreasing. As a result, QoS is increasingly essential to maintain revenue stream and to differentiate in the highly competitive market place. We expect research and development of QoS for heterogeneous networks to intensify in the next few years, and the results will become increasingly critical and rewarding with time.

Since a significant portion of the mobile internet traffic is over-the-top with origination outside of operator's network, such as Skype and YouTube, a heterogeneous network today mostly serves as best-effort data pipe. There are two basic approaches to introduce QoS support in this case. The first one is to expand the traditional QoS model we discussed earlier in this sector into the third-party application service providers. The MOSAP development in 3GPP (interworking between Mobile Operators using the evolved packet System and data Application Providers) can be used as a framework [28]. We can call this a control-plane-centric approach since it requires the control entities of the mobile operator network and the data application provider system to exchange application requirements and policy information with each other so that the mobile operator can properly set up the data pipe for QoS support. However, since many over-the-top applications are already deployed and are working well in the field, many data application providers may not be willing to make additional investment in this capability and negotiate a business agreement with mobile network operators. In addition, the required interoperability testing may be challenging to scale up the solution. Finally, the question of how to handle QoS within a heterogeneous network remains the same.

Another very different approach is to introduce advanced signal processing on the data plane, thereby shifting some of the complexity from the control plane to the data plane. A classic example is the application of deep packet inspect (DPI) at both the base station network and the mobile device. By performing DPI, the data plane integrates an autonomous QoS handling capability, which can remain intact when a radio link is handed over between different layers of networks in a heterogeneous network. For the base station network, the data plane packet processing can be integrated with other advanced content distribution technologies including caching, cross-layer optimization and adaptive media reformatting to address the network overloading conditions. With the convergence of information technology and the telecom in the cloud network, data plane processing is becoming increasingly feasible and attractive. For mobile devices, shifting complexity from the control plane to the data plane has the added potential benefits of minimum additional signal processing. This is because both the applications and the complete protocol stack are implemented within the same physical device, so in practice there is little need to perform packet-by-packet inspection since such information can be easily and directly obtained from the API for each application. In a heterogeneous network environment the mobile devices also have the advantage of knowing its RF environment and the availability of each network. It is natural for the mobile devices to take a greater role in QoS management and decision making.

We expect the wireless industry and the research community to continue to investigate the traditional QoS mechanisms and to look for new technology options in the next few years in the context of IT and telecom convergence and media content distribution for heterogeneous networks. We expect cloud network architecture and multi-communication device to play a major role in QoS over heterogeneous networks. From an economic viewpoint, it is also one of the most important capabilities for mobile network operators since it can offer unprecedented user experience in terms of service quality and consistency to the end users in a highly complex and sometimes confusing heterogeneous network environment, and can also provide a major revenue opportunity for operators who are facing the challenge of becoming a featureless transparent-data-pipe utility provider.

1.4.9 Security and Privacy

Security and device interconnectivity engineered in a heterogeneous network and within the protocol stack of each layer of networks enable consistent and compelling experience for consumers and expand the network applications for machine-type connected devices in a future information society. As devices like smartphones, tablets and smart TVs become more popular, consumers are clamouring for access to applications and data everywhere, but are faced with security barriers of each individual system and therefore disparate user experiences. For machine-type communications such as smart grid and smart transportation, security is paramount in protecting the integrity of strategic infrastructure and for ensuring the normal functioning of an intelligent society and the underlying future embedded internet [2].

As mobile internet penetrates into every aspect of an individual's daily life and becomes the fabric of our modern society, the concerns over security further increase. With the smartphone operating systems becoming more concentrated around four options (the combined market share of Android, Symbian, iOS and Blackberry accounts for more than 95% globally [29]), any major security breach would potentially affect a massive number of users around the world. In addition to the large-scale potential security impacts, the drive towards better user convenience also requires more security safeguards. Features such as single-sign-on (SSO) and unified billing completely rely on secure communications among different systems. Ultimately, convenience does not mean anything unless it is secure. The industry is taking steps to fortify security including embedding security functions directly in the silicon chips [30].

With the proliferation of mobile internet there is also an increased awareness of, or concerns over, privacy. In the heterogeneous network environment in particular, since layers of network need to exchange essential information to ensure the compute continuum, potentially among different operators that own and operate each layer of the network, protecting privacy is an even more critical issue. Since mobile network operators have a strong desire to offer context-aware and location-based services instead of simply providing a featureless data pipe, the collection – and more importantly the exchange – of user and application information sometimes become critical. Taking device-to-device (D2D) for social networking applications as an example, if everyone is concerned with privacy and security, therefore turns off the D2D function, this feature will never become useful. For location-based services, privacy has always been a concern, in particular in a heterogeneous network environment involving a mixed network ownership. We expect mobile network operators to play a bigger role in managing and maintaining security and privacy, and to heavily leverage their subscriber ownership and the cloud infrastructure.

1.4.10 Capacity and Performance Evaluation

The capacity and performance evaluation of heterogeneous networks poses a unique set of challenges and issues. On the one hand, the cellular industry has established an elaborate system capacity modelling and evaluation methodology for traditional homogeneous cellular networks. The focus has been on ubiquitous network coverage (e.g., 95%), overall cell throughput and spectral efficiency, and tail user throughput (e.g., 95% point on the CDF curve). On the other hand, the IT industry uses its own network modelling and evaluation methodology, with focus on enterprise and hotspot traffic density, and peak data rate. For heterogeneous networks, the cellular community has extended its traditional model to cover femto/small-cell deployment scenarios. This includes adding median user throughput, average macrocell area spectral efficiency, and the percentage of total throughput carried by low-power cells. In addition, latency-based metrics were also considered for the case of bursty-traffic evaluations. The cellular model has been further extended to include new types of wireless connectivity and use scenarios, such as closed subscriber group (CSG) and relay, including in-band self-backhauling use cases [31].

Nevertheless, significant research and development are still required for heterogeneous network capacity and performance evaluations. Key focus areas include more accurate modelling of hotspot and hot-area user distribution, including: the impact of network coupling and inter-working performance on subscriber distribution among layers of networks; the inclusion of new types of connectivity in the models such as mobile relay and device-to-device communications; cross-network-layer interference coordination and radio resource management; a unified or hybrid cellular and WiFi evaluation methodology that enables more TCO modelling; and a more accurate modelling of unlicensed band radio environment.

1.5 Future Heterogeneous Network Applications

The initial deployments of heterogeneous networks have been for network capacity and traffic offloading [32]. Future technology development needs to support new capabilities for innovative applications that offer unprecedented new user experience and to deliver a true world of compute continuum.

We can expect that traffic offloading solutions continue to evolve and improve, to address congestion in both access networks and core network. This is achieved through the deployment of both IP flow mobility and local breakout solutions such as LIPA and SIPTO [26]. As backhaul transport becomes a major cost factor, we will see an increasing number of existing and new hybrid solutions using optical fiber, copper cable, microwave, in-band relay, multi-hope relay and mesh networks. Network security and user privacy need to continue to improve as user devices increasingly roam between layers of networks and between coverage areas of different mobile network operators. Finally, we expect the wireless industry to expend significant research and development effort in addressing very dense heterogeneous network design and deployment issues for public venues such as sports stadium and mass public transportation hubs.

Collaborative communication will become a major driving force for heterogeneous network innovations to support new types of connectivity and the associated new business models [33, 34], which include device-to-device communication, client relay and mobile relay, and sensor hub for machine type communication devices. Cognitive radio may also be added to this framework since it may offer an additional connectivity option for collaborative communication.

Future heterogeneous networks are also expected to implement and to take full advantage of the next generation air interface features. A good example is the implementation of carrier aggregation over a heterogeneous network. For a deployment scenario shown on the left side of Figure 1.4, a mobile device may need to handover frequently between macro and small networks without carrier aggregation, which disrupts services and add signalling traffic to the network. With carrier aggregation, it is possible to anchor the primary control channel at the macrocell, preferably operating at a lower frequency band for reliability and coverage, while operating the secondary data channels from small cells at a higher frequency band to deliver high data rate and network capacity in an opportunistic fashion. There is no handover required as long as the mobile device is under the coverage area of the macrocell. This is shown on the right side of Figure 1.4.

Multiple carriers also enable interference management between different power class cells as well as open access and closed subscriber group (CSG) cells. Long-term resource partitioning can be carried out by exclusively dedicating carriers to certain power class cells, while dynamic radio resource management and network load balancing can be applied by sharing those carriers

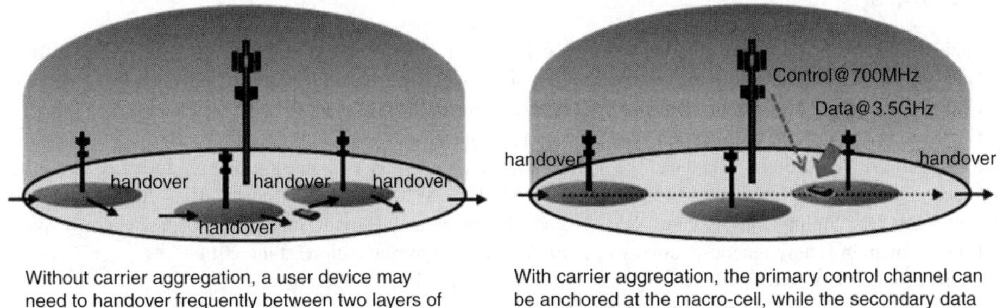

Figure 1.4 Carrier aggregation in a heterogeneous network.

among cells through cross-network-layer coordinated multi-point transmission and beam-forming, application scheduling, power control, inter-cell and inter-network-layer interference coordination and alignment, examples of which include fractional frequency reuse (FFR) and time-domain resource partitioning.

As the wireless technology focus shifts from spectral efficiency to network efficiency, a major network performance improvement is expected to come from media optimization through the application of information and communication technology (ICT) [35]. In particular, caching enables more efficient content distribution for video, local navigation mapping information, social networking content uploading and distribution. Virtualization technology allows a unified signal processing platform for both control and data planes. It also enables dynamic workload balancing between signalling and application processing. The introduction of ICT brings many new possibilities for future heterogeneous networks, including the co-existence of both super-sized cloud base station and small cells, the possible introduction of a super control node that coordinates the operation of different network layers, the proliferation of data plane signal processing for cross-network QoS support, and the further flattening of the core network which facilitates the implementation of cross-layer optimization [36].

Finally, future heterogeneous networks also require advanced multi-mode and multi-band user devices. Consider in the following a rather extreme use case in China: a sales representative drives his car to visit his customer. He is making a GSM call using his Bluetooth headset, while downloading a movie via TD-LTE on his new smartphone, while his smartphone is searching for a WiFi hot spot for offloading the cellular network, while he is listening to the latest MP3 hits, streamed from his smartphone to the car radio via FM radio, while a GPS navigation application on the smartphone is running. And, by the way, the car is running at all only due to the fact the sales representative could identify himself as its owner by the NFC ID feature in his nice smartphone. As the world goes increasingly wireless and increasingly connected, it is no longer uncommon to find a user device with seven or more antennas. The devices themselves become an integral part of the future heterogeneous networks. The tunable RF circuit becomes essential to achieve compact design at low cost. Co-existence technologies based on TDM, FDM, SDM domains become increasingly critical. Together with advanced mobile devices, future heterogeneous networks will bring user experience to an unprecedented new level.

References

1. Panel discussion by Google CEO Eric Schmidt at Techonomy conference in Lake Tahoe, CA, August 2010.
2. Geng Wu, Shilpa Talwar, Kerstin Johnsson, Nageen Himayat and Kevin D. Johnson, 'M2M: From Mobile to Embedded Internet', IEEE Communications Magazine, April 2011.
3. Ericsson, 'More than 50 billion connected devices', February 2011.
4. McKinsey Global Institute, 'Big data: The next frontier for innovation, competition, and productivity', May 2011.
5. Cisco, 'Cisco Visual Networking Index: Global Mobile Data Traffic Forecast Update, 2010–2015', 2011.
6. Shu-Ping Yeh, Shilpa Talwar, Geng Wu, Nageen Himayat and Kerstin Johnsson, 'Capacity and Coverage Enhancement in Heterogeneous Networks', IEEE Wireless Communications, June 2011.
7. China Mobile, 'C-RAN, The Road Towards Green RAN', Version 2.5, October 2011.
8. ABI, Mobile Operator CAPEX Market Data, 4/5/2012.
9. K. Johansson, A. Furuskär and C. Bergljung, 'A Methodology for Estimating Cost and Performance of Heterogeneous Wireless Access Networks', PIMRC '07.

10. K. Johansson, J. Zander, and A. Furuskär, 'Modeling the cost of heterogeneous wireless access networks', *Int. J. Mobile Network Design and Innovation*, Vol. 2, No. 1, pp. 58–66.
11. Caroline Chan and Geng Wu, 'Pivotal Role of Heterogeneous Networks in 4G Deployment', ZTE Technologies, Issue 1, 2010.
12. Jennifer Pigg, 'The Operator as Innovator: Smartphones, Smart Apps and Smart Pipes', Yankee Group Research, February 2011.
13. Aleksandar Damnjanovic et al., 'A Survey On 3GPP Heterogeneous Networks', IEEE Wireless Communications, June 2011.
14. Sayandev Mukherjee, 'UE Coverage in LTE Macro Network with mixed CSG and Open Access Femto Overlay.' 2011 IEEE International Conference on Communications (ICC), Kyoto, Japan, 4 June 2011.
15. R. Balakrishnan, X. Yang, M. Venkatachalam Ian F. Akyildiz, 'Mobile Relay and Group Mobility for 4G WiMAX Networks', 2011 IEEE Wireless Communications and Networking Conference (WCNC), Cancun, Mexico, 28–31 March 2011.
16. Klaus Doppler, Cássio B. Ribeiro and Jarkko Kneckt, 'Advances in D2D Communications: Energy Efficient Service and Device Discovery Radio', 2011 2nd International Conference on Wireless Communication, Vehicular Technology, Information Theory and Aerospace & Electronic Systems Technology (Wireless VITAE), 28 February to 3 March, Chennai, India.
17. Seppo Hämäläinen, Henning Sanneck and Cinzia Sartori, *LTE Self-Organizing Networks (SON): Network Management Automation for Operational Efficiency*, John Wiley & Sons, ISBN 1119970679, 7 February 2012.
18. Small Cell Forum, 'Femto Forum Summary Report: Interference Management in UMTS Femtocells', February 2010.
19. 3GPP TS 23.402, 'Architecture enhancements for non-3GPP accesses'.
20. 3GPP TR 36.938, 'Improved Network Controlled Mobility between E-UTRAN and 3GPP2/Mobile WiMAX Radio Technologies'.
21. 'GSMA, WBA Collaborate on wifi Roaming', http://wirelessweek.com/News/2012/03/gsma-wba-collaborate-on-wifi-roaming/.
22. Julio Puschel, 'Learning from the Femtocell and Wi-Fi Pioneers', Webinar, Informa, 18 May 2011.
23. Qian (Clara) Li, Rose Qingyang Hu, Geng Wu and Yi Qian, 'On the Optimal Mobile Association in Heterogeneous Wireless Relay Networks', IEEE INFOCOM 2012, 25–30 March 2012, Orlando, Florida USA.
24. Rose Qingyang Hu, David Paranchych, Mo-Han Fong and Geng Wu, 'On the Evolution of Handoff Management and Network Architecture in WiMAX', IEEE Mobile WiMAX Symposium, Orlando, USA, pp. 144–149, March 2007.
25. 3GPP TS 23.261, 'IP flow mobility and seamless Wireless Local Area Network (WLAN) offload'.
26. 3GPP TR 23.829, 'Local IP Access and Selected IP Traffic Offload (LIPA-SIPTO).
27. Miikka Poikselkä, Harri Holma, Jukka Hongisto and Juha Kallio, *Voice over LTE (VoLTE)*, John Wiley & Sons, ISBN 1119951682, March 13, 2012.
28. 3GPP TR 23.862, 'EPC enhancements to support Interworking with data application providers'.
29. Gartner, 2011 Q3 Global Smartphones OS Market Share.
30. McAfee, 'McAfee Deep Defender – security beyond the OS to expose and eliminate covert threats', Data Sheet, 2011.
31. 3GPP TR 36.814, 'Evolved Universal Terrestrial Radio Access (E-UTRA); Further advancements for E-UTRA physical layer aspects'.
32. ABI Research, 'Mobile Network Offloading', December 2010.
33. Qian (Clara) Li, Rose Qingyang Hu, Yi Qian and Geng Wu, 'Cooperative Communications for Wireless Networks: Techniques and Applications in LTE-Advanced Systems', IEEE Wireless Communications April 2012.
34. Aamod Khandekar, Naga Bhushan, Ji Tingfang and Vieri Vanghi, 'LTE-Advanced: Heterogeneous Networks', 2010 European Wireless Conference, 12–15 April 2010.
35. Geng Wu, 'Virtualized Next Generation Wireless Network in the Cloud', International Mobile Internet Conference 2010, 13–14 December 2010, Beijing, China.
36. Nokia Siemens Networks, 'Liquid Radio – Let traffic waves flow most efficiently', White Paper, 2011.

Part I

Radio Resource and Interference Management

2

Radio Resource and Interference Management for Heterogeneous Networks

Lingjia Liu,[1] Ying Li,[2] Boon Loong Ng,[2] and Zhouyue Pi[2]
[1] *University of Kansas, USA*
[2] *Samsung Telecommunications America, USA*

2.1 Introduction

The demand of wireless data traffic is explosively increasing due to the increasing popularity, among consumers and businesses, of smartphones and other mobile data devices such as tablets, netbooks and eBook readers. It was predicted that mobile data traffic will grow 26 times between 2010 and 2015 [1]. In order to meet this spectacular growth in mobile data traffic, improvements in radio interface efficiency would be of paramount importance.

The current fourth generation (4G) cellular technologies [2] including LTE-Advanced and Advanced Mobile WiMAX (IEEE 802.16m) use advanced physical layer technologies such as orthogonal frequency division multiplexing (OFDM), multiple input multiple output (MIMO), multi-user MIMO and adaptive modulation and coding (AMC) to achieve spectral efficiencies which are close to theoretical limits in terms of bps/Hz/cell. Continuous improvements in air-interface performance are being considered by introducing new techniques such as carrier aggregation, higher order MIMO configurations, coordinated multi-point (CoMP) transmission and relays. However, based on the current network infrastructure, any further improvements in spectral efficiency would be marginal even in the best case.

When spectral efficiency of a single cell in terms of bps/Hz/cell cannot be improved significantly, another possibility to increase the overall system capacity is to deploy a large number of smaller cells in order to achieve cell-splitting gains. To meet the demand of mobile data

Heterogeneous Cellular Networks, First Edition. Edited by Rose Qingyang Hu and Yi Qian.

traffic, future wireless systems are expected to be heterogeneous networks with base stations of diverse sizes and types.

Compared to traditional homogeneous networks, there are new scenarios and considerations in heterogeneous networks. One of them is traffic offloading from large cells to small cells. As the penetration of wireless devices increases, traffic within a large cell increases. Therefore, the large cell is highly likely to be heavily loaded. Due to the higher deployment density of the small cells, it is beneficial to expand the footprint of the small cells and enable more user equipments (UEs) (otherwise known as mobile stations) to connect to the small cells to take advantage of the higher deployment density (that is, cell-splitting gains). Another new scenario is the support of closed subscriber group (CSG). As the name suggests, the CSG cell only allows its member UEs to access the cell. In the case of co-channel deployment, non-member users may experience strong interference from a CSG cell when they are in close proximity to the CSG cell.

Due to the new deployment scenarios and considerations, traditional interference coordination technologies such as soft frequency reuse (SFR) are not sufficient. Note that it is already a common practice in legacy cellular systems to overlay cells with different sizes using different carriers such that high mobility users dwell in one frequency carrier in large cells and low mobility, users are served in another frequency carrier in small cells. In this way, the handover overhead can be reduced for high mobility users while the system capacity can be improved for low mobility users due to frequency reuse. However, in voice-centric legacy systems, the radio resource management tends to focus on admission control and mobility management, which happens on the order of hundreds of millisecond or longer, while not considering either the target of offloading or the support for the CSG deployment.

In order to meet the challenge of new scenarios and requirements, enhanced inter-cell interference coordination (eICIC) technologies are necessary. There are two categories of ICIC techniques, namely multi-carrier inter-cell interference management and single-carrier (or co-channel) inter-cell interference management.

For multi-carrier inter-cell interference management, frequency can be assigned to large cells and small cells to achieve interference coordination. Joint carrier management and power control among large cells and small cells can be performed to achieve high system performance. Cross-carrier scheduling is introduced for systems deployed with carrier aggregation (CA), which can be applied in heterogeneous networks in which the large cell schedules its UEs using the control channel region in one component carrier while the overlaid small cell can schedule its UEs via the control channel region in another component carrier.

For single-carrier (or co-channel) inter-cell interference management, where the large cells and the small cells are deployed on the same carrier, the time-domain resource partition among large cells and small cells can be used on the subframe level. In some subframes, a cell (e.g., macro, pico or femto) does not transmit data and only transmits the minimum control signals needed to maintain system operation. In this way, the interference to other cells is minimized. To resolve the coverage hole issue created by the CSG, advanced power control or power setting algorithms can be applied to control the transmission power of a CSG cell at a reasonable level, so as not to interfere too much with the macrocell UEs that are not members of the corresponding CSG cell. Other technologies including range expansion can be used to offload the traffic from macrocells to picocells.

Accordingly, this chapter is organized as follows. It first describes the deployment scenarios of heterogeneous networks, including the CSG scenario and the pico scenario. It categorizes the radio resource and interference management based on spectrum usage, that is, the multi-carrier case, and the single-carrier case. Then for each category (multi-carrier and single-carrier), several detailed technologies are presented for inter-cell interference management. The suitability of the technologies for the CSG scenario or the pico scenario is discussed along with the technologies. A summary is presented finally.

2.2 Heterogeneous Networks Deployment Scenarios and Interference Management Categories Based on Spectrum Usage

2.2.1 Heterogeneous Network Deployment Scenarios

In heterogeneous networks, cells with different sizes can be used in a hierarchical network deployment. The type and location of the base stations (also called evolved Node B (eNB) in LTE/LTE-Advanced systems) controlling these cells will play a significant role in determining the cost and performance of multi-tier deployments. For example, indoor femtocell deployments using home eNBs (HeNBs) can utilize the existing backhaul thereby significantly lowering the cost of such deployments. With outdoor picocell deployments through pico eNB, the operator will need to provide backhaul capability and manage more critical spectrum reuse challenges. Other deployment models cover indoor enterprise or outdoor campus deployments that may impose different manageability and reliability requirements.

Figure 2.1 shows a typical heterogeneous network with a macrocell/macro eNB, a picocell/pico eNB and a femtocell/HeNB. Furthermore, the femtocell is a CSG cell which only allows its member UEs to access it. Assuming an operating bandwidth of 10 MHz, a typical configuration of the macro eNB is 46 dBm transmission (Tx) power per cell. Assuming a

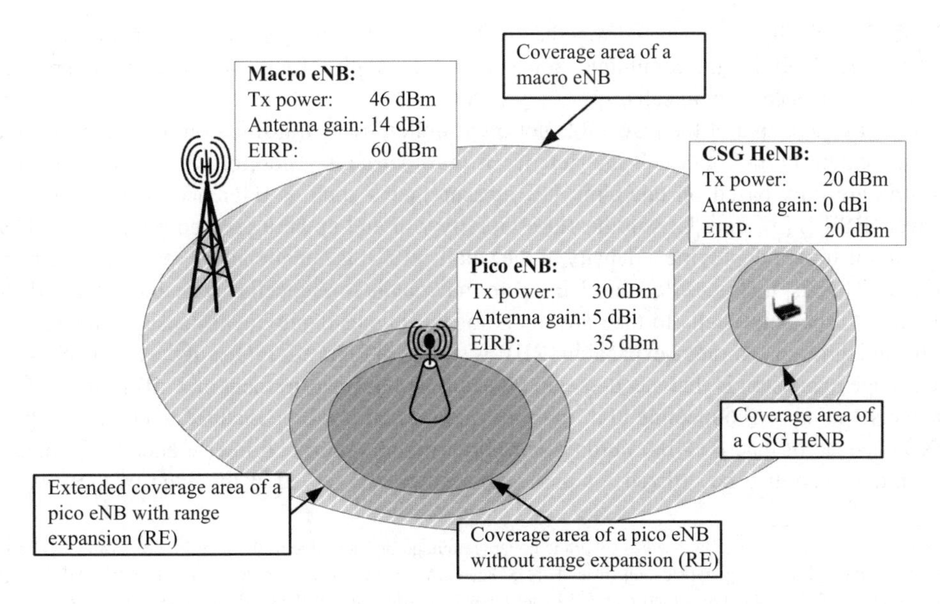

Figure 2.1 A typical heterogeneous network.

14 dBi antenna gain (including feeder loss), the equivalent isotropic radiated power (EIRP) for the macro eNB is 60 dBm. For the heterogeneous network shown in Figure 2.1, the pico eNB only has an EIRP of 35 dBm, which naturally results in a significantly smaller coverage than the macro eNB. On the other hand, the HeNB has the smallest EIRP of only 20 dBm.

2.2.1.1 Closed Subscriber Group (CSG) Scenario

As mentioned, hierarchical networks may be deployed using a variety of lower tier network elements in different locations. The deployment scenario will determine whether access to the lower tier network is available to all users in the network. For instance, in-home femtocells (HeNBs) may only allow access to the users who are part of the household. The terms closed subscriber group (CSG) and open subscription group (OSG) are used to refer to private and public femtocells controlled by home eNBs (HeNBs) respectively. As shown in Figure 2.1, a HeNB can be of a CSG cell.

Due to the fact that the licensed spectrum is limited, CSG cells may be deployed at the same carrier frequency as the macrocell (macro eNB). When a macro UE, which is not a member of the closed subscriber group, is approaching a CSG cell, it cannot handover to this particular CSG cell. On the other hand, the UE may receive strong interference such that it may completely lose the connection with its serving macro eNB. In other words, the CSG cell may generate a coverage hole to the non-member UEs. This problem has to be resolved if a CSG cell is to coexist with other cells at the same carrier frequency.[1]

2.2.1.2 Pico Scenario

Picocells typically are managed together with macrocells by operators. In general, they are open access. Note that the communication technologies applied for picocells can be generalized to other open access small cells.

The coverage area of the picocell is not only limited by its transmission power, but also to a large extent by the inter-cell interference from other cells. Therefore, if the cell selection criteria are only based on downlink UE measurements such as reference symbol received power (RSRP) [3], only UEs in the close vicinity will end up being served by the pico eNB. Due to the higher deployment density of the small cells, it is beneficial to expand the footprint of the picocells, that is, offloading UEs from macrocells to picocells, to enable more UEs to connect to the small cells to take advantage of the higher deployment density. This can be achieved through cell range expansion (RE), as shown in Figure 2.1. One of the approaches for cell range expansion is that a cell-specific bias to the UE measurement of X dB is applied for pico eNB to favour connecting to it. In this way, more UEs will be inclined to connect to pico eNBs instead of macro eNBs. Furthermore, time-domain inter-cell interference coordination techniques can also be utilized for pico users which are served at the edge of the serving

[1] While ICIC techniques can be used to solve the coverage outage problem due to the interference from CSG cells, another solution is to deploy hybrid cells instead of CSG cells. A hybrid cell is accessible as a CSG cell by UEs which are members of the CSG and as a normal cell by all other UEs. Members of the CSG can have a higher priority in access than the non-member UEs.

picocell, for example for traffic offloading from a macrocell to a picocell. This technique will be discussed in details in Section 2.4.1.1.

2.2.2 *Interference Management Categories Based on Spectrum Usage*

Spectrum allocation across multiple tiers is an important aspect of deployment and use of hierarchical architectures. According to the spectrum used, multi-tier cell deployments are possible for the following cases and interference management can be categorized to these cases

- *Multiple carriers case*: The multi-tier cells are deployed on multiple carriers. When multiple carriers are available, choices can be made to enable flexible cell deployment. For example, the macrocell and small cells can be deployed on distinct carriers, or on the same set of carriers while having joint carrier and power assignment/selection to better manage inter-cell interference.
- *Single carrier case*: The multi-tier cells are deployed on a single carrier. This can also be called co-channel deployment.

Large cells such as macrocells and small cells such as femto- or picocells, can be deployed over the distinct set of carriers to avoid inter-tier interference. For cells of the same tier, resource (carrier, power, time, space etc.) allocation can be made for intra-tier interference management. With good design of resource allocation, the coverage hole problem caused by CSG cells to non-member UEs can be resolved. The goal of offloading traffic from macrocells to picocells can be achieved. Details about the multi-carrier interference management can be found in Section 2.3.

Due to the overlying architecture, when macrocell and the overlaid small cells, such as pico- and femtocells, are deployed on the same carrier, interference management across tiers becomes an important design aspect that must be addressed. Advanced interference management solutions for both control channel and data channel are important to support heterogeneous networks, especially for CSG support and pico offloading support. Details about the single-carrier interference management can be found in Section 2.4.

2.3 Multi-carrier Inter-cell Interference Management for Heterogeneous Networks

Control/data channel interference can be mitigated to enable multiple carrier multi-tier deployment, where the cells are deployed on multiple available carriers. Large cells and small cells can be deployed over a distinct set of carriers to avoid interference across various tiers. Large and small cells can also be deployed over the same set of carriers or have carriers overlapped, where resource allocation can be made, to avoid or mitigate the inter-cell interference.

For example, eNBs of multi-tiers may use different carriers based on the measurement of interference from the other eNBs. Different regions of time/frequency resource in multi-carriers can also be allocated to the control channels of each cell for robust transmission. Power control can be jointly used with the resource management in the time or frequency domain, to manage the interferences. In LTE-Advanced, cross-carrier scheduling is introduced for systems deployed with carrier aggregation, which enables dynamic radio resource management across multiple carriers on subframe basis (i.e., 1 millisecond). This scheme can be readily applied in heterogeneous networks in which the macro eNB schedules its users using the control channel region in one component carrier while the HeNB schedules its users via the control channel region in another component carrier.

In this section, we present several multi-carrier inter-cell interference management techniques for heterogeneous networks:

- interference management via carrier partitioning [4,5];
- enhanced carrier reuse with power control [5]; and
- carrier aggregation based inter-cell interference coordination [6].

These techniques are applicable for both downlink and uplink interference management.

Carrier management is particularly challenging for heterogeneous networks due to uncoordinated deployment of cells in the setting. Dynamic carrier management schemes have also been proposed for LTE-Advanced systems (see [7] and references therein). For instance, a system can be deployed with at least one carrier only utilized by the large cells but not the CSG cells (a.k.a. the 'escape carrier') [8].

2.3.1 *Interference Management via Carrier Partitioning*

A simple interference management strategy for heterogeneous networks is to deploy cells of different tiers (that is same power/access class) on different carriers. The carriers involved can be intra-band carriers (contiguous carriers) or inter-band carriers (non-contiguous carriers). Since cells of different tiers are partitioned in frequency, interference between cells of different tiers can be avoided.

An example of carrier allocation by frequency band is illustrated in Figure 2.2, where macrocells are deployed on a frequency band f_1, for example 800 MHz, and picocells (remote radio heads or hot zone cells) are deployed on another frequency band f_2, for example

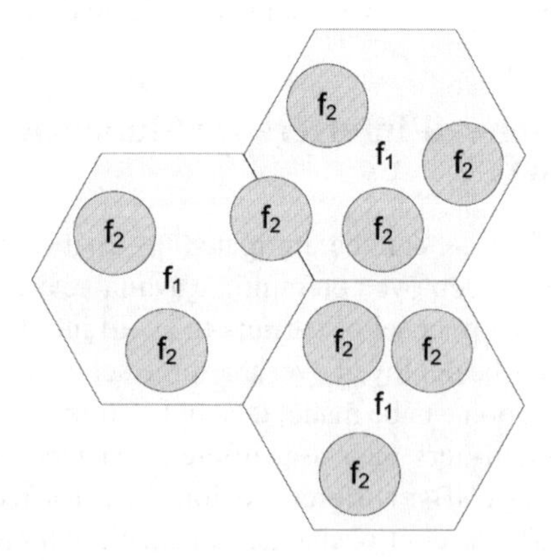

Figure 2.2 Carrier partitioning by frequency band. Macrocells are deployed on frequency f_1 and pico/femtocells are deployed on frequency f_2.

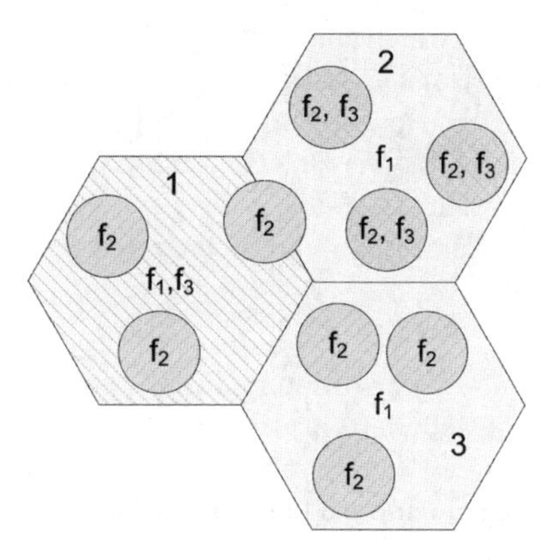

Figure 2.3 Different carrier allocation for different sites.

3.5 GHz. In this example, the picocells are free of interference from the macrocells and vice versa. To offload the traffic from the macrocells to the picocells, the macrocell UEs can be handed over to the picocells if the macrocell UE moves within the coverage area of the picocell.

A similar carrier allocation scheme can be used to avoid interference between open access cells and CSG cells. For example, open access cells are deployed on frequency f_1 and closed subscriber group cells are deployed on frequency f_2. This can avoid the coverage hole problem for the UEs which do not have the membership to the CSG cells.

In LTE-Advanced systems, it is possible to aggregate the carriers from multiple tiers to expand the effective bandwidth for downlink/uplink transmissions. This is termed DL/UL carrier aggregation [4]. In this way, the user's peak/average throughput can be increased. In carrier aggregation, one carrier is configured as the primary component carrier while the remaining carriers can be configured as secondary component carriers. Typically, the carrier with large coverage is configured as the primary component carrier.

An eNB can be allocated with multiple carriers and, depending on the offered traffic, neighbouring eNBs of the same power or access class can have different numbers of carriers allocated [5]. A more sophisticated example of the carrier allocation scheme is shown in Figure 2.3. In this example, macrocell 1 is allocated with carriers 1 (f_1) and 3 (f_3), while macrocell 2 and 3 are allocated with carrier 1 (f_1). The picocell is allocated carrier 2 (f_2), so it is free of interference from both macrocells 1 and 2. However, picocells that are closer to macrocell 2 and further from macrocell 1, can also be allocated carrier 3 without interference from macrocell 1.

Since cells of different tiers do not interfere with each other with strict carrier partitioning, the range of small cells can be extended, for example by increasing the transmission power, so that more users can be offloaded to the small cells when the macrocells are highly loaded. Inter-cell interference management by carrier partitioning involves hard allocation of different carriers to cells of different tiers. Such resource partitioning is typically static or semi-static. To improve carrier reuse among cells of different tiers, power control, fractional frequency reuse (FFR) and frequency/time-domain resource partitioning can be used within the reused carrier [5].

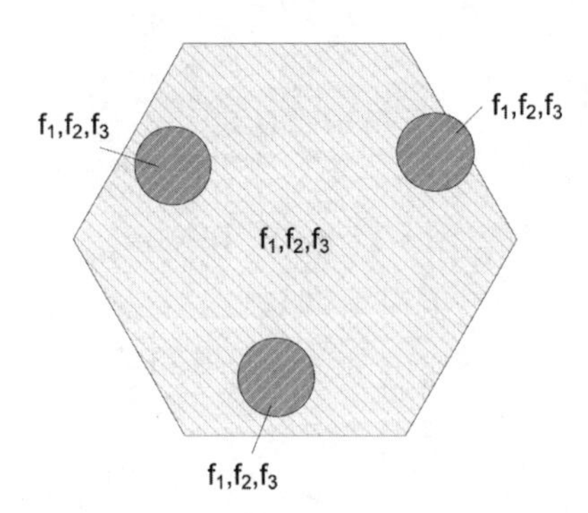

Figure 2.4 Full carrier reuse and full transmission power from the macro.

2.3.2 *Enhanced Carrier Reuse with Power Control*

An alternative to the carrier partitioning scheme detailed in Section 2.3.1 is the full carrier reuse as illustrated in Figure 2.4, where three carriers, f_1, f_2 and f_3, are allocated to both the macrocell and the picocells. In Figure 2.4, the macro and pico transmit on carriers f_1, f_2 and f_3 with full power. However, due to direct inter-cell interference, the coverage of the picocells is limited.

The range of the picocells can be extended with a simple carrier-based power control by the macro [5] as illustrated in Figure 2.5, where the macrocell transmits carriers f_1, f_3 with full power and carrier f_2 with reduced power, whereas the picocell still transmits on carriers f_1, f_2 and f_3 with full power. However, due to the reduced power from the macro on carrier f_2, the coverage range of the picocell on carrier f_2 can be extended.

2.3.3 *Carrier Aggregation Based Inter-cell Interference Coordination*

Carrier aggregation based inter-cell interference coordination (CA-based ICIC) is introduced for the LTE-Advanced system [4] as a means of managing the inter-cell interference on the control channel for heterogeneous networks. This technique assumes that users are capable

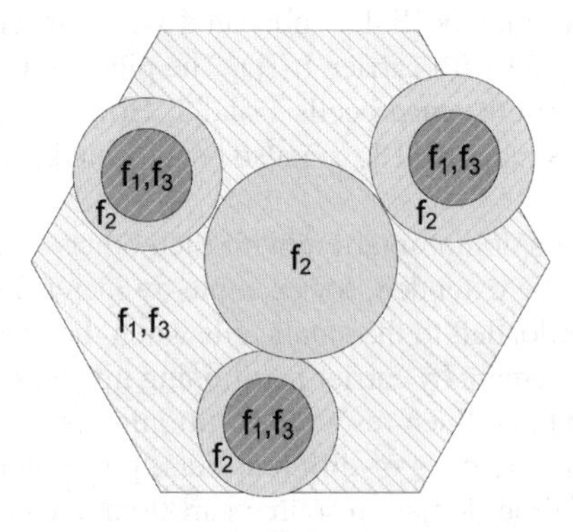

Figure 2.5 Power level of macro carrier f_2 is reduced to enable RE of the pico carrier f_2.

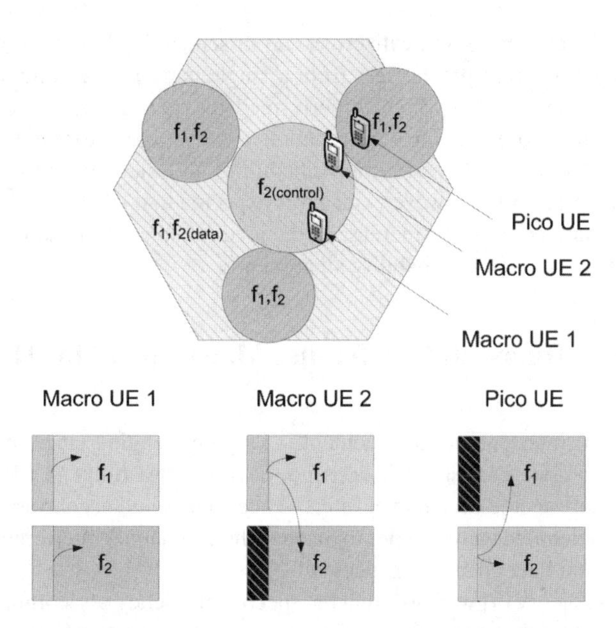

Figure 2.6 Carrier aggregation based ICIC. Cross carrier scheduling is configured for macro UE 2 and pico UE.

of carrier aggregation, that is, capable of simultaneous data reception from multiple carriers. In CA-based ICIC, the set of carriers allocated for an eNB is partitioned into two subsets, where one subset is used for control as well as data transmissions, while the other is for data transmission. Inter-cell interference on the control channel is managed by allocating different carriers used for control channel for cells of different tiers, to avoid inter-tier interference on control channel. Inter-cell interference on the data can be managed using downlink interference coordination techniques.

CA-based ICIC is illustrated in Figure 2.6 where for the macrocells, carrier f_1 is used for control and data transmission, and carrier f_2 is used for data transmission (macro UE 2). For the picocells, carrier f_2 is used for control and data transmission, and carrier f_1 is used for data transmission (pico UE). Note that for UEs that are not capable of carrier aggregation, macro UEs will simply be connected to carrier f_1 and pico UEs to carrier f_2. UEs in idle mode can only camp on carriers with control.

It is noted that carrier 'without control' may not necessarily mean that the control channel of the carrier is completely muted by the network. Instead, the control channel of the carrier may be transmitted with reduced power so as to mitigate the impact of inter-cell interference. This is equivalent to employing a power control technique on the control channels of the heterogeneous nodes. In the example illustrated in Figure 2.6, the data channel of the macro on carrier f_2 is transmitted at full power and the control channel of the macro on carrier f_2 is transmitted at reduced power. Macro UEs close to the centre of the macrocell (macro UE 1) can also receive the control channel on carrier f_2.

To enable scheduling of data on the carrier without control, the control message for data transmission on the carrier can be sent on another carrier, using the so-called cross-carrier scheduling feature. Users configured with cross-carrier scheduling have to monitor and decode the control messages for multiple carriers from a single carrier. The configuration of cross-carrier scheduling is semi-static which is performed by radio resource control (RRC). To indicate the target carrier of the control information, a carrier indicator field is included in the downlink control information. Cross-carrier scheduling will increase the load of the control channel of the scheduling cell. However, since it is expected that UEs with carrier aggregation capability will be relatively few, the increase in the load of the control channel is not expected to degrade the control blocking probability significantly.

2.4 Co-channel Inter-cell Interference Management for Heterogeneous Networks

Co-channel inter-cell interference management is very challenging since the multi-tier cells are deployed on the same carrier, so it does not have the flexibility in the carrier domain. This section first analyses the co-channel interference causes and scenarios in heterogeneous networks, then a few detailed technologies to manage the co-channel interference are presented, on control channel and data channel, respectively.

From information theory [9] we know that the spectral efficiency of a communication system is mainly determined by the signal-to-noise-plus-interference ratio (SINR) at the receiver. In general, a lower SINR corresponds to a lower achievable spectral-efficiency while a higher one corresponds to a higher achievable spectral efficiency. To be specific, the SINR at a receiver can be written as:

$$SINR = \frac{P}{I + N}$$

where P is the received power at the receiver of the transmitted signal, I is the interference power received from other interfering sources and N is the variance of additive white Gaussian noise. Accordingly, low SINR usually happens in either of the two scenarios: *noise-limited* and *interference-limited* [10].

In the noise-limited scenario, the noise-plus-interference ($I + N$) is mainly governed by the noise (N). Therefore, a natural solution to boost the SINR is to increase the received signal power (P). A simple way is to boost the transmission power. More sophisticated methods include utilizing transmit or receive beam-forming and using relay nodes.

On the other hand, in the interference-limited scenario, we have $N \ll I$ and I is on the same order as P ($I \sim P$). In this case, noise power is negligible compared to the interference power and a low SINR is mainly due to the fact that the interference power is large. The interference-limited scenario is the prevailing scenario for cellular networks and the SINR cannot be improved by simply boosting the transmission power from all the cell sites. This is because transmission power boosting may increase the received signal strength, but it will also create stronger inter-cell interference to other cells' mobile stations and hence reduce the corresponding SINRs.

In general, there are many ways to increase the SINR for a target mobile station without boosting the transmission power [11]. One way is to configure heterogeneous networks where low-power nodes such as picocells, femtocells and/or relay nodes are deployed within a macrocell's coverage. In this way, the mobile stations will have better wireless channels linking their destinations since they are closer to the destinations. Furthermore, since low-power nodes are usually effectively only serving mobile stations nearby, cell-splitting gains of heterogeneous networks can be achieved.

However, the heterogeneous networks also introduce additional inter-cell interference since the transmit signals from the low-power nodes will inevitably interfere with macrocell's signals if they are on the same frequency carrier. Since picocells, femtocells and relays are usually using much lower transmission powers than macrocells, the introduced inter-cell interference is usually less severe, but in certain regions, such as the proximity of the small cells, the introduced interference to the macrocell UE can be strong.

As shown in Figure 2.1, despite the relative low EIRP of the HeNB, it still creates a so-called dominance area where UEs served by the macro eNB will experience strong inter-cell interference from the HeNBs. The interference issue will be signified by the deployment of HeNBs with restricted access CSG. Accordingly, the co-channel deployed CSG HeNBs are often said to cause macro eNB coverage holes when no active interference management technique is applied [12]. The interference problem associated with HeNBs is further complicated by the fact that such nodes are deployed in an uncoordinated way by users, resulting in an inherently chaotic interference footprint.

In addition, as shown in Figure 2.1, the pico eNB's coverage area is not only limited by its transmission power, but also to a large extent by the interference experienced from the macro eNB. The service area of the pico eNB can be increased by applying the range expansion (RE) technology. In range expansion, a cell-specific bias to the UE measurement can be applied for pico eNB. In this way, UEs will have a higher chance of connecting to the pico eNB, offloading the traffic in the macro eNB. However, the exact value of the bias applied to the UE measurement may depend on the underlying interference management scheme. For example, in a traditional co-channel scenario without interference management schemes, only small values of the cell bias, say a few dB, should be applied to UEs. Otherwise, the pico UE will experience too much inter-cell interference from the macro eNB. However, when sophisticated interference management schemes are applied, much higher bias can be applied for the UEs to connect to pico eNBs. This will significantly increase the offloading from the macro eNB.

For heterogeneous networks that support HeNB and pico eNB deployment, there are several key interference management technologies for control channel and data channel, as follows.

2.4.1 Control Channel Interference Management

In Releases 8 and 9 (Rel-8/9) LTE systems, inter-cell interference schemes for shared channel (data channel) are designed for homogeneous networks. In principle, similar methods can be used for shared channel in Release 10 (Rel-10) LTE-Advanced heterogeneous networks. However, there are no specific interference management schemes for control channel in Rel-8/9 LTE systems. Furthermore, due to the deployment of low-power nodes within macrocell coverage, inter-cell interference for heterogeneous networks is much more severe than for homogeneous networks. Typical macrocell mobile station SINR distribution under the presence of CSG femtocells can be found in Figure 2.7.

In Figure 2.7, Femto-MS stands for the mobile stations (UEs) that link to the CSG femtocell, while macro-MS stands for the mobile stations (UEs) that are not members of the closed subscriber group and can only connect to the macrocell. In LTE-Advanced systems, a UE

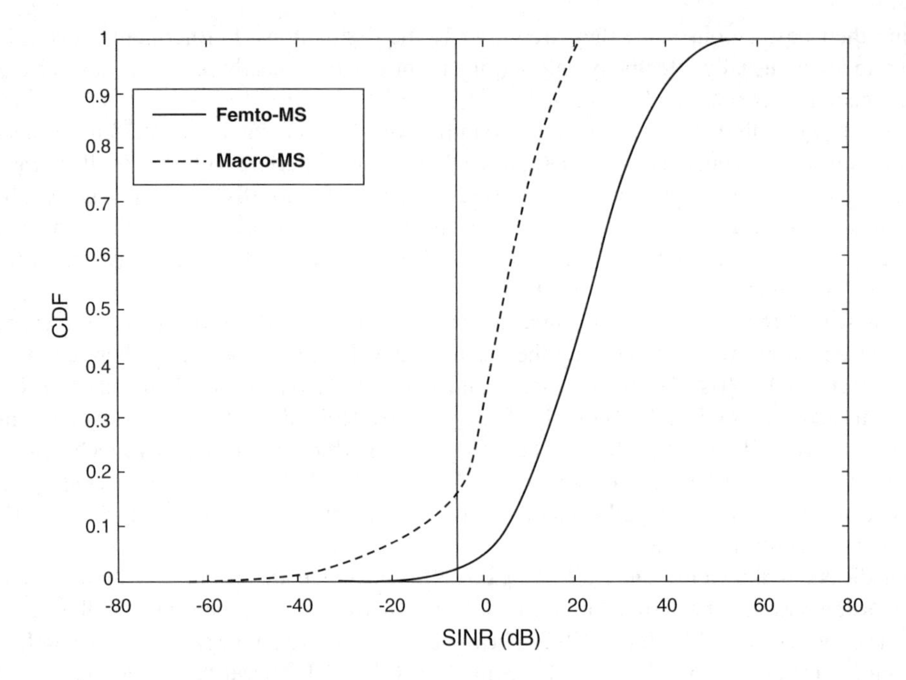

Figure 2.7 UE's SINR without interference mitigation.

cannot decode the control channel if the received SINR of the corresponding channel is below −6 dB. Therefore, we define the outage probability as the ratio of the UEs whose SINRs are below −6 dB to the total number of UEs. For the system performance shown in Figure 2.7, the outage probability of the macro UEs is 16%. This means that CSG femtocells create a large 'dead zone' for the macro UEs that are not members of the corresponding CSGs.

Therefore, one of the major tasks for interference management in heterogeneous networks in LTE-Advanced is to design interference mitigation schemes that provide sufficient protection for the downlink control channels. In Rel-10 LTE-Advanced systems, there are two major methods to coordinate the inter-cell interference for heterogeneous networks: time-domain and power setting solutions. In the following section, we will describe them in detail.

2.4.1.1 Time Domain Coordination

In time domain coordination, the signals from multiple transmission nodes are coordinated in time. In LTE and LTE-Advanced systems, the scheduling granularity in time is subframe where the duration of a subframe is 1 ms. Accordingly, other than the regular subframes, a new type of subframe called an almost blank subframe is introduced in Rel-10 LTE-Advanced for the purpose of inter-cell interference coordination for heterogeneous networks.

In regular subframes, control channels, shared/data channels, as well as reference signals are all transmitted. However, in the almost blank subframe, only the most essential information required for the system to work for legacy Rel-8/9 LTE mobile stations/UEs is transmitted. Thus, during the transmission of almost blank subframes, the signals that are

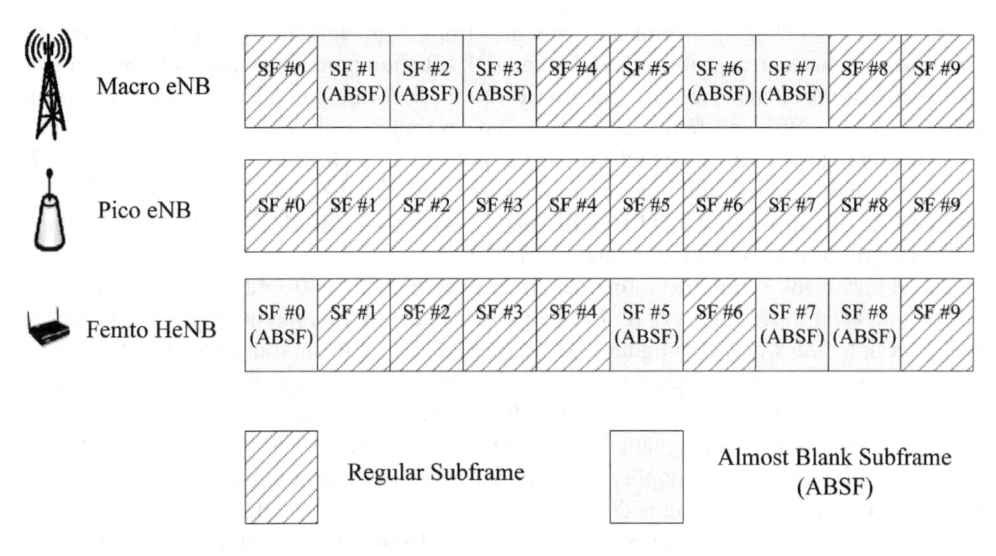

Figure 2.8 Time domain solution ICIC for heterogeneous networks.

mainly transmitted are common reference signals (CRS), as well as other mandatory system information, synchronization channels, and paging channel if configured. Compared to a regular subframe, the average transmission power from an almost blank subframe is therefore often reduced by approximately 10 dB, assuming that base stations use two transmit antennas.

The basic principle of time-domain inter-cell interference coordination (ICIC) is illustrated in Figure 2.8 for a scenario with co-channel deployment of macrocell, picocell and femtocell HeNBs. In the figure, 10 subframes forming a frame (10 ms) are listed where SF #0 stands for subframe number 0 and ABSF stands for almost blank subframes.

The time domain inter-cell interference coordination relies on accurate time synchronization on subframe resolution among all the nodes within the same geographical area. Furthermore, it is often assumed that time-domain ICIC for heterogeneous networks is operated together with advanced UE receivers that are capable of further suppression of the residual interference from almost blank subframes. During subframes where femto CSG HeNBs are using ABSF (for example, SF #0, SF #5 and SF #8 in Figure 2.8), macro-UEs in the close vicinity can therefore still be served by the macrocell eNB in those subframes (obtaining control information), which would otherwise experience too much inter-cell interference during the regular subframes from the femto CSG HeNBs. Similarly, during subframes where the macrocell eNB is using almost blank subframes (SF #1–3, SF #6 and SF #7 in Figure 2.8), there is less interference generated for the users served by picocells and HeNBs. This implies that the picocells and HeNBs are capable of serving UEs from a larger geographical area during those subframes. This essentially means that utilizing almost blank subframes at macrocells makes it possible to increase the offload of the traffic to the low-power nodes. It can be seen that the number of almost blank subframes in a frame can be used to tradeoff the system performance of a macrocell and that of a low-power node. Therefore, the number of subframes configured as almost blank subframes for a cell needs to be carefully chosen to maximize the overall system performance.

In order to obtain performance benefits from time-domain inter-cell interference coordination, the central packet scheduler and link adaption functionality need to be aware of the applied almost blank subframe patterns at various transmitting nodes including macrocell eNBs, picocell eNBs and femtocell HeNBs. For example, a picocell should only schedule users that are close to the macrocell (subject to potentially high interference from macrocell) during the almost blank subframes from the macrocell. On the other hand, macro-UEs that are close to a non-member CSG HeNB should only be scheduled during the almost blank subframes of the corresponding femto CSG HeNB.

The almost blank subframe muting pattern is periodic with a 40-subframe period for FDD mode. For TDD mode, the period of the almost blank subframe muting pattern depends on the exact uplink/downlink configuration. The periodicity of 40 subframes for FDD has been selected to maximize the protection of common channels, including uplink hybrid automatic repeat request (HARQ) performance. The almost blank subframe muting patterns are configured semi-statically and signalled between eNBs over the X2 interface or via the HeNB gateway if X2 interface is not applicable. Since the period of the almost blank subframe muting pattern is 40 ms, X2 signalling is done by means of bit maps of length 40. The exact pattern can be, in principle, configured freely by the network. However, to maximize the performance benefits of the time-domain solution, transmitting nodes of the same type in a given local area are recommended to use the same almost blank subframe muting pattern. That is, clusters of HeNBs within the same geographical area are suggested to be configured with either the same or at least overlapping ABSF muting patterns. Due to the architecture characteristics of HeNBs, it is assumed in Rel-10 LTE-Advanced systems that the ABSF muting pattern for such nodes is statically configured from the network management system.

For other types of transmitting nodes such as macrocell and picocell, LTE-Advanced systems support mechanisms for distributed dynamic configuration of ABSF muting patterns that seek to maximize the overall system performance. For example, as illustrated in Figure 2.8, for the case of macro-pico scenario, it is the macrocell eNB that is expected to use the almost blank subframes. Usually, the macrocell is acting as the master, deciding which subframes it wants to configure as almost blank subframes. In Rel-10 LTE-Advanced system, the macro eNB has various sources of information for it to decide the number of almost blank subframes and its associated pattern. For example, the macrocell eNB can decide the ABS pattern based on the quality of service requirements of its own serving UEs. Furthermore, the macrocell eNB can decide the ABSF pattern collaboratively with the picocell based on the load information it receives from the picocell through the X2 interface. In the load information, the picocell and macrocell could share their intended transmitting power level and coordinate on the almost blank subframe pattern to be used. In this way, the time-domain inter-cell interference coordination scheme could perform well for heterogeneous networks.

The ABSF patterns configured by the serving cell and the neighbouring cells around the UE create a complex and dynamic interference observed by the UE that changes from subframe to subframe. This can have a significant impact on radio resource management (RRM)/radio link monitoring (RLM)/channel state information (CSI) measurement and procedure performed by the UE. For example, if a pico UE performs RLM measurement of the picocell for subframes which include the non-ABSF subframes which are heavily interfered by the macrocell, radio link failure may be declared by the pico UE due to the significant interference observed by the UE even though the UE can still be connected with the picocell using the resources protected by the ABSF pattern of the macro. Similarly, the accuracy of RRM and CSI measurement performed by the UE can be compromised if the ABSF configuration by the network is not taken into account. To resolve this problem, the UE can be configured with 'measurement resource restriction' patterns for RRM/RLM/CSI measurement purposes. In LTE-Advanced, there are three types of measurement restriction patterns that can be configured to the UE:

- Pattern 1: A single RRM/RLM measurement resource restriction for the serving cell,
- Pattern 2: A single RRM measurement resource restriction for an indicated list of neighbour cells operating in the same carrier frequency as the serving cell, and
- Pattern 3: Resource restriction for CSI measurement of the serving cell.

2.4.1.2 Power Setting Solution

Power setting solution is specifically for the scenario of macro-femto network. The basic principle of the power setting solution is to reduce the transmission power of CSG femto HeNB so that the interference from CSG femto HeNB to macro-UE can also be reduced.

In general, the purpose of the power setting at the femto CSG is twofold:

- Mitigate the experienced interference of macrocell UEs
  $\Rightarrow$ Reduce outage probability of macrocell UE (MUE)
- Maintain the coverage and throughput of HeNB UEs
  $\Rightarrow$ Maintain throughput and outage probability of HeNB UE (HUE)

Therefore, in this section we will show the performance of various power setting/control schemes for mitigating MUE interference and maintaining HUE throughput. The corresponding specification and signalling support will also be discussed.

Related Power Setting Schemes

In general, depending on whether there is information exchange between the victim MUE and HeNB, there are two types of the power setting (PS) schemes: with information exchange and without information exchange.

- In PS type 1, there is no information exchange between the victim MUE and the interfering HeNB. The HeNB controls the power setting based only on its own measurement from the macro eNB and the measurement report from its serving HUEs. Therefore, there is no additional interface that needs to be defined between the victim MUE and the interfering HeNB.
- In PS type 2, there is information exchange between the victim MUE and the interfering HeNB. Accordingly, the HeNB controls the power setting not only based on its own measurement from the macro eNB and the measurement report from its serving HUEs but also based on the measurement report from the victim MUE. Therefore, an additional interface has to be in place for this kind of power setting scheme.

There are many proposed power setting schemes in these two categories.

PS type 1: PS Schemes without Information Exchange (MUE → HeNB)
The most prevailing schemes in this category is:

Power setting method 1 (**PS1**): $P_{tx} = median\ (P_{max},\ P_{min},\ \alpha P_M + \beta)$ [13],

where P_{tx} is the power setting of the CSG femtocell, P_{max} is the maximum, P_{min} is the minimum allowed power setting value, P_M denotes the femtocell's received power from the strongest macrocell and α and β are predefined system parameters for the corresponding CSG femtocell. It can be seen from PS1 that the transmission power of a CSG femtocell depends on the relative distance to its nearest macrocell. This is because P_M is a monotonic decreasing function of the distance between the femtocell and its nearest macrocell. PS1 suggests that when a femtocell is further away from a macrocell, it should use lower transmission power. This is because in the vicinity of the corresponding CSG femtocell, the received signal strengths of those non-member macro UEs are usually low. This power setting method can efficiently achieve the goal of mitigating inter-cell interference, but it does not help to improve the femtocell's coverage and throughput.

On the other hand, we can construct a power setting scheme based on maintaining the coverage and throughput requirement of home UEs. To be specific, power setting scheme could be constructed as follows:

Power Setting Method 2 (**PS2**): $P_{tx} = median\ (P_{max},\ P_{min},\ P_{HUE\_received} + x + PL)$ [14],

where $P_{HUE\_received}$ is the received interference plus noise power at the HUE, x is the target SINR at the HUE and PL is the reported path loss between the HUE and the corresponding HeNB. In this method, the transmission power of the HeNB is set based upon the received SINR at the HUE.

From PS1, we can see that controlling the transmission power of HeNB based on the received power from the macro eNB to HeNB (P_M) can efficiently achieve the first goal (mitigating interference from HeNB to non-member macro UE), but it does not help the second goal (improving home UE's throughput). On the other hand, we know that schemes similar to water-filling may help us increase the system throughput [9].

Therefore, we can linearly combine the following two terms in different ways to design the power setting algorithms:

1. P_M, the received power from the eNB to HeNB,
2. P_H, the received power from the HeNB to HUE.

On top of these parameters, we also add the path loss (PL) between the HUE and the corresponding HeNB as an offset to the power setting equation.

Power Setting Method 3 (**PS3**): $P_{tx} = median\ (P_{max},\ P_{min},\ \gamma P_M + (1-\gamma)P_H + PL)$ [15].

In PS3, γ is a scalar between 0 and 1 to balance the two effects:

- The first effect relies on P_M which can help to achieve the first goal.
- The second effect relies on P_H which can help to achieve the second goal.

Table 2.1 Parameters for performance evaluation

	PS1		PS2	PS3	
Parameter	α	β	X	α	β
Value	1	70 dB	−4 dB	1	20 dB

Actually, performing power setting based on P_H has the flavour of performing water-filling in the sense that HeNB will transmit higher power for high geometry HUEs while transmitting lower power for low geometry HUEs.

PS type 2: PS Schemes with Information Exchange (MUE → HeNB)

In [16], a power setting scheme is proposed to achieve very good performance. To be specific, the power setting scheme is described as follows:

$$\text{Power Setting Method 4 } (\textbf{PS4}):\ P_{tx} = median\,(P_{max},\ P_{min},\ \alpha P_{SINR} + \beta)\ [16],$$

where P_{SINR} is the SINR between the macro eNB → MUE and closest HeNB → MUE.

Evaluation of the Power Setting Schemes

We can evaluate the MUE and HUE performance of the related power setting schemes. The corresponding system parameters for evaluating different power setting schemes are listed in Table 2.1.

Accordingly, the outage and throughput analysis for MUE and HUE can be found in Table 2.2.

From Table 2.2, it can be seen that there is a clear tradeoff between the MUE performance and the HUE performance. To be specific, compared with baseline scheme where no additional power setting equations are supported (each HeNB transmits at P_{max}), all the power setting schemes reduce the average throughput of the HUE.

We also note the following:

- PS1 achieves a good balance of all the performance metrics (outage of HUE, outage of MUE and average HUE throughput).
- In PS2, since HeNB is transmitting power according to some target SINR received at the HUE, it minimizes the outage for the HUE and MUE at the expense of the average HUE

Table 2.2 Performance evaluation of current power settings

		PS type 1			PS type 2
	No power setting	PS1	PS2	PS3	PS4
Outage for HUE (%)	1.8	7.29	0.25	5.84	5.25
Outage for MUE (%)	15.8	7.20	4.83	5.63	4.72
Average HUE throughput (bps/Hz)	4.17	2.47	1.68	2.07	3.77

throughput. In this power setting scheme, HeNB is extremely conservative to only barely maintain the links between HeNB and HUEs.

- PS3 outperforms PS1 in outage of HUE and MUE with similar average HUE throughput. Also PS3 achieves a very good performance tradeoff of the three performance metrics.
- PS4 outperforms PS1 in all aspects at the expense of additional measurement reporting from MUE to the interfering HeNB.

In Rel-10 LTE-Advanced systems, it has been decided that no information should be exchanged between the victim MUE and HeNBs for simplicity, so power setting type 2 is not supported. However, it can be seen that once there is a communication link between these two entities, the performance can be improved and smart algorithms can be implemented to improve the overall system efficiency.

2.4.2 Data Channel Interference Management

The interference management methods introduced in Section 2.4.1 can also be used to mitigate data channel interference. In this section, we will introduce additional interference management methods specific for the data channel in heterogeneous networks.

Unlike the control information which is broadcast to all the UEs within the coverage area over the whole bandwidth, the data is usually UE-specific, occupying a subset of the whole system bandwidth. Accordingly, the data for different UEs is usually multiplexed in the frequency domain. This gives us a different domain for operating interference coordination. To be specific, data channel interference could be mitigated in the frequency domain in a similar to the time domain.

In the downlink of LTE/LTE-Advanced systems, macrocells and picocells are connected through the X2 interface. Accordingly, they could coordinate their transmission power in different subbands to mitigate the inter-cell interference. To be specific, a bitmap called relative narrowband TX power indication, RNTP(n_{PRB}), is defined to exchange between cells through the X2 interface. The determination of reported $RNTP(n_{PRB})$ is defined as:

$$RNTP\left(n_{PRB}\right)=\begin{cases} 0, & \text{if } \dfrac{E_A\left(n_{PRB}\right)}{E_{\max\_nom}^{(p)}} \leq RNTP_{threshold} \\ 1, & \text{if no promise about the upper limit of } \dfrac{E_A\left(n_{PRB}\right)}{E_{\max\_nom}^{(p)}} \text{ is made} \end{cases}$$

where $E_A\left(n_{PRB}\right)$ is the maximum intended energy per resource element (EPRE) of UE-specific data channel resource elements in OFDM symbols not containing reference signals in the considered future time interval; n_{PRB} is the physical resource block number in the frequency domain; $RNTP_{threshold}$ takes on one of the following values [dB]:

$$RNTP_{threshold} \in \{-\infty, -11, -10, -9, -8, -7, -6, -5, -4, -3, -2, -1, 0, +1, +2, +3\}$$

An exemplary RNTP bitmap can be seen in Figure 2.9. In this example, there are altogether 12 physical resource blocks in the frequency domain and a cell communicates this RNTP

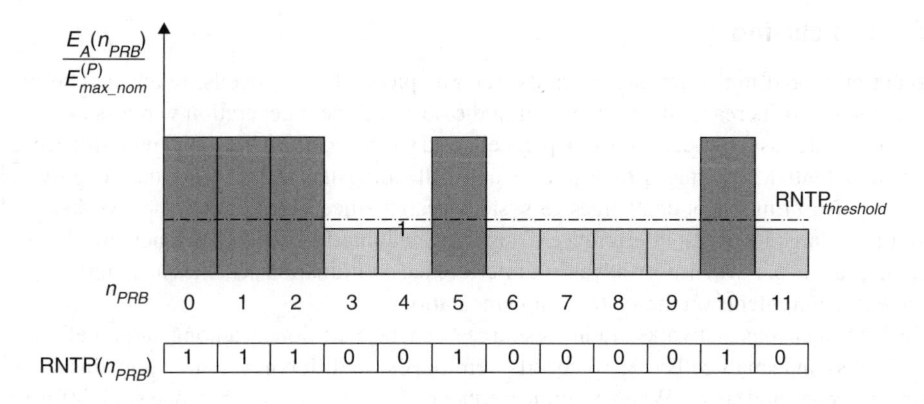

Figure 2.9 RNTP bitmap for downlink inter-cell interference coordination.

bitmap together with the four-bit $\text{RNTP}_{\text{threshold}}$ to another cell for the purpose of inter-cell interference coordination. Upon reception of the information, a cell may have a better understanding of the downlink inter-cell interference and can schedule its UEs to avoid the high interference in the frequency domain.

Similarly, the data channel interference can also be mitigated in the spatial domain. This is because the data traffic can be transmitted to different UEs by applying different precoders at the transmitter using multiple antennas. In this way, the data traffic for different UEs can be multiplexed in the spatial domain using the same time and frequency resource.

In homogeneous networks, coordinated multi-point (CoMP) transmission is going to be supported in Rel-11 LTE-Advanced systems. Similarly, this concept can be used in heterogeneous network for beyond LTE-Advanced systems. In CoMP, multiple cells cooperate to serve multiple UEs simultaneously to combat inter-cell interference [17]. Depending on whether the targeted UE will receive data from multiple cells, CoMP is classified into two categories: coordinated beam-forming/coordinated scheduling and joint transmission. These methods could potentially bring large gains for both the cell-average spectral efficiency and the cell-edge spectral efficiency.

In coordinated beam-forming, different cells coordinate to use different transmit beamforming vectors taking the inter-cell interference into account. By doing so, the received signal power of a target UE would reduce, but the inter-cell interference would also be reduced. For cell-edge UEs where the low performance is mainly caused by large inter-cell interference, the received SINR may, in general, increase. In coordinated scheduling, multiple cells adjust their scheduling decision to reduce the inter-cell interference by avoiding the beam collision from coordinated cells. The gain of coordinated scheduling is essentially multi-user scheduling gain. On the other hand, in joint transmission, multiple cells jointly transmit the same data streams to target UEs and, in this way, the interference becomes a useful signal and can be coherently added over the air. It is expected that the joint transmission could potentially bring the largest performance gain. However, it also imposes high demands for the system. For example, joint transmission requires that the data streams be available at all the coordinated cells, which largely increases the backhaul traffic and load.

2.5 Conclusion

Having more and multi-tier base stations, such as picocells, femtocells, relays and so on in cellular systems increases the momentum in the design of next generation wireless networks. Small cells such as femtocells, hotspot picocells and so on become a hot topic, partially because of their potential advantages of low cost and offloading the traffic. This new deployment scenario also brings new challenges to system design since heterogeneous networks would suffer from large inter-cell interference if sophisticated interference management mechanisms were not in place. Furthermore advanced interference coordination schemes also impose more stringent requirements on the system implementations.

In heterogeneous networks, radio resources can be partitioned among large cells (e.g., macrocells) and small cells (e.g., picocells, femtocells) in different resource domains such as time, frequency and space. When multiple frequency carriers are available, one straightforward way for radio resource management in heterogeneous networks is to use different frequency carriers for macro eNBs and HeNBs. Otherwise, different time slots can be assigned to macro eNBs and HeNBs that use the same frequency carriers to mitigate interference. Advanced coordinated beam forming using multiple antennas can be used in the spatial domain for interference cancellation. In addition, power control and interference cancellation techniques can also be applied in a variety of scenarios in heterogeneous network deployment to mitigate interference and improve the quality of service of the wireless links.

In both single-carrier and multi-carrier cases, advanced interference mitigation scheme needs to be implemented to achieve cell throughput enhancement. Without those techniques, system performance of heterogeneous networks could even be degraded compared to homogeneous networks.

Note that effective radio resource management among large cells and small cells also comes with certain requirements. For example, in order to support cross-carrier scheduling or time-domain ICIC, symbol level synchronization among large cells and small cells is required. The resource partitioning among large cells and small cells in time and frequency in heterogeneous networks creates different interference scenarios across carriers and subframes. As a result, a UE may need to monitor the channel quality not only in different carriers, but also in different sets of subframes. All these impose additional implementation cost at the UE.

References

1. White_paper_c11-481360, 'Cisco visual networking index: Forecast and methodology', June 2011.
2. ITU, Report M.2135: 'Guidelines for evaluation of radio interface technologies for IMT-Advanced', 2008.
3. 3GPP TS 36.214 V10.1.0 (2011–03).
4. 3GPP TS 36.300 V10.5.0 (2011–09).
5. R1-092062, Carrier Aggregation in Heterogenous Networks, Qualcomm, May 2009.
6. 3GPP TR 36.814 V9.0.0 (2010–03).
7. R1-093898, Mechanism for Cell Specific Component Carrier Usage, Nokia Siemens Networks, Nokia, Qualcomm Europe, China Unicom, October 2009.
8. R1-101924, Macro+HeNB performance with escape carrier or dynamic carrier selection, Nokia Siemens Networks, Nokia, April 2010.
9. T. M. Cover and J. A. Thomas, *Elements of Information Theory*. Wiley-Interscience, 1991.
10. D. Tse and P. Viswanath, *Fundamentals of Wireless Communication*. Cambridge University Press, 2005.
11. L. Liu, J. Zhang, Y. Yi, H. Li and J. Zhang, 'Combating Interference: MU-MIMO, CoMP, and HetNet', Journal of Communications, 2012 (Invited).

12. López-Pérez, *et al.*, 'OFDMA Femtocells: A Roadmap on Interference Avoidance', in *IEEE Communications Magazine*, pp. 41–48, September 2009.
13. 3GPP TR 36.921 V9.0.0, Home eNodeB (HeNB) Radio Frequency (RF) requirements analysis.
14. R1-102671, 'Evaluation of R8/9 Power Control and Enhancements for DL Interference Coordination in Macro-Femto', CATT.
15. R1-104626, 'Discussion on DL Power Setting for Heterogeneous Networks', Samsung.
16. R1-104102, 'Performance Evaluation for Power Control Based on Femto Deployment', Alcatel-Lucent Shanghai Bell, Alcatel-Lucent.
17. L. Liu, J. Zhang, J.-C. Yu and J. Lee, 'Inter-cell interference coordination through limited feedback', *International Journal of Digital Multimedia Broadcasting*, vol. 2010, February 2010.

3

Capacity and Coverage Enhancement in Heterogeneous Networks

Shilpa Talwar,[1] Shu-ping Yeh,[1] Nageen Himayat,[1] Kerstin Johnsson,[1] Geng Wu,[1] and Rose Qingyang Hu[2]
[1] *Intel Corporation, USA*
[2] *Utah State University, USA*

The exponential growth in mobile traffic due the introduction of sophisticated client devices (e.g., smartphones, notebooks, tablets) into the market is driving up traffic load and straining the capacity of current wireless networks. At the same time, network providers face the challenge of meeting this demand with relatively flat revenues per bit. Radical innovations in mobile broadband system design will be required to deliver the projected capacity demand in a cost-effective manner. Heterogeneous multi-tier network architecture is one of the most promising paradigms to offer a significant areal capacity gain and indoor coverage improvement at low cost. We describe a heterogeneous network architecture composed of a hierarchy of tiers (that is footprints), multiple types of infrastructure elements and one or more radio access technologies. Together, these network elements address a multitude of capacity, coverage and quality of service concerns. Of particular importance are the scenarios of single-frequency multi-tier deployment and multi-radio multi-tier deployment, since they offer the highest capacity at a relatively low cost. As a result, we focus on these use cases, outlining the challenges and presenting some promising interference mitigation solutions. This chapter also describes industry trends, standardization efforts and future research directions for this rich area of investigation.

Heterogeneous Cellular Networks, First Edition. Edited by Rose Qingyang Hu and Yi Qian.

3.1 Introduction

The growing adoption of rich multimedia devices and services has resulted in an explosive demand for wireless data capacity. Recent studies indicate that wireless traffic has grown at a rate that is approximately an order of magnitude higher than the spectral efficiency (that is capacity) enhancements needed to meet demand. This gap will only increase as the number of devices per person increases and as newer devices enable the consumption of richer multimedia content [1–4]. In fact, it is estimated that a smartphone consumes 30 times more capacity than a simple cell phone, while connected computing devices, such as laptops and tablets, can consume 450 times more capacity. As laptop and netbook adoption continues to rise, these devices will further fuel the mobile broadband data growth mega-trend. Despite this increase in traffic, operators are facing flattening revenues per bit due to largely flat-rate data-centric plans. Hence, it is imperative that operators find ways to add capacity at a significantly lower cost per bit.

With traditional network design strategies, mobile service providers essentially have three primary capacity expansion tools, which include cell splitting, advanced air interface and newer spectrum. Increasing macrocell site density can be costly, since each cell split potentially requires regulatory approval, new sites and civil work. Technology upgrade to OFDMA-based 4G technologies WiMAX and LTE can provide a three- to fourfold increase in capacity. However, 4G air interfaces have been optimized to the point that their spectral efficiency is close to system capacity limits. Acquiring new spectrum can be expensive, limited by availability and subject to government regulatory timelines.

As a result, the focus of the wireless industry is shifting from solely increasing spectral efficiency to improving network efficiency. Here, new heterogeneous networks (HetNets) are a promising paradigm that can cost-effectively improve system coverage and capacity [5–8]. Recent work in the area of heterogeneous networks has primarily focused on multi-tier techniques that offload traffic to smaller cells as much as possible. While the gains from this approach are promising, they represent only a starting point. We envision heterogeneous networks playing a central role in the evolution of mobile wireless broadband, and serving as a platform and enabler for disruptive technology innovations.

The primary benefit of heterogeneous networks is that they adapt the infrastructure elements (both in power/range and cost) to the environment (see Figure 3.1). For example, the cheapest, lowest-power device in the network, which is the client, can cooperate with other clients to *locally* improve access and throughput to higher layers. On the other hand, pricey, high-powered macro base stations (MBSs), provide blanket coverage and seamless mobility across large geographic areas. In between these extremes, pico base stations (PBSs) and relay stations (RSs) enhance coverage and capacity with smaller cells of radius on the order of several hundred metres; while femto access points (FAPs) shrink cell radius to below 100 metres and bring coverage indoors. By shrinking the transmission range and increasing the spatial reuse of spectrum, low-power network elements (FAPs, PBSs, RSs and clients) can achieve significant improvement in coverage and areal capacity. To boost capacity further, heterogeneous deployments are well-suited to leverage spectrum across different radio access technologies, such as WLAN, based on their multi-tier nature. Integrating WLAN (e.g., WiFi) in a cellular network allows operators to offload significant amounts of traffic to unlicensed bands [9, 10]. In order to exploit this alternate spectrum, both infrastructure and consumer devices need to integrate multiple radio technologies, which is becoming a trend.

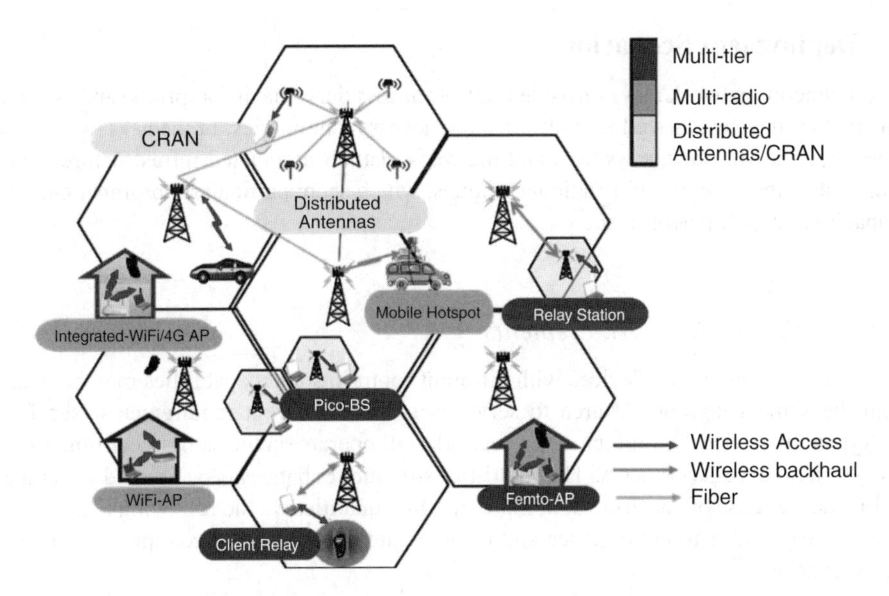

Figure 3.1 Heterogeneous network architecture.

In addition to the capacity increase, there is a cost advantage to heterogeneous networks. Tactically, addressing coverage and capacity needs with inexpensive network elements saves the provider a significant amount of money compared to buying more spectrum or deploying denser macro/microcells. Not only are these elements cheaper, but in many cases, site acquisition and backhaul costs are reduced if not eliminated altogether [11,12]. In addition to all these, the judicious use of free, unlicensed spectrum lowers the cost further.

In this chapter, we provide an overview of the heterogeneous network architecture and associated protocol solutions. We then describe the problems related to the most difficult, yet the most cost-effective means to add capacity with multi-tier deployments, which is to maximize the utilization of scarce spectrum through reuse of the same carrier across the multiple tiers in the network. We show that the promised multi-tier capacity and coverage gains with this deployment can only be realized with proper interference management techniques that mitigate cross-tier interference. Furthermore, we show that adding multi-radio capability to infrastructure and client devices can further boost capacity of cellular networks by taking advantage of unlicensed spectrum.

This chapter is organized as follows. In Section 3.2, we outline the infrastructure elements and deployment scenarios that comprise the heterogeneous network architecture. In Section 3.3, we describe the interference issues related to single-carrier multi-tier networks and promising interference mitigation (IM) solutions. In Section 3.4, we show performance gains from multi-radio heterogeneous deployments. Section 3.5 summarizes the status of standards efforts on heterogeneous networks as well as highlighting potential research directions. Finally, we conclude the chapter in Section 3.6.

3.2 Deployment Scenarios

A heterogeneous network may consist of multiple access tiers (that is footprints) and associated infrastructure devices, as well as multiple radio access technologies. In the next two sections, we describe the various access tiers and the roles of their associated infrastructure devices. We then describe several multi-radio techniques, which seem particularly promising in terms of capacity and QoS performance.

3.2.1 Multi-tier Network Elements

In future mobile networks, devices with different footprints and capabilities may be overlaid within the same geographical area to serve users with different requirements (see Figure 3.1). Typically, devices within the same network will operate on the same spectrum, utilizing the same radio access protocols. The multi-tier structure enhances capacity and coverage by enabling dense reuse of spectrum and improving link qualities. In the following list, we define the role of each infrastructure device and its associated tier and give examples of applicable usage scenarios.

- *Macrocells/microcells*: In current cellular networks, macro/micro base stations (MBSs) are deployed for wide area coverage. The MBS footprint varies depending on traffic demand. Macrocells with more than 500 m site-to-site distance cover rural or suburban areas, while urban areas require microcells with smaller cell radii. In heterogeneous networks, macro- and microcells provide essential wide-area coverage. They also have the advantage of being able to support high mobility users without requiring frequent handovers.
- *Picocells*: Picocells serve cells with inter-site distance on the order of 100–300 m. Pico base stations (PBSs) can be deployed as hotspots in capacity-starved locations such as shopping malls, airports and stadiums. PBSs are basically simplified MBSs with reduced power and cost. They are typically part of the operator's public infrastructure and, therefore, open/accessible to all network customers. The deployment of PBSs is usually carefully planned by the operators.
- *Relays*: Relay stations serve areas similar in size to those of PBSs. They provide coverage extension and throughput enhancement to macro/microcells by forwarding traffic between BSs and mobile stations. RSs communicate wirelessly to the BSs, so landline resources are not required. However, this reduces the amount of spectrum available for access. RSs are often deployed in areas traditionally covered by PBSs when wired backhaul is unavailable or difficult to access.
- *Femtocells*: Femto access points serve small areas on the order of 10–50 m (e.g., in home or apartment). Unlike MBSs or PBSs, which connect to the network through operator-owned backhaul, FAPs reduce infrastructure costs by utilizing existing residential backhaul such as DSL or cable. In addition, FAPs are usually privately owned and more efficiently deployed by a user based on his needs, for example where there is poor indoor coverage. FAP access is often restricted to a specific subscriber group (closed subscriber group), and when an FAP shares the same spectrum with other network tiers, this restriction may cause significant interference to subscribers that are in the FAP's vicinity but not members of its access group.

- *Client relays:* Client relays create yet another tier in the heterogeneous network architecture. This tier is between clients and, therefore, the only tier that is not anchored by infrastructure elements. This tier comprises very short range links on the order of 10–20 m. Client cooperation (CC) utilizes the good link between a cooperating-client and a BS (could be macro, micro or pico) to buttress the link of an end-client whose link to that BS is poor, thereby increasing its probability of successful transmission. In effect, CC improves the virtual link quality of users in poor locations (that is cell-edge users), resulting in a significant reduction in the amount of channel resources and battery power they consume and in the amount of interference they cause to other cells. Studies show that CC can improve average network throughput by anywhere from 80–200% [13].
- *Distributed antennas:* In addition to this range of infrastructure devices and capabilities, there are new radio technologies that create new tiers in the heterogeneous network architecture. A primary example of this is the distributed antenna system (DAS) [14]. DAS spatially separates the antennas of a conventional BS and connects them to a common processing unit via a fast transport medium such as optical fibre. It effectively creates a macrocell out of a collection of smaller cells. DAS allows the operator to replace its high-power centralized antennas with multiple low-power antennas that cover the same cell area. The advantage of DAS is that less power is needed, since (near) line-of-sight channels are often present, resulting in better link qualities and improved coverage and capacity.

China Mobile's C-RAN [15] is a promising evolution of the DAS architecture. Through optical fibre connections, a C-RAN base station aggregates the baseband signals from several hundred sectors/cells to a server platform for centralized signal processing. This architecture creates a super base station with distributed antennas that can support multiple protocols and dynamically allocate signal processing resources to accommodate the varying traffic load within its coverage area.

3.2.2 Multi-radio Techniques

Given that an increasing number of clients in the network are equipped with multiple radio interfaces (such as WiFi and 4G), an operator can exploit unlicensed bands to add low-cost capacity and improve coverage and QoS of existing cellular networks. In the following list, we present some promising multi-radio technologies.

- *WiFi offload*: This is a technique where the operator judiciously offloads low-priority and/or delay-tolerant traffic to WiFi access points to add capacity at a much lower cost without compromising QoS requirements. It requires that operators have some level of control of the WiFi AP.
- *Virtual carrier*: This technique calls for the synergistic use of licensed (e.g., 3G/4G) and unlicensed (e.g., WiFi) spectrum to improve network capacity and user QoS. Different levels of synergy are possible, depending on whether the multi-protocol client connects to distinct licensed and unlicensed access points or to an access point with integrated licensed/unlicensed capabilities. New network devices, such as the integrated WiFi/4G femto AP depicted in Figure 3.1, can implement more dynamic use of the spectrum available from licensed and unlicensed bands. Some of the virtual carrier techniques available to multi-radio

clients are carrier aggregation, interference avoidance, diversity or redundancy transmission and QoS/load balancing. As an example, more details on multi-radio interworking between 802.16 and 802.11 are available in [16].

- *Mobile hotspots*: A mobile hotspot is a multi-radio, portable device with both cellular and LAN/PAN capabilities, which connects non-cellular devices within short range to the cellular network, and routes their traffic to/from the BS. These devices create a mini WiFi hotspot, connecting consumer devices, such as netbooks, cameras or printers, to the internet via cellular backhaul. Mobile hotspots allow WiFi-only consumer devices to access the internet wherever cellular service is available.

3.3 Multi-tier Interference Mitigation

As discussed in the previous section, the features critical to achieving the promised capacity and coverage gains from HetNets are (1) the multi-tier structure with a distributed hierarchy of infrastructure devices, and (2) efficient use of multiple radio access technologies (RATs). In this section, we focus on the problem of efficient spectral reuse across multiple tiers of a single-RAT system, while the next section will show performance gains from a multi-RAT system. While additional network capacity may be added by utilizing the available spectrum across multiple RATs, intelligently reusing scarce spectral resources across the multiple tiers of a single RAT system is also essential for extracting maximal HetNet capacity.

3.3.1 Multi-tier Spectral Reuse Scenarios

Several spectral reuse patterns are possible in a multi-tier network. For example, devices on different tiers can implement 'full spectral reuse' and operate with the same frequency across all tiers, or operate with 'no reuse' and use distinct frequencies across tiers. While sharing the same spectrum promises the highest increase in capacity, it also introduces significant cross-tier interference, which may be avoided by using orthogonal frequencies across tiers.

For spectrum-rich operators who own more than one frequency carrier, the simplest technique for spectrum planning is to assign distinct operating frequencies for use by first- and second-tier devices, such as MBSs and FAPs respectively, thereby avoiding any cross-tier interference. If additional spectrum is available, second-tier devices such as PBSs and FAPs can also choose from more than one frequency, and interference from same-tier devices can also be avoided. However, most network operators are spectrum constrained and cannot afford to deploy separate carriers across tiers. Hence, intelligent single frequency multi-tier deployments, where a single carrier is judiciously used across tiers, are critical to the success of multi-tier architectures.

3.3.2 Cross-tier Interference

When all first-tier MBSs and second tier FAPs, PBSs or RSs are operating simultaneously using the same spectrum, the resulting co-channel interference can severely limit network performance. The inter-tier interference problem is even more prominent with closed subscriber group (CSG) FAP deployments, due to the additional coverage holes created by FAPs for users without access permission. As FAP deployment density increases, a growing proportion of macro-users are driven into outage due to the increased interference, leading to unsatisfactory macro performance.

The interference problem in single carrier multi-tier networks is illustrated by the first two entries in Table 3.1, which shows simulation results demonstrating the potential coverage and capacity improvement with femtocell overlay networks. Comparing the performance of macro-only deployments with those with the femtocell overlay, we observe that although FAPs provides greater indoor coverage and over 100× areal capacity gain, a significant portion of outdoor users are driven into outage due to co-channel interference from FAP transmissions. Therefore, interference mitigation techniques are essential for enabling co-channel deployments at a scale where reasonable density of FAPs may be supported.

The remainder of this section is focused on the problem of cross-tier interference mitigation. We start by describing requirements for multi-tier network synchronization, which are critical for enabling the use of standard interference management (IM) solutions to mitigate cross-tier interference. We also investigate several IM techniques and address issues relevant to applying IM in heterogeneous networks. Our focus is on downlink interference management. Simulations results comparing the performance of various IM schemes are also included. The evaluation is carried out in the context of an overlay CSG FAP deployment as it presents the most challenging interference environment. However, the techniques are also applicable to other types of heterogeneous networks deployments, for example a picocell overlay network.

3.3.3 *Network Synchronization for IM*

Synchronization between network devices is critical for most interference management schemes to work without incurring excessive complexity and performance loss. In the context of OFDMA systems, this requirement implies that an interfering signal must be received within the cyclic prefix of the OFDMA symbol boundary, otherwise an asynchronous interferer can cause significant inter-carrier as well as inter-symbol interference, which are difficult to mitigate with standard IM schemes.

In [17], it is shown that femto networks may be synchronized with the overlay macro network through existing over-the-air synchronization capabilities used in the overlay system. However, it is also seen that simply aligning the timing reference of each tier to a global reference is insufficient to guarantee synchronous interference across tiers. Here it is also important to maintain the relative time alignment between devices, such that transmissions from devices across tiers are received synchronously at the receiver. This requirement is analogous to how uplink transmissions from devices in a single tier network are adjusted based on their distance from the base station, so that they may be aligned at the receiving base station. Similarly, time alignment between devices across tiers must also be maintained to guarantee synchronous interference in multi-tier networks. Further details are available in [17].

3.3.4 *Overview of Interference Mitigation Techniques*

3.3.4.1 Power Control

Power control is an important IM technique as it does not require network level synchronization for its implementation. Here, a FAP's transmission power level can be adjusted to control its

Table 3.1 Performance of femtocell overlay networks (max FAP TX power: 10 dBm, 50 FAPs/sector ~231 FAPs/ km^2). Adapted from S.-H. Yeh, S. Talwar, G. Wu, N. Himayat, and K. Johnsson, "Capacity and Coverage Enhancement in Heterogeneous Networks", IEEE Wireless Communications magazine, Vol. 18, Issue 3, pp. 32–38, June 2011

Scenarios/Interference mitigation scheme		Center Cell Throughput (Mbps)	Coverage[a] (%)		50% user rate (Mbps)	
			Indoor users	Outdoor users	Indoor users	Outdoor users
Macro-only (No FAP deployment)		258	57.8	74.2	0.03	0.06
Power Control	Fixed FAP Power Level	45120 (~175x)	99.7	46.4	25.0	0
	Macro-user QoS	32168 (~125x)	96.2	68.5	14.3	0.06
	Femto-user QoS	7059 (~27x)	99.94	71.4	3.6	0.08
Time/ Frequency Planning	Separate channels for MBS & FAP (50% each)	26400 (~102x)	99.94	74.2	21.4	0.05
	With Femto-Free Zone (FFZ ratio= 27.25%)	32929 (~128x)	99.88	74.2	18.2	0.06

[a]For data, coverage is defined as SINR >= −0.8 dB (SE = 0.5).

coverage range and the amount of interference it generates in the network. Although higher transmission power can provide wider coverage and better signal quality, it can, at the same time, cause tremendous interference to the surrounding users. Therefore, properly balancing the FAP transmission power level can help manage the interference from FAPs to macro-users, while maintaining femtocell performance.

In [18], three different power control schemes are developed and evaluated, including: (1) fixed power level operation, (2) Femto-QoS power control, which adjusts the femto power to maintain a minimum specified data rate for the femto user, and (3) Macro-QoS power control, which controls the FAP power to guarantee minimal degradation to macro user's QoS. Referring to the simulation results for power control schemes shown in Table 3.1, we observe that although FAP power control helps to reduce the outage caused to macrocell users, the resulting degradation in FAP signal quality causes substantially reduced data rates femto users. Nevertheless, power control can still be applied for control channels, where low data rates are sufficient for transmission.

3.3.4.2 Cell Association

When FAP power control is applied, mobile stations may need to perform cell re-association after power back-off is performed at their serving base stations. Here, users may adopt the same rule as the one used in the initial association process, and select from the maximum received signal level, for example. Different rules, such as performing re-association only when the user is in outage, may also be used to ensure that sufficient traffic is offloaded to low-power devices while improving coverage.

More advanced interference mitigation may involve joint operation of power control and cell association. For example, for a given geographical distribution of users, a clustering optimization process may be performed to group users based on the potential to minimize interference through proper control of power settings at the base stations.

3.3.4.3 Low Duty Cycle

Even with lowered FAP transmission power, increasing device density can still cause significant aggregated interference in heterogeneous networks. For example, in an urban environment with multi-story buildings, there can be close to a thousand FAPs installed per sector within a 500 metre cell. Since current standards require all FAPs, both active and inactive, to transmit control signalling messages at the same time as the macro-BSs, an intolerable level of interference on control channels is inevitable at high FAP densities, which prevents wide adoption of FAPs in urban areas.

One possible approach to reducing control channel interference is to transmit FAPs control channels with a low duty cycle, and to require neighbouring FAPs to take turns in transmitting their control channels. This approach 'virtually' reduces the equivalent FAP density. The Femto-QoS power control scheme can also be used together with this low duty cycle operation to achieve sufficient coverage under reuse-1 setting [18].

3.3.4.4 Advanced Time-frequency Planning

Time-frequency planning is another possible approach for interference mitigation in multi-tier networks. For example, in femtocell overlay networks, CSG FAPs can cause severe interference

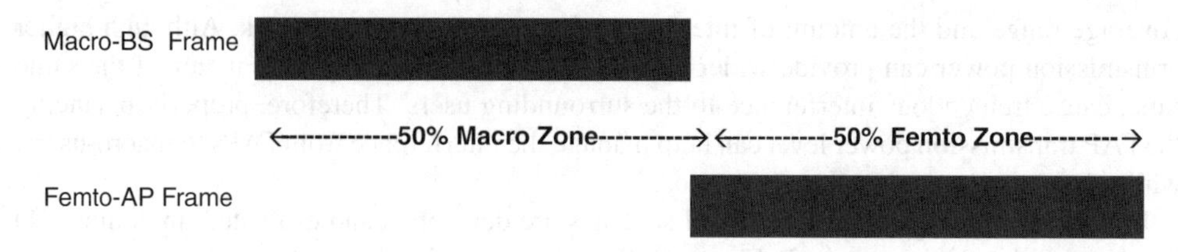

Figure 3.2 Time-frequency planning scenario with orthogonal resource allocation.

to macro-users, and may cause outage for a large number of macro users. A simple approach to avoiding cross-tier interference is to assign distinct time-frequency zones to first- and second-tier devices, such as MBSs and FAPs respectively. Alternatively, it may be more efficient to designate a time-frequency allocation wherein FAPs do not transmit, and hence the resource block is free of FAP interference. In such a femto-free zone (FFZ), macro users suffering from femto interference can be scheduled. Figure 3.2 illustrates the case where orthogonal time-frequency resources are used between a macro-BS and femto-AP, and Figure 3.3 shows femto-free zoning. Performance of the two approaches is compared in Table 3.1, showing good coverage and data throughput for both schemes. However, it is seen that FFZ yields a better outdoor user rate and cell throughput, at the expense of slightly lower indoor user rate. Both these approaches can be applied, together with other advanced frequency planning approach like fractional frequency reuse (FFR). The size of FFR and FFZ partition can be chosen based on the geographical distribution of users across the multiple tiers in the system.

3.3.5 Performance Comparison of IM Schemes

Table 3.1 compares the performance of several IM schemes described in this section. Performance is compared across several metrics, including outage and the 50% throughput performance for both indoor and outdoor users, in addition to the aggregate cell throughput. Results are based on simulation methodology described in [8]. In particular, outage is defined as SINR < -0.8 dB (SE $= 0.5$ bits/s/Hz), which means that a user's SINR is insufficient to support the minimum modulation and coding scheme.

The results show that time-frequency planning techniques deliver huge areal capacity gains, while achieving the best tradeoff across indoor and outdoor user performance, when compared to power control based IM techniques. Therefore, advanced time-frequency planning is a promising solution to mitigate interference for data channel transmission. Time-frequency

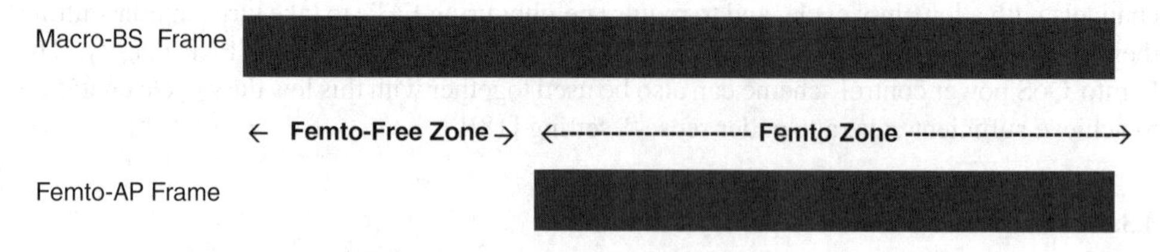

Figure 3.3 Time-frequency planning scenario with femto-free zone.

planning solutions are also desirable for control channel transmission, but require air interface changes that may only be implemented in future standards.

3.4 Multi-radio Performance

In this section, we estimate the performance gains with multi-tier WiFi offload, as well as assessing the performance gains when integrated multi-radio small cells (e.g., picocells) are used for combined transmission across LTE and WiFi carriers (virtual carrier WiFi). Due to the growing popularity of WiFi hotspots, we focus on the outdoor hotspot deployment model as described in 3GPP 36.814. An event-based multi-tier, multi-RAT simulator is used to accurately model the contention-based WiFi MAC protocol. For complexity reasons, the results are based on average-power-based channel statistics; fast channel fading and frequency selective scheduling are not modelled. Further details on methodology are available in [19].

Table 3.2 compares the performance of multi-tier WiFi offload with the multi-tier LTE-pico offload, with and without IM. A macrocell radius of 1.5 km is assumed with two picocells deployed per sector. A 50% ratio of macro-free zone is used, wherein macro-BS does not transmit (analogous to the femto-free zone), to control macro-to-pico interference. As in the previous section, it is seen that interference management across multiple tiers of the same RAT (macrocell and picocell) can significantly boost combined throughput. In this case, macrocell throughput is halved due to macro-free zoning, while picocell throughput is significantly boosted due to lower interference from the macrocell(s). Now, if a WiFi AP is used instead of a pico BS with approximately same range, it can be seen that offloading to 20 MHz of unlicensed WiFi spectrum can restore macrocell throughput by avoiding the need for a macro-free zone, as well as offering the higher throughput of WiFi radio access. Hence, WiFi offloading is a very cost-effective means for operators to achieve higher network capacity.

Further improvement in performance can be obtained if both WiFi and LTE spectrum are aggregated through the use of an integrated WiFi and LTE base station/AP device. Table 3.3 illustrates these performance gains. It is seen that greater than two times gain in per sector throughput can be obtained with LTE+WiFi integrated devices, when compared to the LTE-pico only offload. Hence, integrated infrastructure devices offer the benefit of combining multi-tier and multi-RAT performance gains in a single network, which can make it easier for operators to manage their network as well as lower operational costs.

It is to be noted that the performance estimates in this section assume standard WiFi technology with 54 Mbps of PHY capacity, no coupling between WiFi and LTE interfaces, and are based on long-term channel statistics. We expect the performance gains to improve

Table 3.2 Throughput performance of with multi-tier pico deployments and multi-tier WiFi offload

	Shared spectrum LTE 10 MHz				Additional spectrum	
	LTE-Pico (no IM)		LTE-Pico (with IM)		WiFi (20 MHz)	
Scenarios	Macro	Pico	Macro	Pico	Macro	WiFi
Macro only	2.3	0				
Multi-tier	3.4	9.8	1.7	30.1	3.6	47

Table 3.3 Throughput performance of with integrated WiFi+LTE infrastructure device

	Shared spectrum (LTE 10 MHz)		WiFi (20 MHz)	
Scenarios	Macro	Pico	WiFi	Total
Macro + LTE pico with IM	1.6	30		31.6
Macro only	2.3			2.3
Integrated WiFi+LTE	1.6	30	34.5	66.1

significantly with dynamic channel models and scheduling with an advanced radio-resource management scheme exploiting tighter coupling across WiFi/LTE. Further gains are possible if additional WiFi spectrum is exploited.

3.5 Standardization and Future Research Directions

In recent years, the potential benefits from heterogeneous networks have drawn a great deal of attention from both academia and industry. In the following two sections, we first give a brief overview of the wireless standards efforts focused on heterogeneous networking and then describe some of the promising research directions in the field.

3.5.1 Status of Wireless Standards

Wireless standard bodies are actively pursuing 'enabling' technologies and working to expedite the deployment of heterogeneous networks.

3GPP has made substantial progress in this area over the past few years. Topics including network architecture, security and RF requirements have been examined. For example, new features have been added to the standard to enable femtocell deployments (H(e)NB in 3GPP terminology). Currently, there are work items in release-10/11 to enable non-carrier aggregation interference mitigation in heterogeneous networks, advanced MIMO techniques for interference management (known as coordinated multi-point transmissions – CoMP) and mobility enhancements for heterogeneous networks [20, 21]. There are also efforts to enable LTE relays and WLAN offloading.

IEEE 802.16 has formed a new study group focusing on hierarchical networks (multi-tier and multi-RAT networks) [22]. This group is developing a hierarchical networks study report, which will describe potential use cases, key features and requirements and example network architecture, as well as the prescribed simulation methodology.

Heterogeneous network technologies are still at an early stage of development. New applications and services will continue to add tougher design requirements. Advanced signal processing techniques are expected to enable new features and architectural options for more cost-effective implementation.

3.5.2 Future Research Directions

3.5.2.1 Vertical and Horizontal Optimization

We expect that future heterogeneous networks will be optimized in both vertical and/or horizontal directions. Vertical optimization consists of cross-layer protocol optimization and/or the further flattening of the network architecture. This approach is essential for reducing network latency and improving end-to-end QoS. Horizontal optimization involves cooperative signal processing across multiple sectors/cells (these can be from the same or different tiers). For example, a super base station can serve as a coordination point managing interference across multiple tiers. Since the X2 interface [23] between base stations is replaced by an internal connection in the super base station, it may be a cost-effective way of implementing CoMP and carrier aggregation in future heterogeneous networks.

3.5.2.2 Service on Network Edge

Future heterogeneous networks need to fully leverage new technologies such as virtualization and multi-core processor platforms. Virtualization enables multi-core processors to support different operating environments simultaneously, and allows dynamic computing resource allocation that adapts to the network's ev-er changing service requirements.

Latency reduction continues to be a priority for wireless broadband systems. As a result, we expect services to continue moving towards the network edge. The significance of this trend is twofold. First, it offers service providers new revenue opportunities, for example, user access to local content, localized radio resource management for improved user experience and value-added data processing at the network edge. Second, it may significantly reduce network congestion in the core network which is increasingly a risk due to the proliferation of media-intensive applications.

3.5.2.3 Device-centric System Design

One continuing challenge for heterogeneous networks is the coordination and management of multiple access networks. Traditional cellular networks employ a centralized or 'network-centric' model, where inter-system interfaces ensure a mostly seamless operation of client devices across different networks. The main limitation of this approach is that it does not scale well in a heterogeneous network composed of multiple access networks.

We expect future heterogeneous networks to employ a distributed or 'device-centric' model. Intelligent multi-mode devices with sophisticated connection managers are a key technology goal. We expect that more radio link management decisions will be made at the client to simultaneously minimize network complexity and improve user experience.

3.5.2.4 Heterogeneous Networks for Machine-to-machine Communications

As the deployment of WiFi APs, FAPs and PBSs continues to increase, the coverage of heterogeneous networks will continue to improve, particularly inside buildings. This creates opportunities for new services, such as those based on machine-to-machine (M2M) communication, to connect devices in locations that were traditionally difficult to reach by cellular (e.g., basement of buildings). Furthermore, we expect that some M2M networks will be integrated into future heterogeneous networks. If this is the case, M2M aggregation points can serve both their intended M2M function as well as provide 'hotspot' access [24].

3.5.2.5 Regulatory Considerations

As heterogeneous networks become part of the mainstream wireless communication infrastructure, special attention must be paid to address existing and future regulatory requirements. Lawful intercept support will become a more complex issue in a heterogeneous networking environment. Also, new services (e.g., m-Health) carried by the network may increase service providers' liability. Future research must consider and address these new requirements.

3.6 Conclusion

Wireless broadband data traffic driven by consumer demand for rich mobile internet services is now the primary driver for both consumer purchases and network operator deployments. Mobile subscribers desire the same rich content availability as on their fixed broadband internet connections – including streaming video. Mobile M2M connections are also expected to increase wireless data consumption. As wireless broadband continues to be rolled out worldwide, the demand for wireless data shows every sign of accelerating. Heterogeneous networks hold great promise for meeting consumer demands, while providing optimum total cost of ownership for the network operators. However, these networks have many technical challenges at the air interface and network layers, which the wireless community is working to address in standards forums. We expect that heterogeneous networks will be a key to enabling cost-effective wireless broadband to every corner of the globe.

References

1. Cisco, 'Cisco Visual Networking Index: Global Mobile Data Traffic Forecast Update, 2009–2014', February 2010.
2. ABI Research, 'Mobile Network Offloading', Updated Research Report, 12 August 2010.
3. Rysavy Research, 'Mobile Broadband Capacity Constraints', 24 February 2010.
4. Sue Marek (editor), 'Managing Wireless Network Capacity', Fierce Wireless eBook, 22 May 2012.
5. V. Chandrasekhar, J. Andrews and A. Gatherer, 'Femtocell Networks: A Survey', IEEE Communications Magazine, September 2008.
6. S. Yeh, S. Talwar, S. Lee and H. Kim, 'WiMAX Femtocells: A Perspective on Network Architecture, Capacity and Coverage', IEEE Communication Magazine, October 2008.
7. A. Damnjanovic, J. Montojo, Y. Wei, et al., 'A Survey on 3GPP Heterogeneous Networks', IEEE Communications Magazine, June 2011.
8. S. Yeh et al. 'Capacity and Coverage Enhancement in Heterogeneous Networks', IEEE Communications Magazine, June 2011.
9. W. Wang, X. Liu, J. Vicente and P. Mohapatra, 'Integration Gain of Heterogeneous WiFi/WiMAX Networks.' *IEEE Transactions on Mobile Computing*, Vol. 10, no. 8, August 2011.
10. A. Atayero, E. Adegoke, A. Alatishe and M. Orya, 'Heterogeneous Wireless Networks: A Survey of Interworking Architectures', *International Journal of Engineering and Technology*, Vol. 2, no. 1, January 2012.
11. Johansson et al. 'A Methodology for Estimating Cost and Performance of Heterogeneous Wireless Access Networks', PIMRC '07.
12. H. Claussen, L. Ho and L. Samuel, 'Financial Analysis of a Picocellular Home Network Deployment', IEEE ICC '07.
13. K. Johnsson, O. Oyman, P. Kuo, P. Ting and Y. Chen, 'Cooperative HARQ', IEEE C802.16m-09/1380r1, July 2007.
14. IEEE C802.16–10/0018, Jiwon Kang, Bin-Chul Ihm, Wookbong Lee, 'Distributed antenna system for future 802.16', LG Electronics, March 2010.
15. China Mobile, 'C-RAN – Road towards Green Radio Access Network', C-RAN International Workshop, http://labs.chinamobile.com/focus/C-RAN, Beijing, April 2010.
16. N. Himayat, et al., 'Heterogeneous Networking for Future Wireless Broadband Networks', IEEE C80216-10_0003r1, January 2010.

17. N. Himayat, et al., 'Synchronizing Uplink Transmissions from Femto AMSs', IEEE C80216m-09_1348r1, July 2009.
18. S. Yeh, S. Talwar, N. Himayat and K. Johnsson, 'Power Control Based Interference Mitigation in Multi-tier Network', IEEE 1st Workshop on Femtocell Networks, Globecomm, December 2010.
19. N. Himayat, G. Wu et al. 'Exploiting multi-radio heterogeneous networks', presentation to WWRF-27, October 2011.
20. 3GPP TR 36.814, 'Evolved Universal Terrestrial Radio Access (E-UTRA); Further Advancements for E-UTRA Physical Layer Aspects', Release 9.
21. 3GPP TR 36.912, 'Feasibility study for Further Advancements for E-UTRA (LTE-Advanced)', Release 10.
22. R. Kim and N. Himayat, 'Study Report on Hierarchical Networks', IEEE C802.16ppc-10/0008, July 2010.
23. 3GPP TS36.420, 'X2 general aspects and principles'.
24. G. Wu, S. Talwar, K. Johnsson, N. Himayat and K. D. Johnson, 'M2M: From Mobile to Embedded Internet', IEEE Communication Magazine, April 2011.

4

Cross-tier Interference Management in 3GPP LTE-Advanced Heterogeneous Networks

Wenkun Wen, Zichao Ji, and Ming Jiang
New Postcom Equipment Co. Ltd., China

4.1 Introduction

Interference management (IM) techniques are one of the key enablers in next-generation mobile networks, such as the evolved universal terrestrial radio access network (E-UTRAN), also known as the third generation partnership project (3GPP) long-term evolution (LTE). In broadband wireless communication systems under a single-frequency network (SFN) deployment, the achievable system capacity is usually limited by the inter-cell interference (ICI). The ICI problem is even more severe in heterogeneous networks (HetNet), since the interference environment is more complicated than that in the homogenous networks (HomoNet) due to the coexistence of different kinds of E-UTRAN nodeBs (eNB) in HetNet deployments. More specifically, the ICI in HetNets is typically a combination of different cell tiers, for example the high-power node (HPN) tier and the low-power node (LPN) tier. An HPN is typically a macro eNB (MeNB), while an LPN can be a pico eNB (PeNB), or a femto eNB (FeNB), which is also often referred to as a home eNB (HeNB).

Aiming at mitigating the ICI problem, especially in HetNet, significant improvements have been made in the evolved version of LTE, namely the LTE-Advanced system (also known as LTE Release-10) during the 3GPP standardization process. In this chapter, we will discuss the interference management issues in the context of LTE-Advanced HetNet scenarios.

Heterogeneous Cellular Networks, First Edition. Edited by Rose Qingyang Hu and Yi Qian.

4.1.1 Heterogeneous Network Deployments

A cost-effective mechanism to offload the heavy traffic at hotspot scenarios, such as shopping malls, airports and railway stations, is to deploy LPNs in addition to the conventional macro base stations. Such a mixed deployment is known as a HetNet. Although such HetNet deployments were used as early as the deployment of 2G networks, the LPNs are typically deployed on carriers different from that designated for the macro tier, in order to eliminate the mutual interference between them. Obviously, this kind of traditional interference management strategy results in low spectral efficiency, especially in mobile broadband systems, such as LTE and LTE-Advanced networks, where data-demanding services requiring higher bandwidths are expected to be dominant.

For the sake of achieving high spectral efficiency, operators strongly prefer to adopt a frequency reuse factor of one in network planning in their next-generation cellular network deployments. In this case, co-channel interference inflicted from one tier on another may significantly outweigh the abovementioned cell-splitting gain, if no appropriate measures are taken for tackling the ICI. For example, deployment of uncoordinated closed subscriber group (CSG) mode femtocells will severely interfere with the cell-edge user equipments (UEs) served in a macrocell that is in close proximity to the femtocells.

Generally, there are two typical scenarios of LPN deployment in HetNet, namely the open subscriber group (OSG) and the CSG modes. Naturally, the statistical characteristics of the interferences associated with the two scenarios are different, which will be detailed in the sequel.

4.1.2 OSG Scenario

The conventional cell selection/reselection strategy in LTE is based on the received power strength, or more explicitly the reference signal received power (RSRP). Thus, in an OSG deployment scenario, where LPNs including relay nodes, picocell nodes, and remote radio head (RRH) nodes, are open for access from all UEs, the UEs would likely decide to camp on the cell, providing the strongest RSRP. However, the cell association strategy based purely on a UE's received power may result in imbalanced downlink (and uplink) coverage and service quality, due to the asymmetric transmission power levels from different cell tiers. Unlike the HPNs, the lower transmission power of LPNs implies that their downlink coverage is poorer, which also constrains their UL services.

Furthermore, the abovementioned strategy may create undesirable cell association ratios. As shown in Figure 4.1, a macrocell UE (MUE-1) camps on the cell with strongest received power (corresponding to the macrocell) instead of the cell with the lowest path loss (corresponding to the LPN cell), due to the higher transmission power from the macrocell in comparison to the picocell. Therefore, the association ratio of LPN may be unnecessarily low, thus degrading the expected achievable cell-splitting gain. Moreover, the undesirable access from UEs to the macrocell also introduces significant ICI to the LPN UEs on the uplink.

One of the solutions to offload the UEs to the cell with lower path loss is to apply the so-called cell range expansion (CRE) mechanism, which tends to associate the UEs with a neighbouring LPN by carefully adjusting the LPN's cell individual offset (CIO). As a result, a promising cell-splitting gain may be achieved with the uplink signal-to-interference-plus-noise ratio (SINR) maximized, yielding a significant improvement on uplink coverage and capacity. More specifically, by absorbing more macrocell-edge MUEs to a LPN, the uplink interference

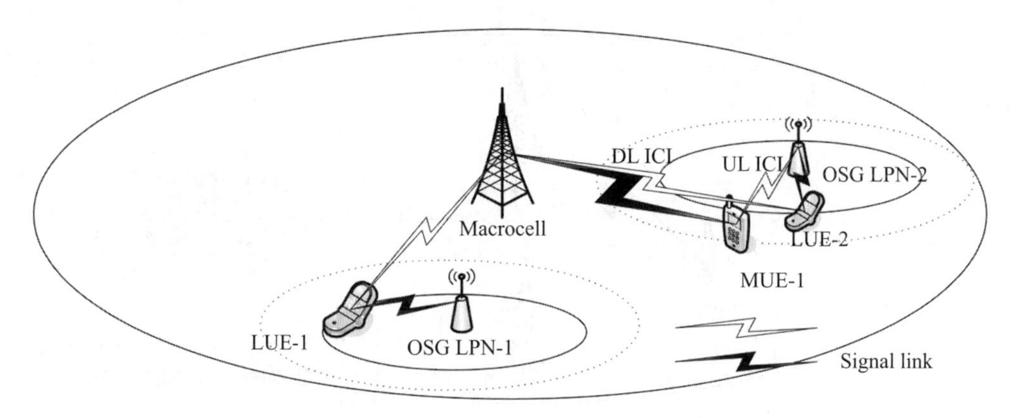

Figure 4.1 HetNet deployment in the OSG scenario.

inflicted by the high transmission power of these macrocell-edge MUEs on the neighbouring LPN UEs (LUEs) would be largely mitigated, since as newly immigrated LUEs, they would now only need to apply an uplink transmission power much lower than that needed in the macrocell scenario, due to the fact that the coverage of an LPN is typically much smaller than that of a MeNB.

However, since the LUEs in the CRE-enabled area would no longer camp on the cell with strongest downlink received power, they may suffer severe downlink interference from the macrocell, as demonstrated in Figure 4.2. More details of CRE-related issues and the potential solutions will be discussed later.

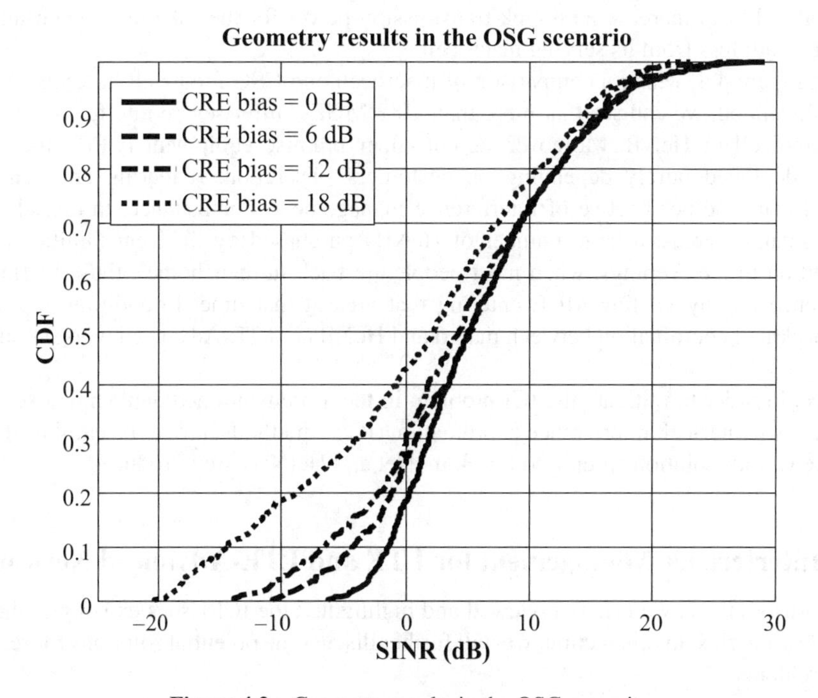

Figure 4.2 Geometry results in the OSG scenario.

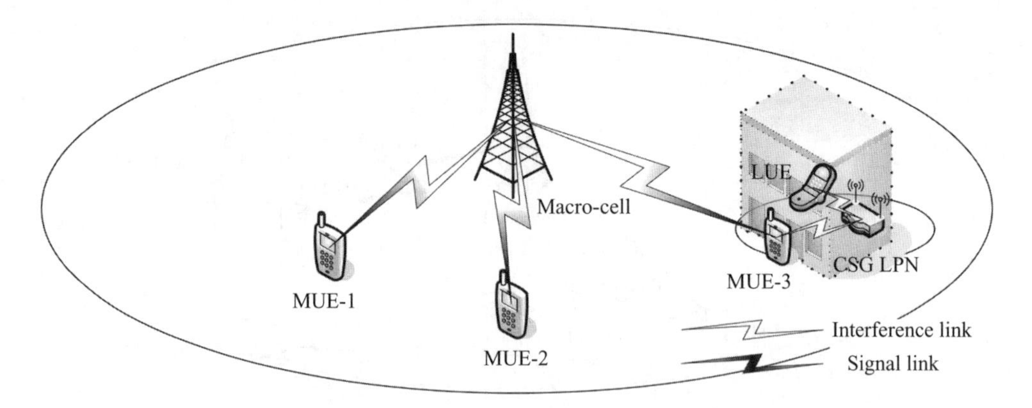

Figure 4.3 HetNet deployment in the CSG scenario.

4.1.3 CSG Scenario

Compared to OSG, the interference characteristics in the CSG scenario are significantly different. More explicitly, the cell that is served by a CSG LPN (typically a femtocell, or an HeNB in the LTE system) and configured to CSG mode, is only accessible by authorized users referred to here as the CSG users. Consequently, a macrocell user (e.g., the MUE-3 in Figure 4.3) visiting the CSG LPN cell area would suffer significant downlink interference from the LPN, since it is not allowed to access the CSG cell, which therefore becomes an interferer. Furthermore, on the uplink, the CSG users would also be interfered by the visiting MUE, as the latter has to increase its uplink transmission power for the sake of compensating the increased path loss from its serving macrocell.

From Figure 4.4, where a comparison of macrocell and CSG femtocell in terms of geometry is illustrated, we can see that more than 10% MUEs suffer severe interference from the CSG femtocell or HeNB. Moreover, as consumer premise equipment (CPE), the HeNBs may be deployed purely depending on customers' discretion, rendering an even worse situation from the perspective of interference management. For instance, in densely populated residential areas, a large number of HeNBs purchased by different families may be switched on in the evenings, when most people are back in their homes, thus creating high interference on any visiting MUE entering that area at that time. In addition, the lack of direct backhaul coordination between macro and HeNBs in LTE-Advanced worsens such ICI issues.

Hence, in order to mitigate the ICI problem in the aforementioned deployment scenarios, cost-effective inter-cell interference coordination (ICIC) methods have to be employed. In the next, the various solutions proposed for HomoNet and HetNets are introduced.

4.2 Interference Management for LTE and LTE-Advanced Networks

In previous sections, we briefly reviewed and highlighted the ICI issues existing in the OSG and CSG scenarios. In this section, we will further discuss the potential solutions for resolving these problems.

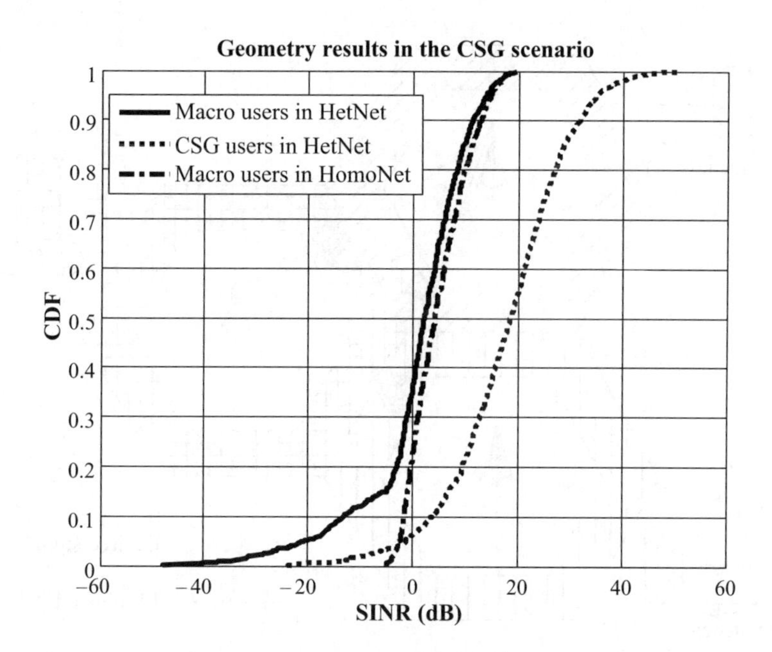

Figure 4.4 Geometry results in the CSG scenario.

4.2.1 Interference Management Methods for Homogenous Networks

As a popular ICIC solution, fractional frequency reuse (FFR) techniques are often exploited in wideband wireless systems under HomoNet deployments, such as in LTE Release-8 networks [1, 2]. One of the typical FFR implementation scenarios is shown in Figure 4.5. In this example, the base station assigns a specific band for cell-edge UEs, who are allowed to use a high transmission power for coverage enhancement. Furthermore, within the cell-edge band, a few orthogonal subbands are allocated for the use of adjacent neighbour cells, such that the ICI generated by the high-power cell-edge UEs from different cells can be avoided. In other words, in FFR-enabled mobile networks, ICI only exists between cell-edge UEs and the corresponding neighbour cells' non-cell-edge UEs. Since the abovementioned two UE groups are usually not too geographically close, the resultant ICI is expected to be low. Hence, the employment of FFR techniques can help to improve the performance of cell-edge UEs in HomoNets, while still maintaining good performance of non-cell-edge UEs.

Two mechanisms are needed to facilitate the FFR function in LTE HomoNets:

1. identification of cell-edge UEs,
2. sharing and coordination among adjacent neighbour cells on information of the specific bands reserved for cell-edge UEs.

These mechanisms have been supported in LTE since Release-8. By exploiting the reported RSRP or power headroom report (PHR) from the UE, the eNB can derive the downlink path loss of the UE. As shown in Figure 4.5, the UE with a sufficiently large path loss is typically

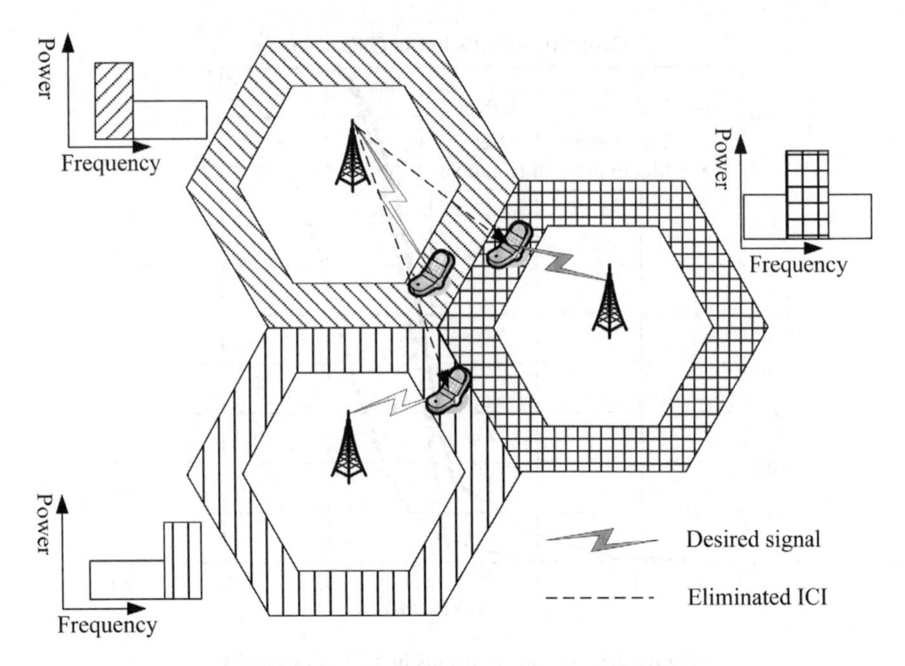

Figure 4.5 Traditional FFR-based ICIC techniques.

far from the eNB, and therefore may be identified as a cell-edge user. Moreover, through the X2 interface specified in LTE for MeNB (Section 4.2.2 in [3]), neighbour eNBs in HomoNet are capable of scheduling their respective cell-edge UEs according to the corresponding coordination on the allocation of FFR bands.

However, the aforementioned two mechanisms are not sufficient for ICIC in HetNet scenarios. Firstly, unlike the typically fixed location of MeNBs, the location of the LPNs in HetNet may be randomly distributed, depending on customers' discretion. Particularly, a CSG LPN may even be powered up at the cell centre, thus creating a so-called coverage hole from the macrocell's perspective, as discussed in the previous section. Therefore, the macrocell may no longer be able to identify the MUEs to be protected by the ICIC mechanism through path loss calculation only, since in this case the interfered MUEs do indeed have a low path loss as they are located at the centre of the macrocell. Moreover, as there is no standardized X2 interface between HeNBs and MeNBs in LTE-Advanced specifications, no backhaul coordination is available for tackling the ICI issue that arises from the HetNet scenario. Apparently, FFR-based ICIC solutions alone are incapable of handling the ISI issues in HetNets. Another drawback of FFR-type ICIC schemes is that they can not be applied to physical downlink control CHannel (PDCCH) in the LTE system (Section 6.8 in [4]). More specifically, PDCCH is interleaved and spread all over the frequency band and hence cannot benefit from the FFR solutions, which require splitting the cell-edge band into a number of orthogonal subbands.

In the next section, the standardized solutions for LTE-Advanced are discussed, which are designed to resolve the ICI issues existing in HetNets.

Table 4.1 Comparison of spectral efficiency (bps/Hz) between HomoNet and HetNet assuming a 2 × 2 LTE system where each macrocell serves 25 active UEs

	Cell average	Cell edge
HomoNet	1.34	0.017
HetNet	1.84	0.048

4.2.2 Interference Management for Heterogeneous Networks

First we will look at the various issues in CRE-enabled heterogeneous networks, where the fast-increasing traffic in mobile wireless networks due to the explosion of today's markets of smartphones and their substantial applications, leads to heavy demands of much higher data rates and much wider bandwidths than the needs in legacy networks. As discussed in previous sections, cellular network operators may prefer dense deployment of LPNs on top of the macrocell tier for the sake of exploiting the cell-splitting gain and maximizing the spectral efficiency. Simulation results of Table 4.1 show that a significant gain may be achievable in HetNet as compared to HomoNet.

As mentioned in Section 4.1.2, an effective mechanism to exploit the cell-splitting gain and at the same time to reduce the impact of the abovementioned downlink and uplink coverage imbalances is to adopt flexible CIOs, namely to apply CRE. The rationale of the CRE mechanism is to intentionally bias the ranking of cell measurement results by adding a CIO to the received power strength of each LPN cell, as indicated by this formula:

$$S_i = Q_i + CIO_i, \tag{4.1}$$

where Q_i is the measurement result of cell i and CIO_i is the CIO of cell i. The UE is associated with the cell with maximum S. By setting a proper CIO for different cells, the cell-splitting gain can be effectively attained.

However, in a realistic system, the eNB communicates with the UE with the aid of the download control channel, but this may not be decodable when the received SINR at the UE is lower than a certain threshold. For instance, for a typical LTE terminal, the demodulation threshold of not exceeding a BLock error ratio (BLER) of 1% for PDCCH is −4 dB [5]. Specifically, the use of CIO may increase the probability of PDCCH outage. The example provided in Table 4.2 shows that significant downlink SINR performance degradation can

Table 4.2 The SINR performance of cell-edge users (5 percentile) in HetNet downlink with a large CRE bias of 20 dB

SINR (dB)	without CRE	with CRE
HetNet UEs	−1.59	−18.52
LUEs	−2.44	−19.59
MUEs	−0.82	−1.41

occur if large CIO bias values are applied, which lead to outage of more than 5% of users due to their corrupted control channel under the large CIO bias setting. Hence, the CIO bias should be carefully selected to strike for an optimum ICIC performance in HetNet scenarios.

4.2.3 Time Domain Based ICIC Schemes

To solve the abovementioned ICI problems, including the control channel issues occurring in HetNets, an enhanced ICIC (eICIC) mechanism is introduced in the LTE-Advanced specifications.

The eICIC technique works in the time domain with the aid of the so-called almost blank subframes (ABS). The ABS has the following characteristics, as defined in [6, 7]:

- UEs can assume the following:
  - All ABSs carry CRS,
  - If PSS/SSS/PBCH/SIB1/Paging/PRS/CSI-RS coincide with an ABS, they are transmitted in the ABS (with associated PDCCH when SIB1/Paging is transmitted),
  - No other signals are transmitted in ABSs,
  - If ABS coincides with an MBSFN subframe not carrying any signal in data region, CRS is not present in data region,
- MBSFN subframe carrying signal in data region shall not be configured as ABS.

From the definition of ABS, we can see that an ABS is a subframe provided by the aggressor cell (i.e., a macrocell in OSG deployment or a CSG cell in CSG deployment) with reduced transmission power (including zero power transmission) on some physical channels [8]. Obviously, the overall ICI is therefore effectively reduced in ABSs. However, to maximize the backward compatibility, in LTE-Advanced networks some essential control signals are allowed to be transmitted within ABSs. These include control signals, for example primary synchronization signals (PSS) and secondary synchronization signals (SSS), and reference signals, such as the cell-specific reference signal (CRS) and the channel state information reference signal (CSI-RS), as well as system information like the master information block (MIB), system information block (SIB), etc. According to [6], no CRS from the serving cell is transmitted in the data region of the multi-cast/broadcast over a single frequency network (MBSFN) subframes, which are usually utilized to provide the multimedia broadcast and multi-cast service (MBMS) or loCation service (LCS), can naturally be exploited as ABSs.

It is clear that the ICI issue in HetNet can be mitigated by time-domain multiplexing (TDM) style scheduling coordination between the neighbour cells, such that the interfering MUEs and LUEs are orthogonally scheduled in different subframes, namely ABSs and normal subframes, on the same carrier. Therefore, when eICIC is activated, the low-geometry HetNet UEs heavily interfered as discussed in previous sections can have a hugely increased probability of accessing an LPN cell enabling CRE, thus benefitting from the cell-splitting gain. An example of LTE-Advanced eICIC mechanism is portrayed in Figure 4.6, where the aggressor cell and the victim cell coordinate their scheduling operations by utilizing ABSs.

Moreover, since the aggressor cells have statistically reduced-power or little transmissions in the control channels, namely the physical control format indicator CHannel (PCFICH), physical hybrid automatic repeat reQuest (ARQ) indicator CHannel (PHICH) and physical

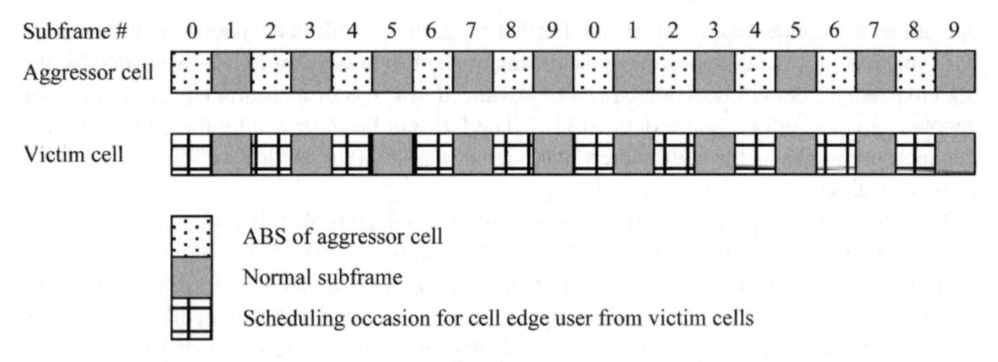

Figure 4.6 Illustration of the eICIC technique adopted by LTE-Advanced.

downlink control CHannel (PDCCH), the interference imposed on these control channels of the victim cell from the aggressor cell is minimized, especially for the LUEs in CRE-enabled cell area.

Nonetheless, configuring MBSFN subframes as ABSs may not be a sufficiently flexible solution due to the limited choices of the configurable periodicity. More explicitly, only subframes 1, 2, 3, 6, 7 and 8 can be configured as MBSFN subframes (Section 6.3.1, [9]). On one hand, the optimum ratio of ABSs against normal subframes depends on a number of factors, for instance the HetNet's load, the number of cell-edge UEs, etc. A straightforward example is that when the aggressor cell has a light load, it may set aside more ABS resources for the use of the victim cell in order to improve the latter's service for an increased number of UEs, especially when CRE is enabled by the victim cell. On the other hand, frequent changes of ABS patterns would require excessive coordination efforts among the neighbour cells, which might outweigh the benefits provided by eICIC. For example, when an ABS pattern is selected for a pair of neighbour cells, their respective neighbour cells may also have to change their ABS patterns such that the various ABS patterns are compatible with one another for avoiding ping-pong-like unnecessary pattern reconfigurations. Thus, configuration of ABSs should allow sufficiently flexibility and be semi-statically adapted based on the varied ICI status within the HetNet. More details about ABS pattern design can be found in Section 4.2.3.3.

The design concept of ABS – namely to enforce reduced-power or little transmissions within selected subframes – is nothing new and has been able to be used since LTE Release-8, but there are some additional impacts to the system that need to be resolved for facilitating the eICIC operations. In the next few sections, we will further discuss the relevant design considerations from the perspectives of measurements, CRS interference handling and pattern design options.

4.2.3.1 Restricted Measurement

The implementation of ABSs requires that the subframes are categorized into two groups, namely normal subframes and coordination subframes (that is ABSs). To be aligned with this strategy, the link adaption (LA) technique adapting the modulation and coding scheme (MCS), the multiple-input multiple-output (MIMO) mode and channel-dependent scheduling configuration, etc. should be performed, according to the type of subframes where these operations are conducted, respectively. Furthermore, the downlink channel state information (CSI) measured within the normal and coordination subframes, respectively, is needed for the eNB to make proper LA decisions. In LTE-Advanced, two sets of measurement configurations denoted as $C_{csi,0}$ and $C_{csi,1}$ are defined [10]. The UE will be informed by the eNB about the two measurement sets through higher-layer signalling, and thus report two sets of CSI to the eNB accordingly.

The intention of introducing ABS is to minimize the outage probability of the cell-edge UEs in the CRE area, by serving them in ABSs only. This suggests that when ABS is configured, it is of little meaning for the UEs to perform radio link monitoring (RLM) in normal subframes, which is instead considered as a valid operation in LTE Release-8/9 networks. Otherwise, unnecessary radio link failures (RLF) may be triggered due to the strong ICI occurring in normal subframes in HetNet scenarios. In order to enforce the UEs to monitor the *relevant* subframes only, ABSs or more precisely a subset of ABSs associated with a subframe pattern, should be provided to the UE for assisting in its proper RLM operations. Similar to RLM, radio resource management (RRM) measurement is also restricted to the subframes associated with a configured ABS pattern in order to have the measurement report and the handover operation handled correctly.

However, there are some differences between RRM measurement and RLM/CSI measurement. Firstly, concerning the measurement target, RLM/CSI measurement is performed only on the serving cell(s), while RRM measurement is performed not only on the serving cell(s) but also on the neighbour cells. This is reasonable, since RRM measurement is designed to support handover operations. Secondly, the output of RRM measurement is RSRP and reference signal received quality (RSRQ), while the output of RLM/CSI measurement is the received SINR. More specifically, measurement of RSRP only requires estimation of the received power strength of the CRS from the target cell, without the interference being taken into account. By contrast, RLM/CSI measurement does consider the ICI aspect, through dividing RSRP by the received signal strength indicator (RSSI), which is the total wideband received power strength including both the serving and the interfering cells.

As required by the implementation of eICIC, not only should the users served by the victim cell perform restricted measurement on their serving cell as similar to the RLM measurement, but also the users served by the aggressor cell should perform restricted measurement on the victim neighbour cell(s) for handover purposes.

When measuring the victim cell, for instance the victim picocell in the OSG macro-pico deployment or the victim macrocell in the CSG macro-femto deployment, the UEs protected by ABS should restrict their RRM measurement to ABSs only. Otherwise, due to the severe interference imposed on the victim cell's CRS from the aggressor cells, the UEs' RSRP estimation for the victim cell may become highly inaccurate, which in turn degrades the detection of the victim cell. Similarly, if the RRM measurement is not restricted to ABSs only, the RSRQ measurement results for the victim cell may also become over-pessimistic. This is because the interference from the transmission in non-ABSs of the aggressor cells is incorrectly taken into account, while the cell-edge UEs within the CRE zone may only be served in ABSs, where the ICI from the aggressor cell occurring on physical downlink shared CHannel (PDSCH) is absent. Consequently, if the given measurement is not restricted to ABSs only, these CRE users may be handed over from the victim cell to the aggressor cell, which is undesirable from the perspective of exploiting the cell-splitting gain or potential offloading UEs to LPNs.

Therefore, in LTE-Advanced networks, two subframe patterns – instead of only one as defined in LTE RLM measurement specifications – can be provided to UEs for RRM measurement on victim serving cell and neighbour cells, respectively.

It should be noted that the abovementioned CSI, RLM and RRM measurement restrictions are only applicable to connected mode LTE-Advanced UEs. For idle mode UEs, the legacy LTE cell ranking and selection/reselection procedures are applied. It is because, unlike with the connected mode UEs, there is no cell-splitting gain or uplink interference issue for idle mode UEs. Nonetheless, as different cell association mechanisms are applied for UEs of different modes, it may lead to imbalanced coverage for connected mode and idle mode UEs, which may result in an increase of unnecessary handovers. More explicitly, UEs may be directed to undesired cells in radio resource control (RRC) reestablishment process during RLF recovery, which affects the mobility robustness optimization (MRO) and/or mobility load balancing (MLB) functions.

Another remaining issue in LTE-Advanced concerning restricted measurement is that it is only applicable to the carrier associated with the primary cell in the context of carrier aggregation (CA). This is due to the fact that the time domain ICIC function designed for LTE-Advanced assumes only non-CA-based deployment.

The abovementioned limitations in LTE-Advanced measurement for eICIC are under considerations in 3GPP and are expected to be further enhanced in future LTE releases. For instance, for the CA-based deployment, other further enhanced ICIC schemes such as cross scheduling techniques among different carriers, that is, carrier-based ICIC schemes [11], etc. may be considered.

4.2.3.2 Interference Mitigation for CRS

Although the introduction of eICIC in LTE-Advanced helps to significantly mitigate the overall ICI from the aggressor cell within ABSs, the interference inflicted by the aggressor's CRS remains unimproved. According to the definition of ABS [6], the aggressor cell should always transmit CRS in ABSs (except in the data region of empty MBSFN subframes) in order to maintain backward compatibility for legacy UEs' measurement operations. Considering the dense distribution of CRS-carrying resource elements (RE) in each resource block (RB) of each subframe, such ICI brings a rather extensive and severe impact on the system performance, which largely depends on the quality of received reference signals, as shown in Figure 4.7.

CRS interference leads to large performance degradation in a HetNet employing CRE, especially when a large bias value is applied, as shown by the simulation results in Figure 4.8 and Table 4.3. However, if interference cancellation (IC) techniques are used, the CRS ICI can be effectively reduced, as also exhibited by Figure 4.8 and Table 4.3. More specifically, we can derive from Table 4.3 that there will be more than 5% cell throughput loss and 10∼45% cell edge throughput loss, if no CRS IC is employed, resulting in reduced achievable gains of the time-domain ICIC function designed for LTE-Advanced.

However, in typical macro-pico or macro-femto deployments, only a few dominant interferers exist, and some statistics of the interference (e.g., the CRS sequence used) may be derived through specific a priori information provided by the aggressor cell, such as the physical cell ID (PCI), the number of CRS ports and the CP type. Based on such information, it is possible to cancel the CRS interference at the victim UE or at the victim eNB side.

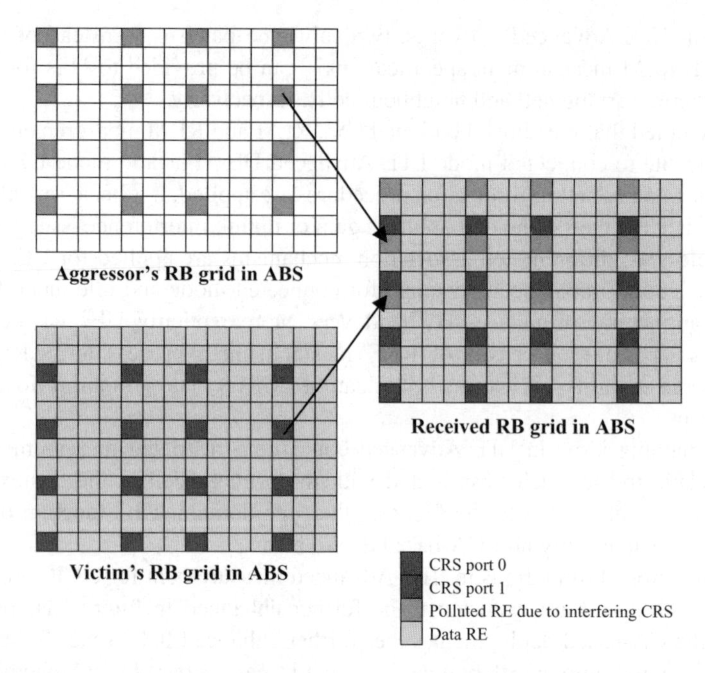

Figure 4.7 Interference from CRS in a RB of LTE or LTE-Advanced subframes.

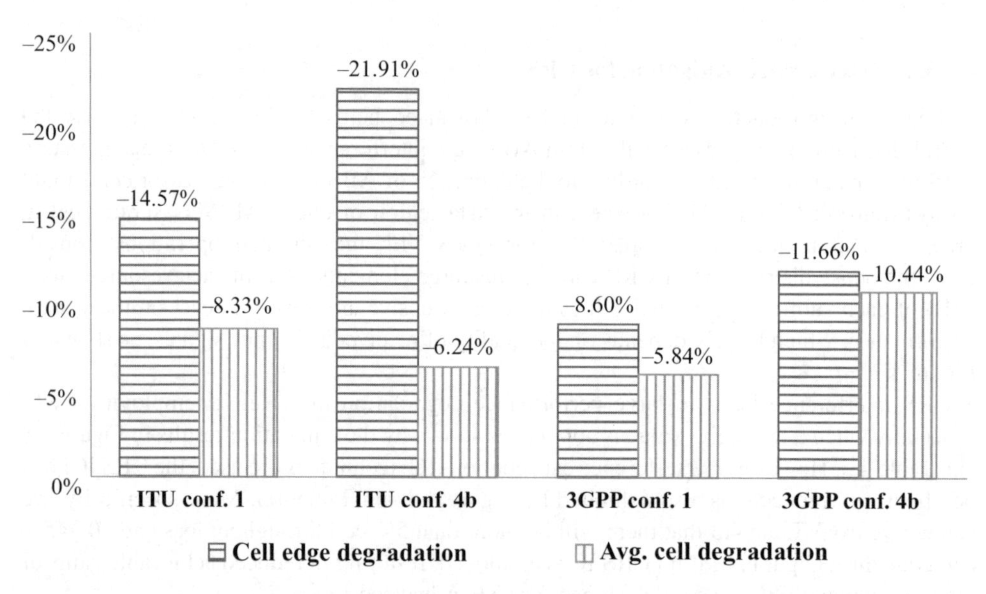

Figure 4.8 Performance loss due to CRS interference (in comparison to perfect CRS IC). 'ITU conf.1' means the general simulation assumptions align with ITU cases (Section A.2.1.1 [14]), and using configuration 1 in Table 4.7 for placing the LPNs and UEs. In the same way, 'conf.4b' means configuration 4b in Table 4.7 is used for placing the LPNs and UEs. The '3GPP' prefix means the general simulation assumptions align with cases defined by 3GPP (Section A.2.1.1 [14]).

Table 4.3 Spectral efficiency of CRS IC techniques

	ITU conf. 1		ITU conf. 4b		3GPP conf. 1		3GPP conf. 4b	
	Cell avg.	Cell edge	Cell avg.	Cell edge	Cell avg.	Cell edge	Cell avg.	Cell edge
Benchmark (no eICIC)	8.256	0.043	8.959	0.058	5.838	0.031	7.538	0.034
Perfect CRS IC	8.427	0.091	9.695	0.117	6.156	0.053	8.597	0.074
Non CRS IC	7.803	0.087	9.090	0.092	5.796	0.049	7.699	0.065

More explicitly, there are mainly two kinds of CRS IC schemes: the receiver-based (Rx-based) IC and the transmitter-based (Tx-based) IC. Either way requires that the aggressor cell informs the victim cell of the CRS-related a priori information through backhaul signalling and/or operations, administration and maintenance (OAM) configurations. For the Rx-based CRS IC methods, the victim eNB needs to forward the abovementioned system information to the victim UEs for facilitating their IC operations.

Moreover, in Rx-based methods a successive interference cancellation (SIC) like processing is usually employed, as illustrated in Figure 4.9. Under the LTE-Advanced eICIC assumptions, the power level of interfering cells is usually strong or even higher than that of the serving cell. Thus, the estimation of the interferers' CRS is to some extent expected to be reliable, and effectively facilitates the removal of the interference from the received signal, which helps to achieve a good SIC performance.

In order to demodulate the aggressor cell's CRS, the UE needs to know the subframe index of the aggressor, such that the occurrence of ABSs can be known too. This means that subframe-level time synchronization between the aggressor and victim eNBs is needed, which is a constraint for network deployment under certain circumstances. Furthermore, Rx-based IC methods require advanced SIC receivers, which increases the implementation complexity at the UE side. Nonetheless, since the main IC functionality is UE implementation-specific, Rx-based IC techniques impose minimum impact on LTE-Advanced specifications.

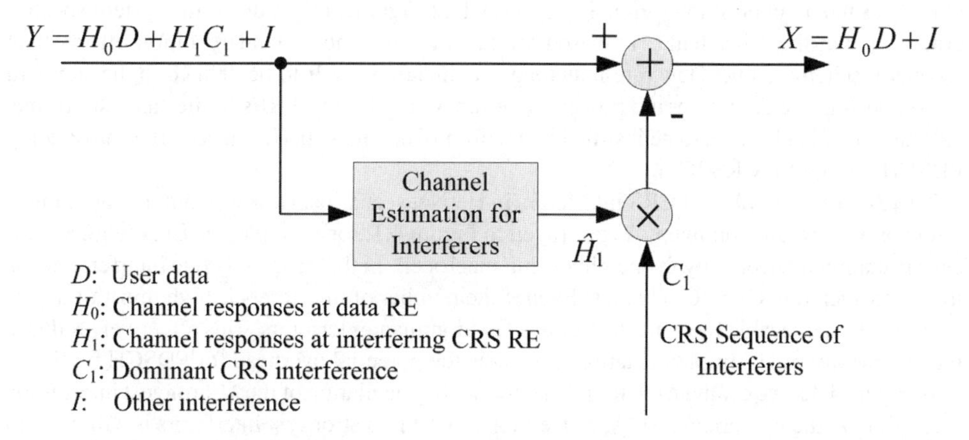

Figure 4.9 Diagram of receiver-based CRS IC techniques.

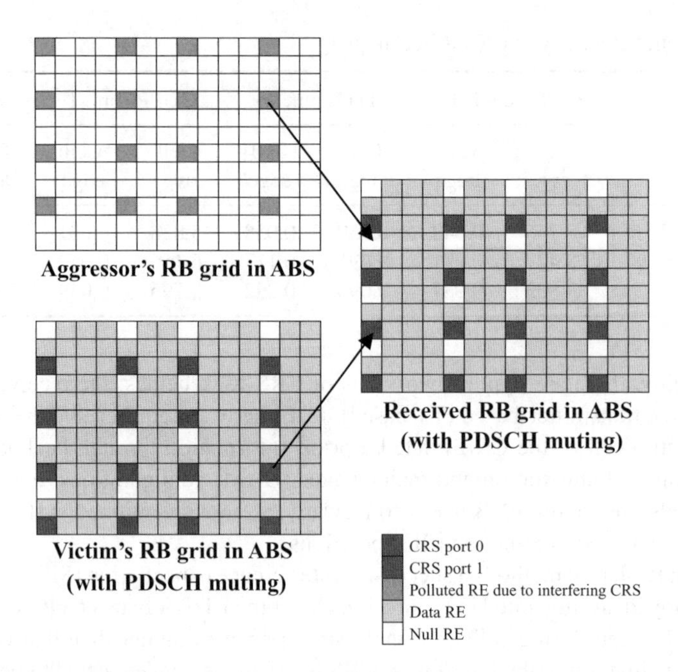

Figure 4.10 Tx-based CRS IC schemes employing PDSCH muting.

The Tx-based IC methods can reduce the relatively high complexity required at UEs by shifting some processing efforts to the eNB side. As shown in Figure 4.10, the victim cell may preclude the polluted REs interfered by aggressor's cell CRS, and thus completely avoid the CRS interference at the victim UEs. Such a method is called PDSCH RE muting in LTE-Advanced systems.

The PDSCH muting scheme could be transparent to UEs, but the muted REs can no longer be used for data transmissions and thus increase system overhead. Moreover, as the density of CRS within a subframe is quite high, the overhead introduced by PDSCH muting increases quickly as the number of interferers increases, which significantly reduces the system spectral efficiency. On the other hand, Tx-based methods require more standardization efforts than Rx-based schemes, since new rate-matching specification needs to be defined in the standard for supporting the data mapping procedure at remaining PDSCH REs in the same subframe. All these drawbacks have to be justified by the improved eICIC performance after introducing PDSCH muting for CRS IC.

There can be a number of CRS interferers in HetNet scenarios. However, among these interferers only a few are dominant. As portrayed in Figure 4.1, for example, the LPN (e.g., a picocell) is mainly interfered by the neighbouring macrocell. In this case, only one interferer needs to be considered in CRS IC schemes. Even if the number of interferers is high, most of the IC gain may be achievable by cancelling only a few dominant interferers. This effectively reduces the complexity of UE implementation as well as the potential overhead for PDSCH muting.

In Figure 4.11, we evaluate the Rx IC method with cancellation of the M strongest interferers. According to the numerical results, cancelling at most two strongest interferers is sufficient to recover the performance degraded by CRS interference.

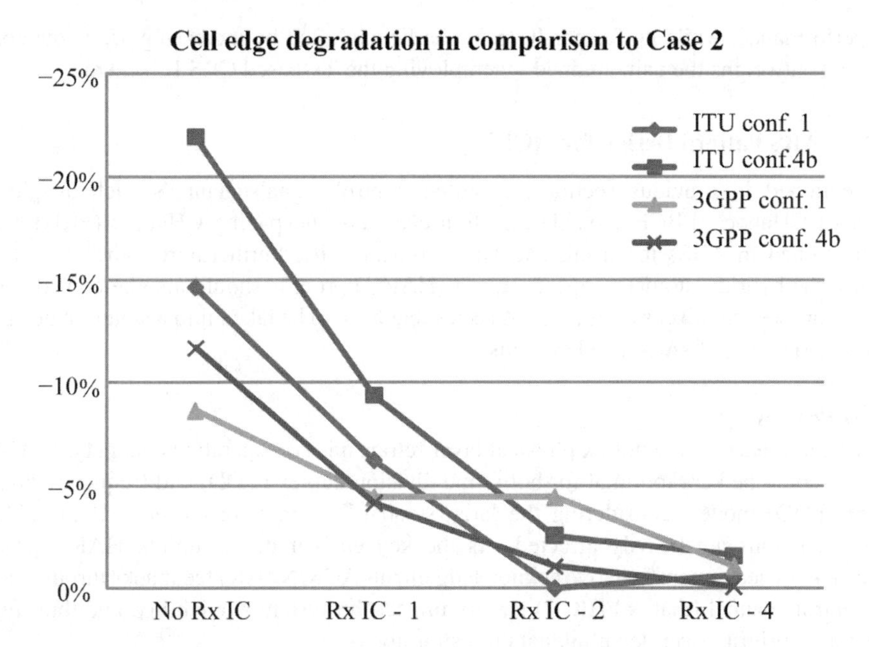

Figure 4.11 Cell edge performance of the Rx-based CRS IC method.

For the Tx-based CRS IC scheme, for example, a total of $12 \times M$ data REs have to be muted in order to avoid the interferences associated with the non-colliding CRSs from two antenna ports. In Figure 4.12, the performance of Tx-based CRS IC scheme is evaluated assuming $M = 1$, 2 non-colliding CRS interferers. The PDSCH muting overhead has been taken into account in the simulation results. As seen from Figure 4.12, the case of $M = 1$ shows the

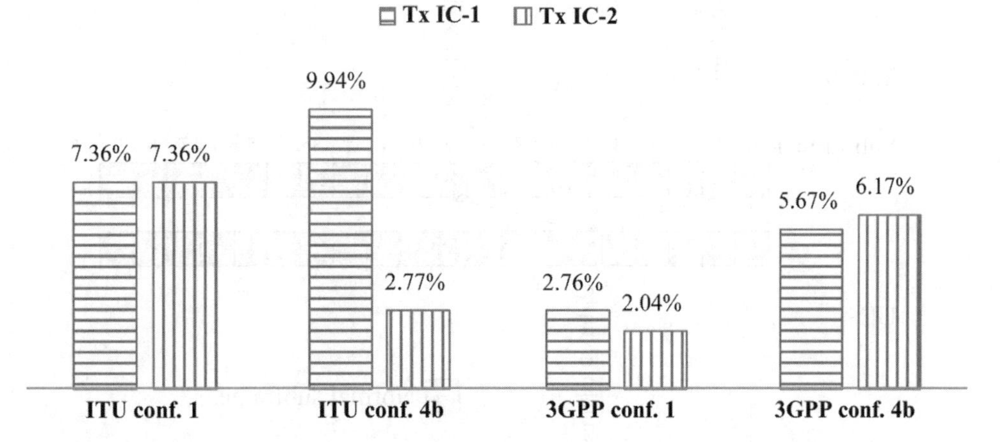

Figure 4.12 Cell edge performance under the employment of the Tx-based CRS IC methods with PDSCH muting.

best performance, while no further IC gain can be exploited by increasing M, as the cost of overhead outweighs the gain attained by employing the Tx-based CRS IC scheme.

4.2.3.3 ABS Pattern Design for eICIC

As mentioned in previous sections, essential control signals/channels such as physical broadcast CHannel (PBCH), synchronization channels and paging CHannel (PCH) should be maintained in ABSs to ensure backward compatibility. Furthermore, other aspects, for instance the hybrid automatic repeat request (HARQ) process should also be considered. In this section, we will discuss a number of issues that have to be taken into account in designing ABS patterns for LTE-Advanced systems.

HARQ Process

First of all, impacts to the uplink physical layer retransmission operations, namely the HARQ process should be kept minimal for both time division duplex (TDD) and frequency division duplex (FDD) modes, considering the large standardization efforts needed if the HARQ-related functions are heavily affected. As the key enabler of the uplink HARQ process, timely acknowledgement/negative acknowledgement (ACK/NACK) feedbacks through downlink control channel (that is PHICH) largely impact the system's reliability, and thus always have higher priority over downlink data transmissions.

Therefore, the design of ABS patterns, including the starting subframe index and the period, needs to be carefully considered to avoid conflict with the downlink ACK/NACK transmissions. More specifically, since ACK/NACKs are transmitted through PHICH, which has to be muted within ABSs as defined in LTE-Advanced, the occurrence of ABS should therefore avoid colliding with the subframes carrying ACK/NACKs. In Figures 4.13 and 4.14, we show an example of potential issues, which may occur if an inappropriate ABS pattern is configured. In

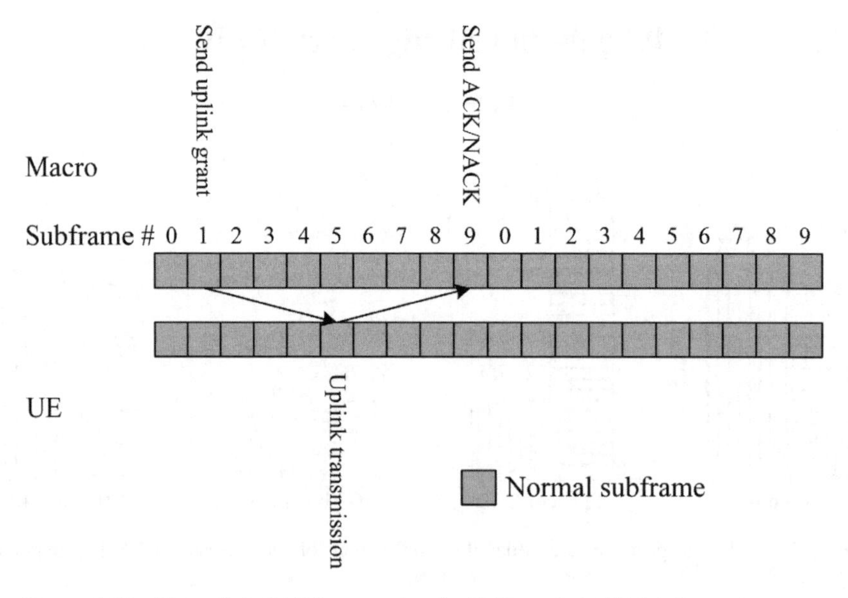

Figure 4.13 The uplink HARQ processing for FDD mode in LTE-Advanced systems.

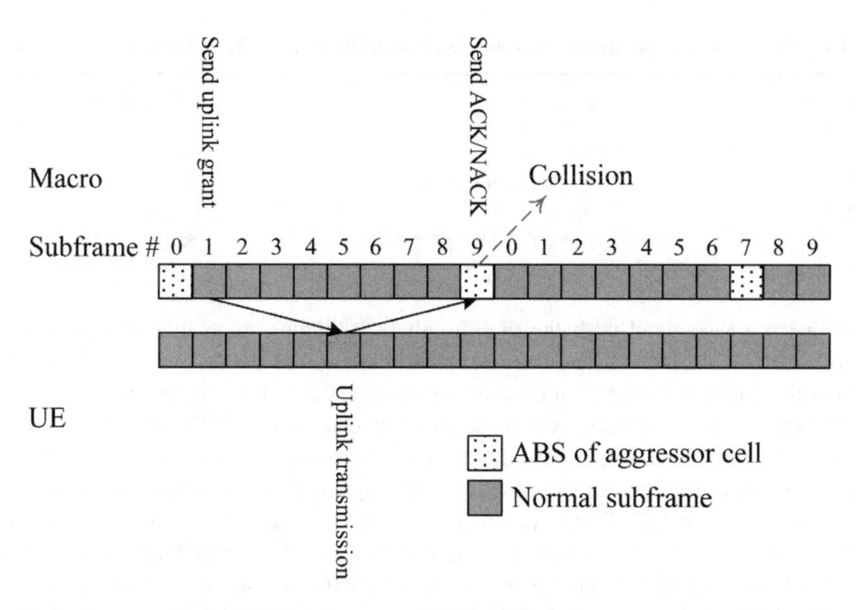

Figure 4.14 Corrupted uplink HARQ processing in LTE-FDD due to inappropriate ABS configuration.

a LTE FDD system exemplified in Figure 4.13, the HARQ period is 8 ms (or duration of eight subframes). In LTE/LTE-Advanced networks, the UE can start uplink transmission only after it receives an uplink grant from the eNB through PDCCH. Hence, a design criterion may be that subframe #x and #(x+8) of a radio frame in the aggressor cell should not be set as ABS, such that the uplink grant can be transmitted in subframe #x and ACK/NACK feedbacks can be transmitted in subframe #(x+8) normally. In this design example, when such constraints are taken into account in ABS pattern design, as long as the eNB avoids scheduling uplink grants in the subframes corresponding to the occurrence of ABS, the associated ACK/NACK feedback transmission will not collide with ABS.

Recall that the duration of a LTE/LTE-Advanced radio frame is 10 ms. Thus, the period for the ABS pattern for FDD can be set to the least common multiple (LCM) of 8 and 10 (i.e., 40 ms), which takes into account both the radio frame duration and the HARQ period in LTE-FDD. Otherwise, collision with ACK/NACK transmissions may occur, as shown in Figure 4.14, which then corrupts the uplink HARQ process in the aggressor cell. As a simple example, assuming subframe #n is a proper candidate ABS, one can set subframes #n, #n+8, #n+16, #n+24 and #n+32 as ABSs (which constitutes an ABS pattern period of 40 ms) in order to avoid similar collisions demonstrated in Figure 4.14.

For TDD, on the other hand, similar considerations on the HARQ process were also taken into account when LTE-Advanced was standardized. The finally agreed ABS pattern period options for TDD are 20 ms for downlink/uplink configurations #1~#6, 70 ms for configuration #0, and 60 ms for configuration #7, respectively.

Common Channels and Other Considerations

The design of ABS pattern also depends on the UEs' association ratios of the aggressor and victim cells, the traffic type and the protection priorities for the essential control signals/ channels mentioned above. It is straightforward to apply an ABS ratio of 10%, if the number

Table 4.4 Subframe configurations for common channels in a LTE/LTE- Advanced radio frame

Common channel	FDD Mode	TDD Mode
PBCH	#0	#0
SIB-1	#5 per even radio frame	#5 per even radio frame
PSS/SSS	#0, #5	#1, #6
PCH	Subset of {#0, #4, #5, #9}	Subset of {#0, #1, #5, #6}

of UEs that are associated with the victim cell is 10% more than that of aggressor cells. Furthermore, to determine which subframes should be used as ABSs, one should consider the level of protection that should be applied to the essential control signals/channels of the victim cells. The subframe configurations of these common channels in LTE/LTE-Advanced, which are summarized in Table 4.4, should therefore be taken into account when designing ABS patterns for maintaining backward compatibility through minimizing their collision with ABSs.

For example, in an LTE-FDD system, if subframes #0 and #5 could be protected, the PSS/SSS and broadcasting channels (that is PBCH and SIB-1) which are related to initial access would be free of ICI. This is a key for improving the success rate of a UE's initial access and thus the first priority should be given to subframes #0 and #5 in ABS pattern design. Moreover, the PSS/SSS channels have a lower priority than PBCH. This is because a UE can still become time-synchronized with the aggressor cell if it fails to synchronize with the victim cell, as the victim cells and aggressor cells are typically synchronized in the time domain when eICIC is enabled. On the other hand, the number of subframes used for PCH depends on the network loads, and therefore subframes #4 and #9 where PCH is possibly carried could be optionally protected. In other words, these subframes should be protected only when necessary. More details on the subframe protection priorities are shown in Table 4.5.

However, since the common channels of Table 4.4 exist in ABSs of the aggressor cell, they inevitably impose ICI on the colliding time-frequency resources of the victim cells deployed on the same carrier. To guarantee the victim cell's performance of the common channels, one solution may be that the aggressor cell applies a subframe offset for mitigating the collision from the particular common channels of aggressor and victim cells, as illustrated in Figure 4.15.

In order to avoid ICI from PBCH, a subframe offset may be applied in LTE-FDD systems by setting subframe #0 of LPN as ABS after subframe shifting. With the HARQ procedure taken into account, subframe #8 in the first radio frame, subframes #0 and #6 in the second radio frame, subframes #0 and #4 in the third radio frame and subframes #0 and #2 in the fourth radio frame need to be set as ABS too. Alternatively, the aggressor cell needs to avoid

Table 4.5 ABS ratios associated with prioritized protection of essential control channels

ABS pattern #	ABS ratio	Protected channels
1	10%	PBCH
2	15%	PBCH, SIB-1
3	20%	PBCH, SIB-1, PSS/SSS
4	30%	PBCH, SIB-1, PSS/SSS and PCH in subframe #9
5	40%	PBCH, SIB-1, PSS/SSS and PCH in subframe #4

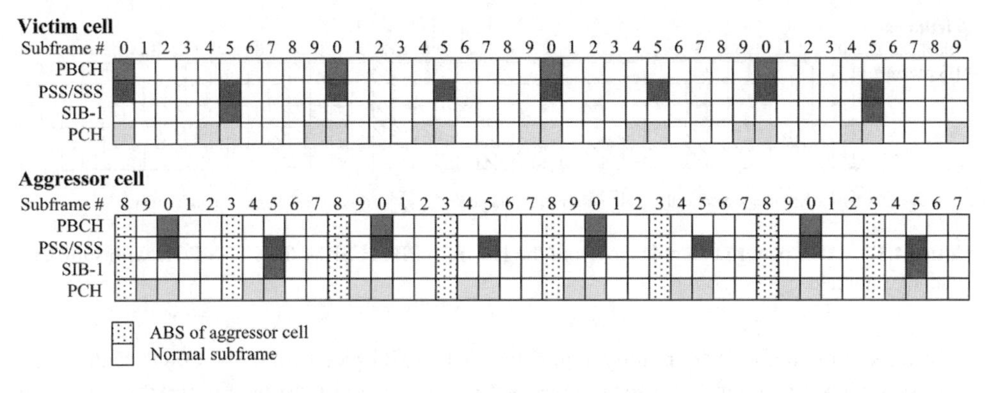

Figure 4.15 Employment of a subframe offset of 2 to avoid ICI on common channels.

scheduling uplink grants in subframe #2, such that no ACK/NACK from the UE needs to be sent at subframe #0 of the next radio frames. An example reflecting this design criterion is provided in Figure 4.16.

Based on the design principle similar to PBCH-specific considerations discussed above, we further exemplify some desired ABS patterns for achieving ICI-free transmissions on other control channels and/or PSS/SSS in Figure 4.17, where subframe offset is also applied. Moreover, the protection priorities summarized in Table 4.5 are also considered. More specifically, in the design example shown in Figure 4.17, different ABS ratios relate to the protection priority given in Table 4.5 for essential control channels. Those subframes that are exempt from uplink grant transmissions can be also set as ABS, if a higher ABS ratio is needed, for instance in a HetNet suffering heavy ICI.

Nonetheless, the subframe-shifting technique discussed above is applicable to FDD systems only, as in TDD systems the shifted subframes would result in undesirable cross-interference between downlink and uplink transmissions. Due to the tight standardization plan of LTE-Advanced, protection mechanisms for PBCH/SIB-1/PSS/SSS/PCH were not standardized and were left for the future LTE releases, that is Release-11.

4.2.4 Power Setting for Femtocells

In addition to the time domain ICIC mechanism introduced in the previous section, other ICIC solutions may also be applied. The adaptive power setting method, which mitigates the ICI by reducing the transmission power of the aggressor cell, is one of the effective ICIC solutions for non-CA-based HetNet deployments. Note that the classical FFR techniques widely used in existing networks can also be considered as one kind of power setting scheme.

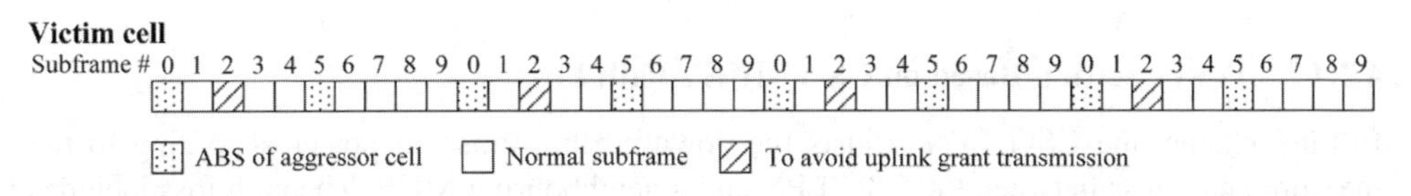

Figure 4.16 An ABS pattern example to avoid the ICI from PBCH with subframe offset of 2 applied.

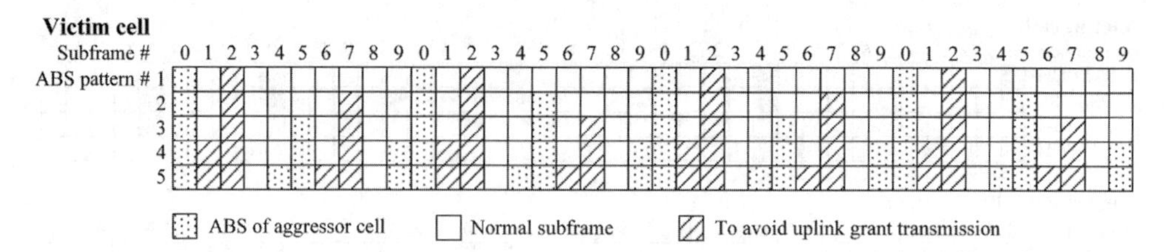

Figure 4.17 ABS pattern examples to avoid the ICI from PBCH/SIB-1/PSS/SSS/PCH with subframe offset of 2 applied.

Power setting methods are mainly applicable to the CSG scenario. In the OSG scenario, the aggressor MeNB can hardly perform power setting for several reasons. Firstly, the macrocell may serve a wide coverage area, which is designed carefully during the network planning stage. Reducing the transmission power of a MeNB may unfavourably and significantly affect the basic network coverage and capacity, and is therefore not reasonable. Secondly, since a large number of UEs are served by the macrocell, it is not optimum to adopt a single power setting scheme, which is only optimum from the ICIC perspective and for some specific UEs, that is those that are close to the LPNs.

By contrast, it is reasonable for a CSG LPN (e.g., FeNB) to reduce its transmission power if any victim MUEs exist in its coverage area, since a CSG LPN usually serves only a few UEs that are readily in the proximity of the CSG LPN. In the sequel, we will discuss several power setting schemes in the context of co-deployment of HPN (e.g., MeNB) and CSG LPNs. Note that the application of most of the power setting schemes assumes that CSG LPNs are equipped with a downlink receiver for facilitating their coordination with HPNs.

4.2.4.1 Power Setting Based on Interference Measurement

In this scheme, the CSG LPN adjusts its maximum downlink transmission power as a function of air interface measurements to avoid interference on MUEs, as described by the following formula [12]:

$$P_{tx} = Median(\alpha \cdot P_m + \beta, P_{\max}, P_{\min})\ (dBm), \tag{4.2}$$

where $P_{\max}$ and $P_{\min}$ are the maximum and minimum CSG LPN transmission power in dB, respectively, while P_m is the strongest received power from the co-channel macrocells measured at the CSG LPN. The parameter α is a linear scalar that allows altering the slope of power control mapping curve. Parameters $P_{\min}$, α, and β are considered to be configuration parameters for CSG LPN, and $P_{\max}$ corresponds to the CSG LPN's maximum transmission power capability.

4.2.4.2 Power Setting Based on CSG MUE's Path Loss

In this scheme, the CSG LPN adjusts the downlink transmission power according to the measured path loss between the CSG LPN and a neighbouring MUE. The path loss includes penetration loss for providing better interference mitigation for the MUE, while maintaining good coverage of the CSG LPN. The path loss based power control method is formulated as [12]:

$$P_{tx} = Median(\alpha \cdot P_m + P_{offset}, P_{\max}, P_{\min})\ (dBm), \tag{4.3}$$

where $P_{\max}$, $P_{\min}$ and P_m have the same meaning as in (4.2) and P_{offset} (dB) is the power offset given by:

$$P_{offset} = Median(P_{inter\_pathloss} + P_{offset\_\max}, P_{offset\_\min}). \tag{4.4}$$

The parameter $P_{inter\_pathloss}$ in (4.4) is a power offset value related to the indoor path loss and the penetration loss between the nearest MUE and the CSG LPN, while $P_{offset\_\max}$ and $P_{offset\_\min}$ are the predefined maximum and minimum values of P_{offset}, respectively [13].

4.2.4.3 Power Setting Based on SINR Measurement

In this scheme, the CSG LPN adjusts its transmission power such that the victim MUE's SINR requirement is satisfied, as denoted by:

$$P_{tx} = \Gamma + \Omega + IoT\ (dBm), \tag{4.5}$$

where Γ is the minimum SINR requirement of the victim UE, Ω is the background noise and IoT represents the interference over thermal noise. In this method, the minimal SINR requirement of all victim MUEs is satisfied at the cost of performance degradation of the LUEs served by the CSG LPN.

4.2.4.4 Enhanced Power Setting

In this scheme, the CSG LPN adjusts its transmission power, taking into account both the interference environment and the serving quality of the UEs, by using the following criterion:

$$P'_{tx} = Median(\alpha \cdot P_m + \lambda \cdot P_{offset} + \beta, P_{\max}, P_{\min})\ (dBm) \tag{4.6}$$

The parameter *P_offset* is the power compensation factor calculated according to the path loss between CSG LPN and victim MUEs:

$$P_{offset} = Midian(P_{inter\_pathloss}, P_{offset\_\max}, P_{offset\_\min}), \tag{4.7}$$

$$P_{inter\_pathloss} = P_{MUE\_tx} - P_{MUE\_rx}, \tag{4.8}$$

where $P_{MUE\_tx} = P0 + PL$ is the estimated transmission power of victim UE, and P_{MUE_rx} is the

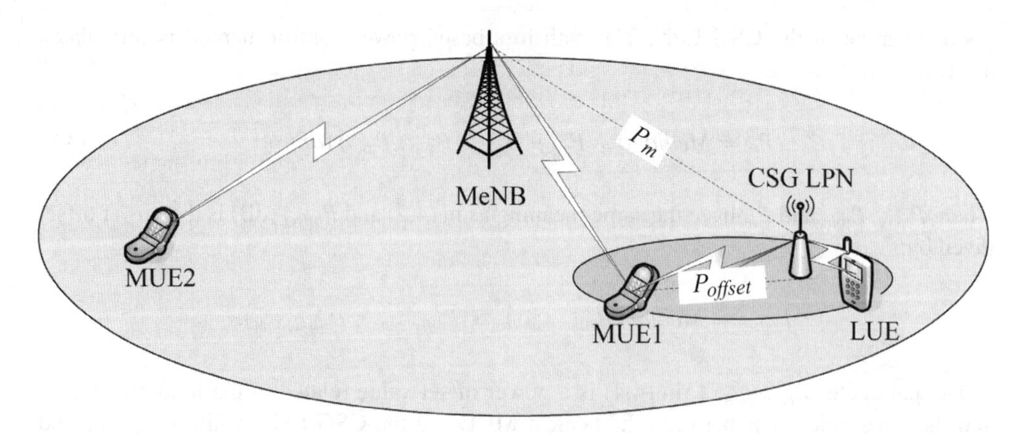

Figure 4.18 Illustration of the enhanced power setting scheme.

uplink received power of victim UE at CSG LPN. Moreover, $PL = P_{M\_tx} - P_m$ is the measured path loss between CSG LPN and the MeNB with strongest downlink received power, while *P0* is the uplink target receiving power of that MeNB, which is same as the *P0* parameter used in uplink power control in LTE [10].

It is worth mentioning that it is not necessary to configure the CSG LPN with a power level lower than the minimum requirement of victim MUEs. Therefore, the transmission power of CSG LPN can be further determined as:

$$P_{tx} = \max(P'_{tx}, \Gamma, \Omega + IoT)\ (dBm) \tag{4.9}$$

The enhanced power setting scheme is illustrated in Figure 4.18. It effectively combines the advantages of (4.1) and (4.2), and therefore can provide a better performance, as confirmed by the simulation results provided in the next section.

4.2.4.5 Performance Comparison

Numeric results of the various power setting methods discussed above are summarized in Table 4.6.

Table 4.6 UE average SINR geometry distribution (dB) under various power setting schemes

	5%-tile (cell edge)		50%-tile (average)		95%-tile (peak)	
Power setting scheme	MUE	CSG LUE	MUE	CSG LUE	MUE	CSG LUE
No power setting	−20.5	−1.4	2.3	18.1	13.5	35.2
IM based	−11.2	−9.6	2.6	14.3	11.8	32.4
Path loss based	−7.0	−6.2	2.9	16.8	14.1	35.0
Victim MUE SINR based	−3.1	−4.4	4.4	−4.0	15.0	8.6
Enhanced	−4.1	−4.0	3.4	12.0	14.6	30.1

Comparing the results of Table 4.6, power setting methods significantly improve the victim MUE's SINR. However, the IM based and path loss based methods still can not meet the SINR requirement of PDCCH (−4 dB), while the method based on minimum SINR requirement of victim MUE meets this target, though at the cost of degradation of all CSG LUEs' performance. More specifically, note that the average LUE SINR is only −4 dB and even the cell-centre LUE can only have a SINR of less than 9 dB. By contrast, when the enhanced power setting method is employed, cell-edge UEs in both macro and CSG cells meet the SINR requirement with the performance of CSG LUEs reasonably maintained.

4.3 Conclusions

In this chapter, the ICI management techniques for LTE-Advanced HetNet deployments are studied. It is concluded that the existence of cross-tier interference invalidates the effectiveness of conventional frequency-domain ICIC methods such as FFR. Therefore, the time-domain-based ICIC solution, also known as eICIC, has been proposed and standardized in LTE-Advanced for tackling the ICI issue in HetNet scenarios.

We also analysed the issues arising from the implementation of ABS in eICIC, such as restricted measurement, RS IC problems and their corresponding solutions. The principles for designing ABS patterns were also discussed in detail.

Besides the eICIC scheme, power setting strategies can also be applied in CSG scenario in HetNet deployments. Numerical results are provided for comparing the gains of various power setting techniques.

Although the eICIC scheme in LTE-Advanced provides an effective way to mitigate ICI in data channels, severe ICI could exist in essential control channels, for example PBCH/SIB-1/PSS/SSS/PCH. Other solutions, such as the subframe-shifting mechanism, the coordinated multi-point coordination (CoMP) technique, etc. need to be investigated for further enhancing the achievable ICIC performance in LTE/LTE-Advanced networks. Our future work will focus on the evolvement of eICIC in next the LTE releases, namely the so-called further enhanced ICIC (FeICIC) solutions.

Appendix: Simulation Models

The simulation models used for HetNet evaluations are defined in 3GPP technical report TR 36.814 [14]. Besides the indoor channel model, the major modelling changes in comparison to the HomoNet scenario are the placement of LPNs and UEs. The simulation results provided in this chapter were generated based on the assumptions in [14]. For readers' convenience, we extract and summarize the relevant information in Tables 4.7 and 4.8.

As an example, the user dropping Configurations #1 and #4b are plotted in Figure 4.19 and Figure 4.20, respectively. In Figure 4.19, we can see that both the MUEs and LUEs are dropped uniformly around the MeNB and LPNs. In contrast, in Figure 4.20 the UEs are dropped in a clustered manner that most UEs are dropped in the close proximity of LPNs, resulting in more LUEs than MUEs.

Table 4.7 Placing of new nodes and UEs [14] *Source:* 3GPP 36.814. Reproduced with permission.

Config.	UE density across macrocells*	UE distribution within a macrocell	New node distribution within a macrocell	Comments
1	Uniform 4.25/macrocell	Uniform	Uncorrelated	Capacity enhancement
2	Non-uniform [10–100]/macrocell	Uniform	Uncorrelated	Sensitivity to non-uniform UE density across macrocells
3	Non-uniform [10–100]/macrocell	Uniform	Correlated**	Cell edge enhancement
4a, 4b	Non-uniform***	Clusters	Correlated**	Hotspot capacity enhancement

*New node density is proportional to the UE density in each macrocell. UE density is defined as the number of UEs in the geographic area of a macrocell.
**Relay and hotzone nodes, often deployed by planning, see Section A2.1.1.4 of [14].
***Clustered UE placement for hotzone cells:

- Fix the total number of users, N_{users}, dropped within each macro geographical area, where N_{users} is 30 or 60 in fading scenarios and 60 in non-fading scenarios.
- Randomly and uniformly drop the configured number of low-power nodes, N, within each macro geographical area (the same number N for every macro geographical area, where N may take values from $\{1, 2, 4, 10\}$).
- Randomly and uniformly drop N_{users_lpn} users within a 40 m radius of each low-power node, where $N_{users\_lpn} = \lfloor P^{hotspot} \times N_{users}/N \rfloor$ with $P^{hotspot}$ defined in Table A.2.1.1.2-5 of [14], where $P^{hotspot}$ is the fraction of all hotspot users over the total number of users in the network.
- Randomly and uniformly drop the remaining users, $N_{users} - N_{users\_lpn} * N$, to the entire macro geographical area of the given macrocell (including the low-power node user dropping area).

Table 4.8 Configuration #4a and #4b parameters for clustered user dropping [14] *Source:* 3GPP 36.814. Reproduced with permission.

Configuration	N_{users}	N	$P^{hotspot}$
Configuration #4a*	30 or 60	1	1/15
		2	2/15
		4	4/15
		10	2/3
Configuration #4b	30 or 60	1	2/3
		2*	2/3*
		4*	2/3*

*Baseline for Configurations #4a and #4b.

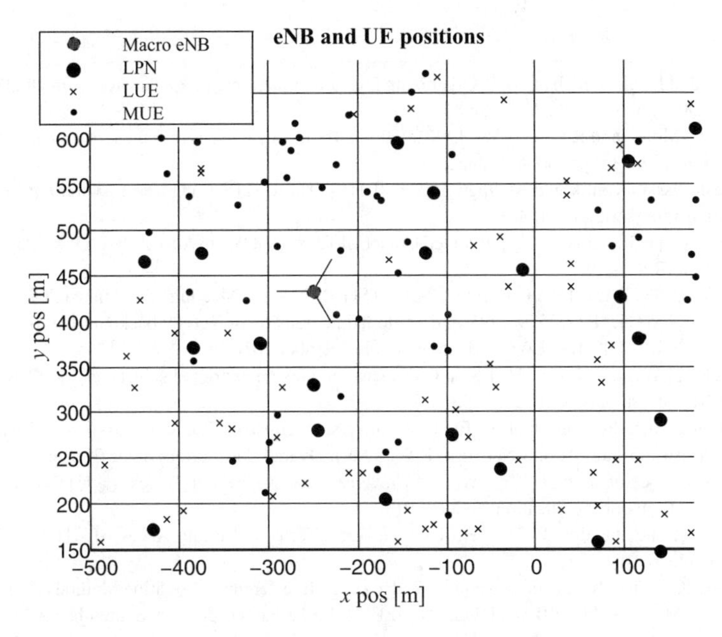

Figure 4.19 User dropping in Configuration #1 (uniform dropping).

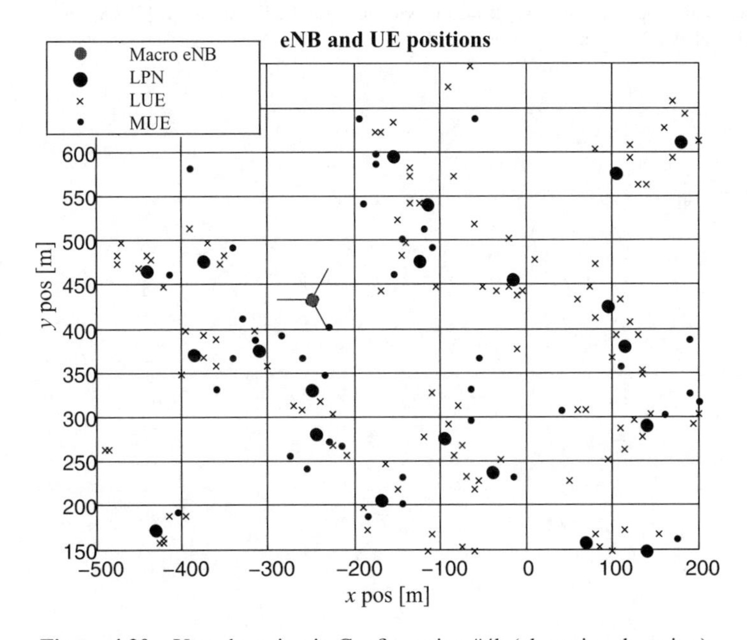

Figure 4.20 User dropping in Configuration #4b (clustering dropping).

References

1. A. L. Stolyar and H. Viswanathan, 'Self-Organizing Dynamic Fractional Frequency Reuse in OFDMA Systems', INFOCOM 2008, pp. 691–699, April 2008.
2. R. Giuliano, C. Monti, and P. Loreti, 'WiMAX fractional frequency reuse for rural environments', *IEEE Wireless Communications*, Vol. 15, pp. 60–65, June 2008.
3. Stefania Sesia, Issam Toufik and Matthew Baker, 'LTE – The UMTS Long Term Evolution: From Theory to Practice', John Wiley & Sons, 2009.
4. 3GPP Technical Specification 36.211, 'Evolved Universal Terrestrial Radio Access (E-UTRA); Physical Channels and Modulation (Release 9)'.
5. Motorola, 'R1-073371, E-UTRA Coverage', 3GPP TSG RAN WG1 Meeting #50, August 2007.
6. 3GPP TSG RAN WG1, 'R1-105793, LS on time-domain extension of Rel 8/9 backhaul-based ICIC for Macro-Pico scenario', 3GPP TSG RAN WG1 Meeting #62bis, October 2010.
7. 3GPP TSG RAN WG1, 'R1-106551, LS on CSI measurements on restricted subframes for eICIC', 3GPP TSG RAN WG1 Meeting #63, November 2010.
8. 3GPP Technical Specification 36.300, 'Evolved Universal Terrestrial Radio Access (E-UTRA) and Evolved Universal Terrestrial Radio Access Network (E-UTRAN); Overall description; Stage 2; (Release 9)'.
9. 3GPP Technical Specification 36.331, 'Evolved Universal Terrestrial Radio Access (E-UTRA), Radio Resource Control (RRC); Protocol specification (Release 9)'.
10. 3GPP Technical Specification 36.213, 'Evolved Universal Terrestrial Radio Access (E-UTRA); Physical layer procedures (Release 9)'.
11. A. Szufarska, K. Safjan, K. I. Pedersen and F. Frederiksen, 'Interference Mitigation Methods for LTE-Advanced Networks with Macro and HeNB Deployments', IEEE VTC Fall 2011, pp. 1–5, September 2011.
12. 3GPP Technical Report 36.921, 'Evolved Universal Terrestrial Radio Access (E-UTRA); Home eNode B (HeNB) Radio Frequency (RF) requirements analysis (Release 10)'.
13. New Postcom, 'R1-105225, Performance of power setting for Macro-Femto co-channel deployment', 3GPP TSG RAN WG1 Meeting #62bis, October 2010.
14. 3GPP Technical Report 36.814, 'Evolved Universal Terrestrial Radio Access (E-UTRA); Further advancements for E-UTRA physical layer aspects (Release 9)'.

5

Inter-cell Interference Management for Heterogeneous Networks

Sai Ho Wong and Zander Zhongding Lei
Institute for Infocomm Research, Singapore

5.1 Introduction

Both 3GPP LTE-Advanced [1] and IEEE WirelessMAN-Advanced have been found to fulfil the requirements [2] defined by IMT-Advanced for fourth generation (4G) cellular mobile networks. Both employ similar technologies, such as OFDMA combined with multiple antennas, channel dependent scheduling and carrier aggregation (CA). It has been shown in [3] that the targeted data rates and spectrum efficiencies can be achieved by employing advanced multiple-input and multiple-output (MIMO) techniques and aggregating multiple component carriers for LTE-Advanced. Commercial operation of LTE and WiMAX networks have started in several countries and deployment for these pre-4G technologies is expected to accelerate rapidly in the next few years. With the spectral efficiency per link of pre-4G technologies already approaching theoretical limits [4], it is important for 4G cellular networks to provide good user experience and quality of service (QoS) everywhere within the cell coverage. One of the enabling techniques is a topology known as heterogeneous networks (HetNets) [5], in which operators leverage on the economy of scale provided by the already-deployed 3G networks and reuse the infrastructure effectively.

In traditional homogeneous cellular networks, the base stations are of similar coverage and are carefully planned by operators with limited overlapping. These base stations support roughly equal number of mobile user equipment (UEs) per cell, and when traffic demand grows, QoS is usually maintained through cell splitting. However, cell splitting gain is limited by the resulting inter-cell interference, which can be an issue in dense deployments.

Heterogeneous Cellular Networks, First Edition. Edited by Rose Qingyang Hu and Yi Qian.

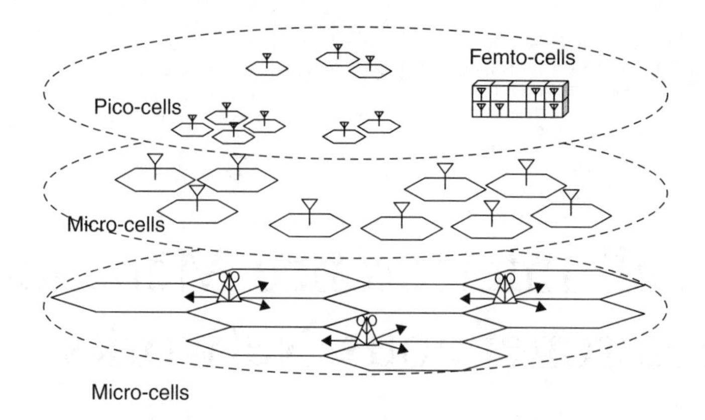

Figure 5.1 Base stations of different coverage overlaid on each other in a heterogeneous network.

Heterogeneous networks on the other hand, consist of layers of low-power nodes (LPNs) overlaid on the traditional planned homogeneous network to achieve the highest user throughput and/or the widest coverage according to the radio environment and required QoS [6]. In decreasing coverage and transmission power, these base stations can be classified into macro-, micro-, pico- and femtocells, as illustrated in Figure 5.1.

The macro layer is similar to a conventional homogeneous network, in which cells are planned and deployed for coverage and seamless mobility. Cells in the macro layer have the highest transmission power compared to other layers.

Relays with wirelessly connected backhaul can be overlaid on the macro layer to tackle the coverage holes that may otherwise exist there. There are many flavours of relays, such as in-band relay, out-band relay and mobile relay. Relays present additional challenges with regards to interference handling. For example, a relay transmitter may cause interference to its own receiver if it is transmitting to provide access to mobile terminals and receiving backhaul connectivity on the same time/frequency resources. Since this chapter deals with inter-cell interference across different heterogeneous layers, such relay-specific interference handling is out-of-scope.

In the micro/pico layer, low-power nodes can be installed by the operator to serve a number of UEs gathered at a certain location, such as in a stadium or a restaurant. Relying on the macro layer alone may not be able to meet the traffic required for such a user concentration. On the other hand with a congregated part of the macro traffic offloaded to the micro/pico layer, load-balancing can be achieved and the overall system capacity and user experience can be improved. As the terms suggest, the main difference of microcell from picocell is in its larger coverage; to a certain extent we can consider relays as a special class of microcells with wireless backhauls. Any macro-pico ICIC techniques described in this chapter are equally applicable to macro-micro deployment.

Lastly in the femto-layer, the cells are set up by the users at home without coordination. In addition, femtocells are usually configured to be operating in closed-subscriber group (CSG) mode, allowing access to a few predetermined UEs for association since it makes use of the users' broadband access at home. This is in contrast with macro-, micro- or picocells in which the location is planned by the operator and access is open to all UEs. With high terminal penetration, growing demand for fixed-mobile convergence and mobile data offloading, femtocells are expected to provide attractive services and data rates in home and indoor environments [7].

Low-power node deployment (e.g., pico- or femtocells) on top of the traditional homogeneous network (i.e., macrocells) causes many interference issues that are not observed in traditional networks. In the upcoming sections, we would like to present enhanced inter-cell interference coordination techniques to address the issue of heterogeneous network interference in a single-radio-access technology (single-RAT) environment, classified into frequency, time, power and spatial domains. While the techniques presented are specific to 3GPP LTE-Advanced, similar principles can also be used in other OFDMA based systems such as IEEE WirelessMAN-Advanced with relevant system-specific tweaks applied. We concentrate on the single-RAT environment since it is more challenging than its multi-RAT counterpart, which can essentially be considered under frequency domain enhanced inter-cell interference coordination.

5.2 Conventional Inter-cell Interference Coordination

The effectiveness of the enhanced ICIC schemes for heterogeneous networks need to be compared with the baseline ICIC techniques already available for homogeneous macro deployment, which will be briefly described in this section before we introduce enhanced ICIC schemes for heterogeneous networks.

Down-tilting – Down-tilting of the base station antennas is a simple static interference reduction technique that improves cell isolation [33]. Figure 5.2(a) shows the simulated antenna gain pattern without any down-tilting for a single site with three-cell macro deployment, and significant radiation out of the cell boundary can be observed. Figure 5.2(b) shows the same antenna setup with an electrical down-tilt of 12°, and the out-of-cell interference is reduced.

Frequency reuse – Another static interference mitigation technique is frequency reuse. Traditional fractional frequency reuse (FFR), as shown in Figure 5.3(a), can completely orthogonalize the available spectrum, but this is done in the expense of a reduction in the total usable bandwidth for both cell-centre and cell-edge UEs and hence decreases the throughput performance. Other variants such as soft frequency reuse (SFR) try to improve on FFR, in which the cell-edge UEs' resource allocation is based on a certain frequency reuse factor, while the cell-centre UEs enjoy the rest of the spectrum that is not used by its own cell edge UEs as illustrated by Figure 5.3(b). FFR and its variants can be supported effectively by scheduling, in which the scheduler may take account of the reported subband channel quality index (CQI) that reflects the interference condition of the reported subband [7]. Cell-edge UEs suffer from large path loss from their serving eNB and high interference from the neighbouring non-serving cells on the downlink, and the schedulers involved in different cells can engage in various FFR schemes for ICIC.

Semi-static ICIC – Interference in LTE is managed semi-statically in the frequency domain, in conjunction with power control by the macro base stations, also known as enhanced nodeBs (eNBs). Coordination is facilitated by inter-eNB signalling through the X2 interface [10].

A bitmap called relative narrowband transmission power (RNTP), which represents the load condition for each physical resource block (PRB), can be exchanged among neighbouring eNBs for the purpose of proactive downlink ICIC [11], in which the eNBs exchange the

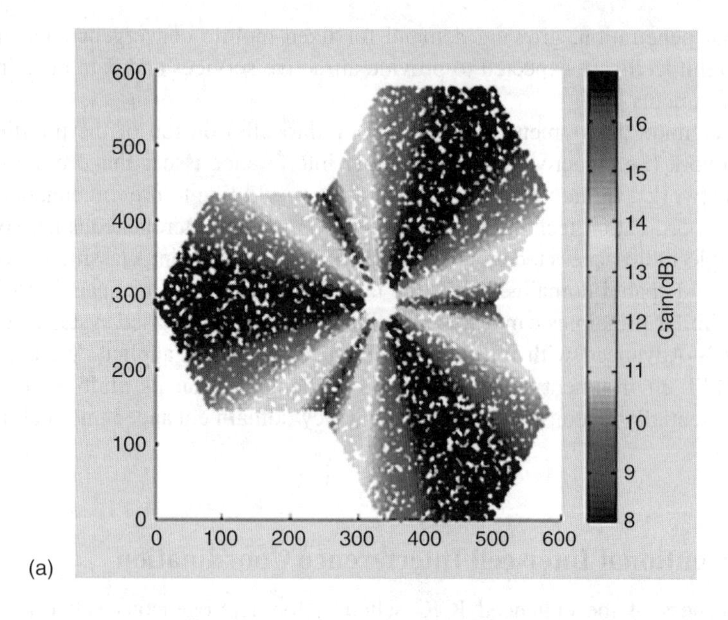

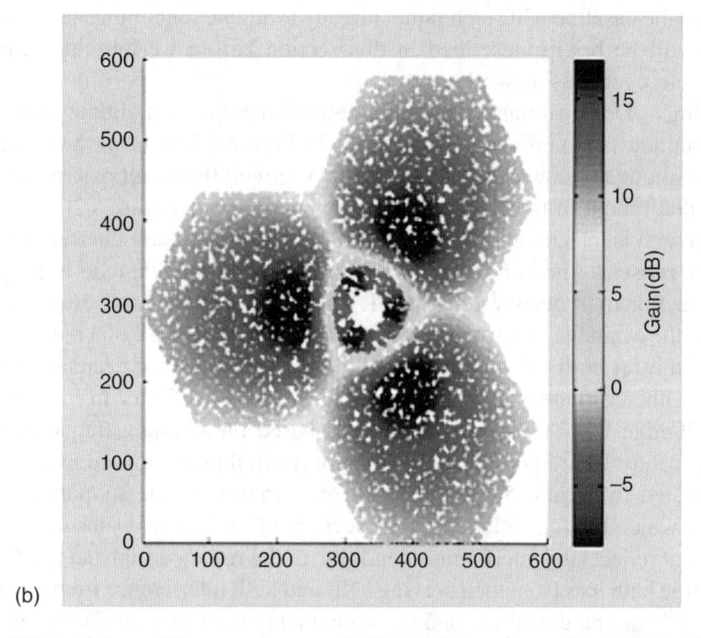

Figure 5.2 Effect of down-tilting on antenna gain G. (a) With no down-tilt (b) With 12° down-tilt.

scheduled transmission information that is going to take place for each PRB. Each PRB is a representation of a group of time and frequency resource elements (REs), and a PRB is defined as 12 subcarriers over 0.5 ms in LTE. For example in Figure 5.4, eNB A sends the RNTP bitmap for each of its six PRBs via the X2 interface to inform eNB B that the transmission power for PRBs 1, 2, 3 and 4 will be below a certain configurable threshold for a configurable period of time, while PRBs 0 and 5 will be above. Similarly, eNB B sends the RNTP bitmap over

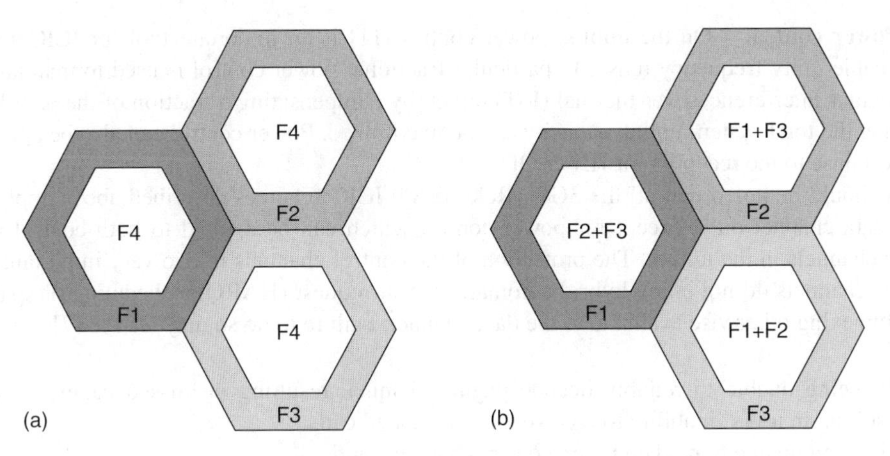

Figure 5.3 (a) Fractional frequency reuse (b) Soft frequency reuse.

the X2 interface to inform eNB A that the transmission powers for PRBs 0, 2, 3 and 5 will be below and PRBs 1 and 4 will be above threshold. As a result, both eNBs may attempt to prioritize the scheduling of physical downlink shared channel (PDSCH) transmission to minimize inter-cell interference in the frequency domain. It should be noted that although RNTP provides information about the expected level of interference in each PRB, the exact behaviour for the eNB scheduler upon receiving RNTP is not standardized.

High interference indicator (HII) is the functional dual of RNTP in uplink (UL), originating from an eNB to inform its neighbouring eNBs that it will be scheduling some UEs with high transmission powers and hence high interference to others at those frequency locations. Besides a proactive indicator such as HII, a reactive indicator called the overload indicator (OI) may be exchanged over X2 for UL ICIC. The OI takes three values, indicating low, medium or high levels of interference received for each PRB in an eNB. Similar to RNTP, the exact behaviour for eNBs receiving HII or OI is not standardized. These semi-static ICIC schemes are not updated more often than every 20 ms, which also helps to reduce the signalling load over the X2 interface.

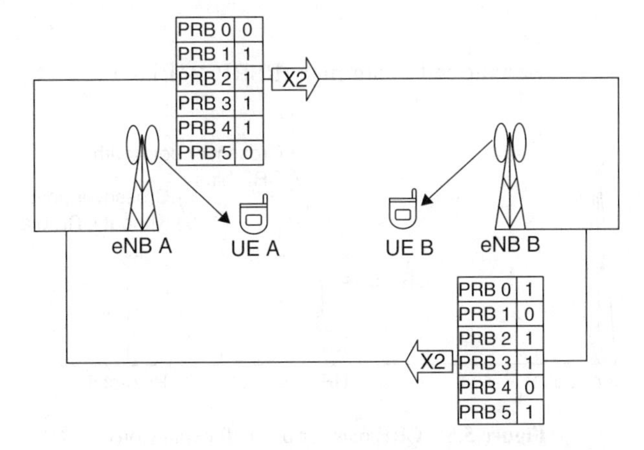

Figure 5.4 RNTP for inter-cell interference coordination.

Power control – On the uplink, power control [11] is an important tool for ICIC so as to enable unity frequency reuse. In particular fractional power control is used to maintain a particular interference-over-thermal (IoT) target, by compensating a fraction of the path loss so that the total system uplink capacity can be maximized. Power control can also be applied in response to the reception of HII or OI.

It should be noted that all the 3GPP Release-8/9 ICIC schemes described above apply to the data channel only, except for power control which can be applied to both control and data channels in the uplink. The protection of the control channels is also very important, as these channels do not enjoy hybrid automatic repeat request (HARQ), scheduling or spatial multiplexing otherwise available to the data channel. Failure to do so may lead to [7]:

- UE being unable to reliably decode paging channel, resulting in missed pages, in turn resulting in users' inability to receive UE-terminated calls,
- UE being unable to read common control channels, and
- Throughput degradation or degraded PDSCH performance.

5.3 Enhanced Inter-cell Interference Coordination

5.3.1 Interference Scenarios in Heterogeneous Networks

5.3.1.1 Macro-pico

Evaluations within 3GPP have shown that without any cell range expansion (CRE) biasing for the picocell, Release-8/9 ICIC schemes described earlier are effective for data and control channels protection in both downlink and uplink [12]. This assumes that the UE associates with the cell that has the strongest received power, through a measurement called reference signal received power (RSRP). As illustrated in Figure 5.5, without CRE bias the RSRP crossover point may occur quite close to the picocell due to its handicap in the transmission power with respect to the macrocell. This may not offer too much traffic offloading from the macro, defeating the purpose of picocell deployment. CRE bias is a value added artificially to the RSRP of the LPN, so that the UEs are biased towards associating with the LPNs, and in effect increasing the coverage of the LPNs. Hence the serving cell is chosen among i cells according to (5.1)

$$\text{Serving cell} = \arg\max_i(\text{RSRP}_i + \text{Bias}_i) \tag{5.1}$$

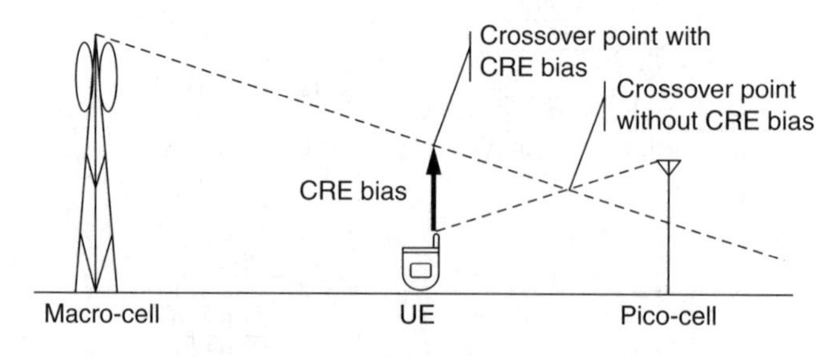

Figure 5.5 CRE bias for picocell expansion.

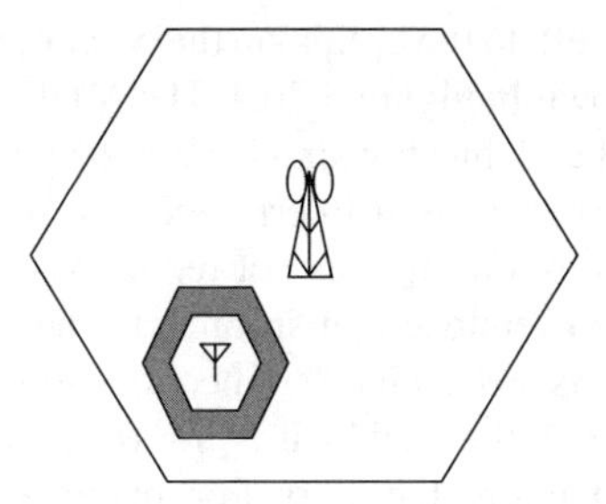

Figure 5.6 UEs in the shaded area suffer excessive interference from the macrocell due to cell range expansion bias.

It has further been shown in [12] that in a typical macro-pico deployment with a small CRE bias, downlink interference is not a problem. However, when a high CRE bias is used, the cell edge pico-UEs suffer serious interference from the macrocell as illustrated in Figure 5.6.

5.3.1.2 Macro-femto

Figure 5.7(a) shows an example of the interference caused by CSG deployment, in which home UEs (HUEs) 1 and 2 are associated with femtocells 1 and 2 exclusively. During downlink (DL) transmission, the macro UE (MUE) sees a relatively strong signal from Femto 2 due to its proximity, but association with Femto 2 is not allowed due to its CSG nature. This strong downlink interference from CSG often overwhelms the signals from the macrocell, resulting

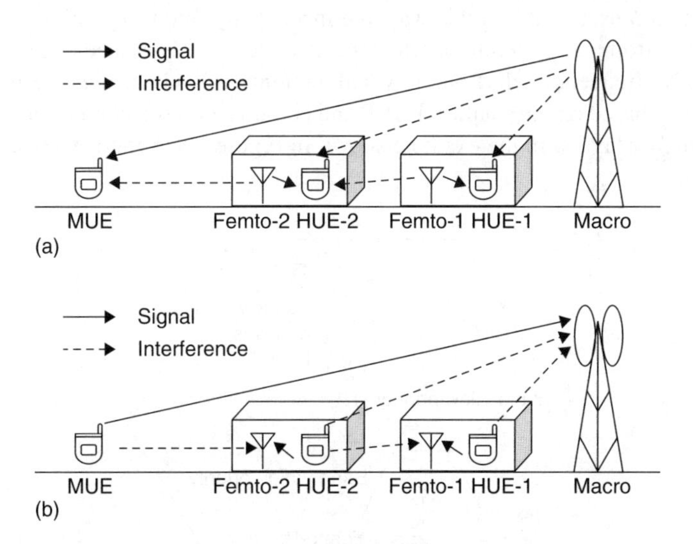

Figure 5.7 (a) Heterogeneous network interference on downlink, based on Figure 3 of [30]. *Source:* I2R, 'R1-104732 eICIC for HeNB UL and MUE DL based on HeNB UL Measurement', Apr 2010. (b) Heterogeneous network interference on uplink, based on Figure 1 of [30]. *Source:* I2R, 'R1-104732 eICIC for HeNB UL and MUE DL based on HeNB UL Measurement', Apr 2010. Reproduced with permission.

in a coverage hole in the macrocell. In the uplink on the other hand, the interference aggressor–victim roles are reversed as shown in Figure 5.7(b). The MUE needs to overcome a relatively large path loss to the macrocell and thus transmits at a large power due to power control from the macrocell. This in turn creates a lot of interference for the HUEs' uplink transmissions nearby, since conventional power-control does not require too much transmission power due to a relatively small path loss for femtocell environment. The interference situation is further aggravated by the fact that there is no backhaul connection between the macro- and femtocells, and thus semi-static ICIC through the backhaul is not operable, and that the deployment of the femtocells is uncoordinated among the users. For macro-femto, it has been shown in [13] that Release-8/9 ICIC techniques are not fully effective in mitigating interference, especially when non-CSG/CSG users are in close proximity of the femto.

5.3.2 Enhanced ICIC Solutions for Heterogeneous Networks

5.3.2.1 Frequency Domain Enhanced ICIC

The previous section has introduced legacy ICIC methods, which are essentially frequency domain ICIC techniques, relying on the scheduler to achieve interference avoidance based on RNTP. As such, these legacy schemes are only useful for channels that can be scheduled, but not for control channels. In this section, we will present further frequency domain techniques for enhanced ICIC.

Resource reservation – One of the simplest forms of resource partitioning between macrocell and LPNs is for them to be frequency division multiplexed (FDM) onto different frequencies. For example, the whole bandwidth can be divided into two bands, each exclusively for macro or femto users only. In this way, the macro-only frequency band is free from any coverage holes introduced by femto interference. Another closely related concept is described in [7] as adaptive frequency selection, in which the femtocell adaptively selects its operating carrier frequency based on the smallest RSRP and certain cell reselection priority information signalled by higher layers. In the example shown in Figure 5.8, Femto-A selects band 2 that

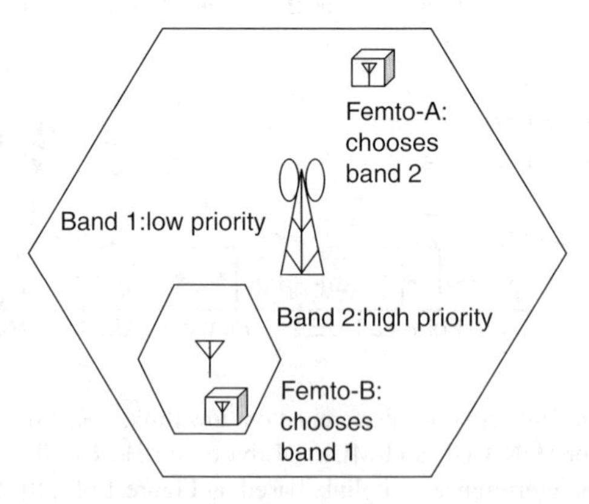

Figure 5.8 Adaptive frequency selection by femtocells, according to RSRP and frequency band priority, based on Figure 7.2.2.1-2 of [7]. *Source:* 3GPP, 'TR 36.921 v10.0.0, FDD Home eNodeB (HeNB) Radio Frequency (RF) requirements analysis (Release 9)', Apr 2011. Reproduced with permission.

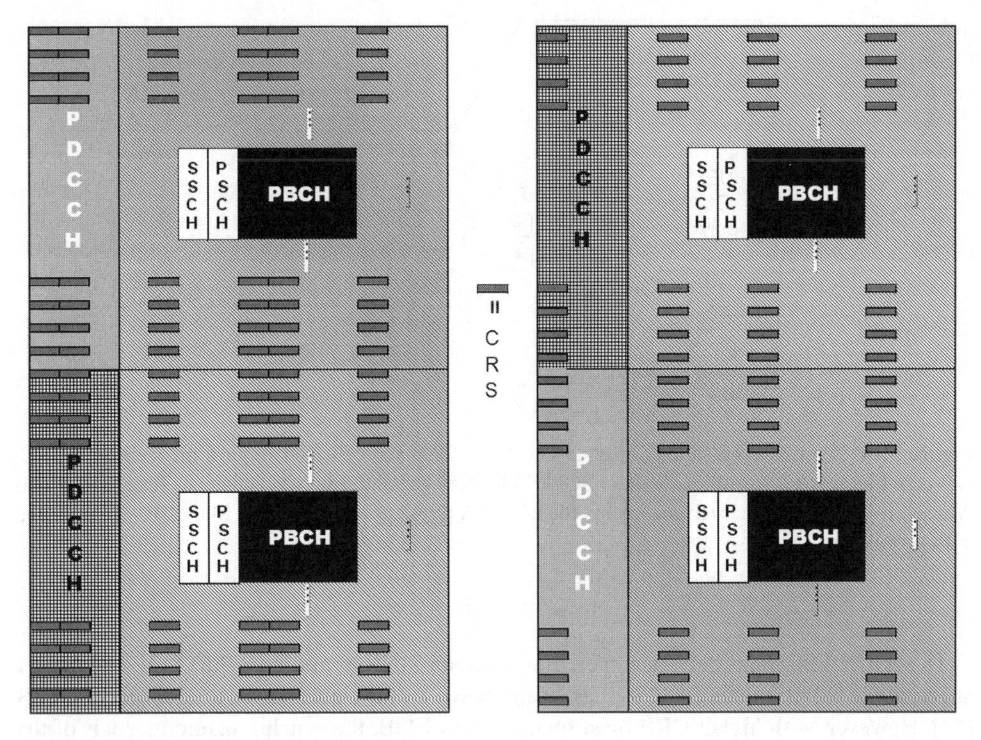

Figure 5.9 Carrier aggregation – PDCCHs of different shades denote possible different transmission power; CRS is shifted between different cells.

corresponds to the band with the smaller RSRP. In case of multiple layers overlaid on each other, as in Femto-B, it selects the lower priority carrier frequency band 1.

Carrier aggregation (CA) – One issue with the resource reservation is that part of the frequency band is possibly under-utilized due to FDM. This can be addressed by cross-carrier scheduling, introduced in Release-10 under carrier aggregation [3]. A component carrier (CC) is a frequency band that is backward-compatible with Release-8/9, with the possibility of standalone operation. Cell-specific reference symbols (CRS) are located in all CCs, with physical broadcast channel (PBCH), primary synchronization channel (PSCH) and secondary synchronization channel (SSCH) in the centre six PRBs of each CC, as shown in Figure 5.9. Referring to Figure 5.10, in which two CCs are aggregated, CC1 can be used for both data and control information of the macrocell, while CC2 is mainly for data. The other cell can complementarily use CC2 for both data and control information and CC1 mainly for data. Backward compatibility is ensured by scheduling Release-8/9 UEs in one of the component carriers while Release-10 UEs capable of CA can be cross-carrier scheduled by the primary component carrier, achieving full bandwidth utilization. Thus when the heterogeneous layers are time synchronized, interference on the physical downlink control channel (PDCCH) can be avoided by FDM without sacrificing the available bandwidth to the data channel. Data channel interference can be handled by other means such as scheduling, and legacy schemes such as RNTP. However, as shown in Figure 5.9, interference of the PSCH, SSCH and PBCH still exists and needs to be handled by other means.

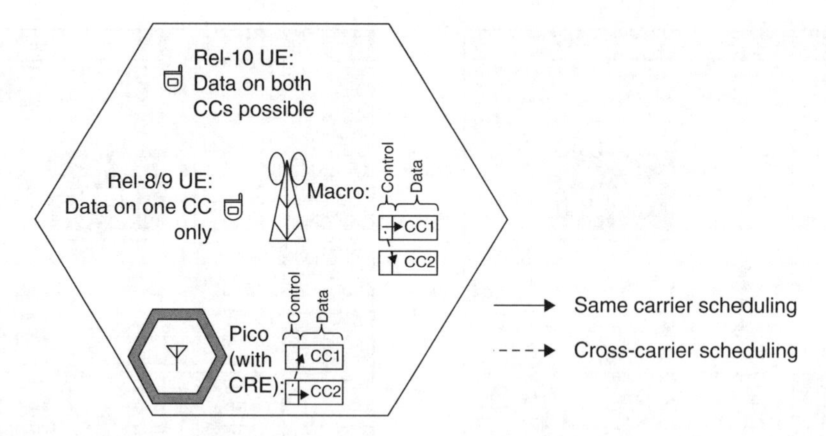

Figure 5.10 Carrier aggregation for enhanced ICIC for macrocell interfering a picocell with CRE, based on Figure 9A.2.1-1 of [3]. *Source:* 3GPP, 'TR 36.814 v9.0.0, Evolved Universal Terrestrial Radio Access (E-UTRA); Further advancements for E-UTRA physical layer aspects', Mar 2010. Reproduced with permission.

The PSCH/SSCH has been designed to work in an SINR of around −9dB [42], and carrier aggregation based enhanced ICIC has been shown to be viable up to a CRE bias of 6dB [43]. However, with higher CRE bias, such as up to 12dB, the synchronization signals of the picocell in the CRE region can be highly interfered by the macro layer, making it difficult for UEs to acquire synchronization and derive the cyclic prefix length from PSCH/SSCH of the victim picocell. A solution to this problem has been proposed in [44], by employing an optional RRC signalling of synchronization and system information of the other aggregated carrier such as the cell ID cyclic prefix length. This may be necessary as UEs operating in the cell range expansion zone may have to solely rely on these signalled parameters to maintain synchronization due to excessive interference to the synchronization channels. An alternative solution is to detect the aggressor's PSCH/SSCH/PBCH for the victim's use, assuming tight synchronization among the cells in the heterogeneous layers [47].

Carrier segment – Another similar enhanced ICIC proposal to CA is known as carrier segment [17], which provides more orthogonality for the control channels, especially for the PSCH, SSCH and PBCH. The differences between carrier segment and carrier aggregation are highlighted in Table 5.1. As CRS does not span over the whole bandwidth, CRS-CRS or CRS-data interference can be avoided. Some energy saving can be achieved as CRS needs not to be transmitted all the time. However, carrier segments are not backward compatible to Release 8/9/10. Other flavours of such new carrier types (NCT) are being investigated in 3GPP Release-12.

Carrier offset – If the difference between the operating bandwidths of the aggressor and victim cells is large enough, carrier offset of more than six PRBs between them can help to mitigate the PSCH, SSCH and PBCH interference [7], as explained in the preceeding paragraph. The aggressor cell can further mute the frequency region that overlaps with the six PRBs belonging to the victim's SCH/PBCH region by not scheduling PDSCH there, as illustrated in Figure 5.11.

Table 5.1 Difference between carrier aggregation and carrier segment

Carrier aggregation	Carrier segment
multiple PDCCHs in whole bandwidth	only one PDCCH, located in a reduced bandwidth, to schedule whole bandwidth
PDCCH region is same as PDSCH region for R8/R10 UE	PDCCH region is same as PDSCH region for R8 UE and smaller than PDSCH for R10
CRS locates in all CCs (whole bandwidth)	CRS share same bandwidth with PDCCH (partial bandwidth)
PBCH/SCH on each CC	PBCH/SCH share same bandwidth with PDCCH (partial bandwidth) and same MAC procedure and system information maintenance as single CC in R8
accepted as part of Release-10 specifications	not accepted as part of Release-10 specifications

A similar concept is to restrict the aggressor's PDSCH transmission within a limited subband, so that the interference to the victim cell can be controlled. This can be done semi-statically for instance, in which the victim cell resource block allocation is dependent on the UE's location. When HeNB gets its own location information, it will know which resource blocks will be assigned to a nearby macro UE and avoid using them, as illustrated in Figure 5.12.

Open-access femtocells [7] – These are femtocells that allow all users to have unlimited access rights. For this case, UEs can associate freely with the cell having the highest RSRP, resulting in a situation similar to macro-pico interference [31]. Its difference from a picocell is summarized in Table 5.2 below. However, this is not expected to be an important deployment scenario.

Hybrid cells [7] – Hybrid cells are in-between open access femtocells and CSGs, in which they provide different service levels to member and non-member UEs, and are supported in Release-9. Hybrid cells have more information about the victim UEs (i.e., non-member UEs) than pure CSG. A hybrid HeNB may reserve some frequency resources for non-member UEs' use, handing over to the macrocell with ICIC such as FFR when these resources are exhausted. In other cases, a hybrid cell may only provide a paging service to non-member UEs that are

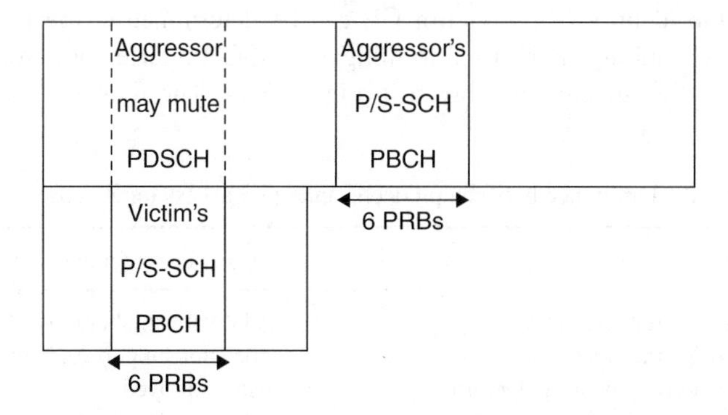

Figure 5.11 Partial bandwidth overlap with carrier offset.

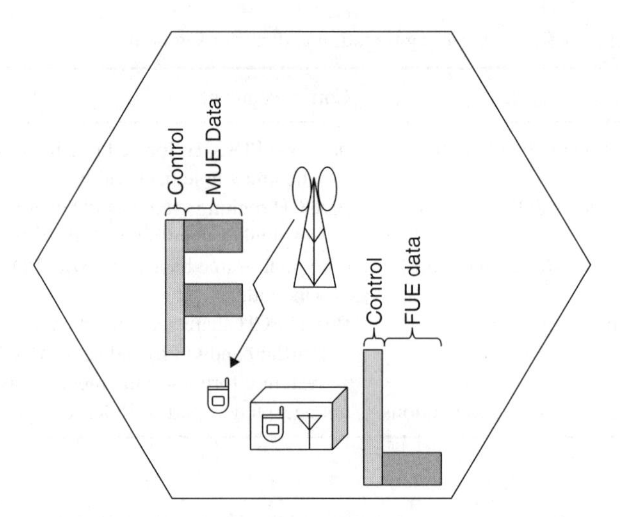

Figure 5.12 Limited subband PDSCH transmission.

allowed to camp and receive paging, but will be handed over to the macrocell if other services are requested. During such handover, the hybrid cell can engage in ICIC, such as transmission power reduction, to minimize interference especially to the macrocell's PDCCH.

Enhanced PDCCH (ePDCCH) [28] – In Release-8/9/10, PDCCH spans the whole system bandwidth, which makes control channel interference coordination for heterogeneous networks in frequency domain difficult. In Release-11, ePDCCH has been introduced, which allows FDM of different ePDCCHs over the same system bandwidth as shown in Figure 5.13. In addition, ePDCCH may make use of the UE-specific reference symbols (UERS) for demodulation to reap precoding and frequency-dependent scheduling gains that is otherwise unavailable to PDCCH. However, Release-8/9/10 UEs cannot benefit from this new design.

Victim detection [7] – By detecting the uplink transmission of the MUEs in the vicinity of the HeNB, for example by changes in the IoT or by the uplink RS characteristics, femtocells are able to ascertain the presence of MUEs so as to avoid activating enhanced ICIC unnecessarily. If backhaul is available, victim detection can be based on MUE reports being sent to the interfering femtocell via the backhaul.

For example, the identity of the victim UE can be determined at the macrocell side so that coordinated scheduling can be used to mitigate PDSCH interference from the femtocell [30]. The steps for this scheme are shown in Figure 5.14. The femtocell performs sensing

Table 5.2 Difference between picocells and open-access femtocells

Picocells	Open-access femtocells
relatively larger coverage	relatively smaller coverage
support X2 interface	does not support X2 interface
planned deployment by operator	user-deployed
relatively lower density	relatively higher density

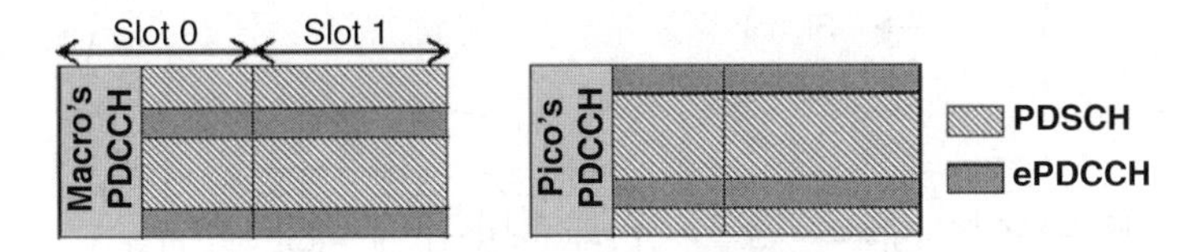

Figure 5.13 FDM of ePDCCHs of different layers.

of MUE uplink transmission and feeds back the interference pattern to the macrocell. The macrocell compares this pattern with its own uplink grants, and identifies the victim MUE. The macrocell then sends the downlink scheduling information to that MUE, and also forwards this information to the interfering femtocell so that it will avoid using those resource blocks.

Another non-backward-compatible method for victim UE detection is for the UEs under strong interference to send a 'distress bit' [29]. For example, CSG cells that detect this bit may activate their enhanced ICIC mechanisms, like that shown in Figure 5.15. This distress bit can ensure that these mechanisms are only turned on when necessary.

5.3.2.2 Time Domain Enhanced ICIC

Time domain solutions provide an additional 'interference-free' dimension and are identified as part of the baseline solutions for non-carrier-aggregation based (i.e., non-frequency-domain-based) enhanced ICIC in Release-10 [8]. However, since a FDD/TDD common solution is desirable [8], the impact of their application on TDD systems needs to be carefully studied due to the possible UL-DL interference. Time domain solutions assume that time synchronization across all participating heterogeneous layers can be achieved.

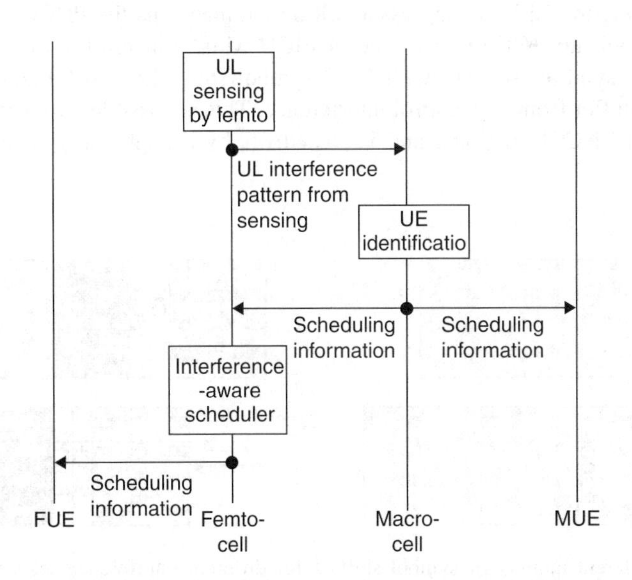

Figure 5.14 Victim identification with coordinated scheduling, based on Figure 4 of [30]. *Source:* I2R, 'R1-104732 eICIC for HeNB UL and MUE DL based on HeNB UL Measurement', Apr 2010.

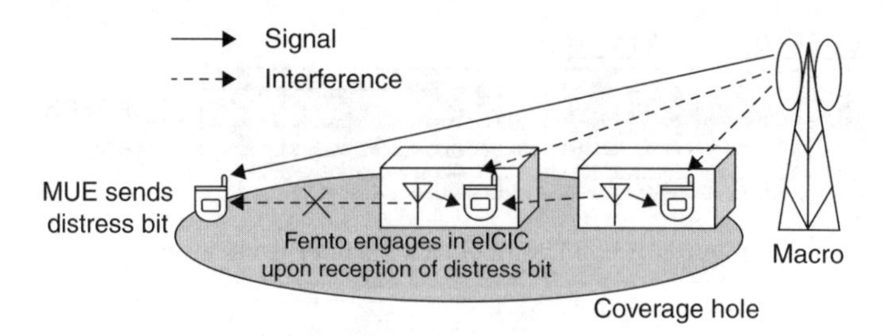

Figure 5.15 MUE sends distress bit to interfering femtocells.

Symbol shifting – Since PDCCH occupies the whole bandwidth, a straightforward approach to prevent direct control-control collision is to allow k symbols of offset between the aggressor and victim cells. Symbol shifting also helps in controlling CRS interference. Depending on the antenna configuration of the heterogeneous layers, various values of k may be necessary to provide interference avoidance to different control channels. Taking Figure 5.16 as an example, $k = 3$ and 4 are required for PDCCH and BCH protection respectively.

Symbol shifting is compatible with FDD since its UL and DL transmissions are on different frequency bands. For TDD, however, symbol shifting may cause excessive UL-DL interference between the shifted symbols [21].

Symbol level shifting can also be used to protect the first OFDM symbol, on which the physical control format indicator channel (PCFICH) and physical HARQ indicator channel (PHICH) is carried together with the PDCCH. The PCFICH resides on the first OFDM symbol, indicating the symbol span of the PDCCH, while the PHICH is transmitted either on the first one or three symbols.

As shown in Figure 5.17, the aggressor cell over-dimensions the PDCCH span by setting control format indicator (CFI) to 3 on the PCFICH while transmitting PDCCH only on the first two OFDM symbols so that, with $k = 2$ symbol shift, the first OFDM symbol on the victim does not suffer from any control interference. This is possible only with lightly loaded cells with small PDCCH regions, and is traded off by a slight drop of efficiency in the aggressor cell.

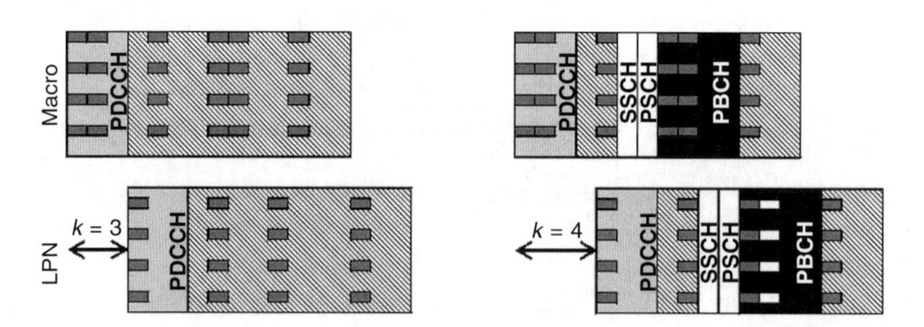

Figure 5.16 Different number of symbol shifts k for different interference scenarios. Macrocell is assumed to have four transmit antennas while the picocell has two. (a) PDCCH Protection (b) BCH Protection.

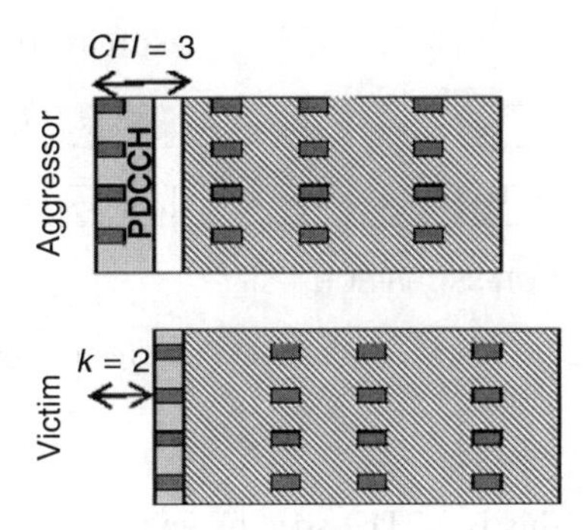

Figure 5.17 Protection of the first OFDM symbol [19].

An alternative construction involves an over-dimensioned PCFICH and a lightly loaded PDCCH. With CFI = 3, the lightly loaded PDCCH is located sparsely in all the three OFDM symbols, so that some of the resource elements are nulled [18] for interference control. PCFICH and PHICH can borrow power from the nulled REs if necessary. Legacy UEs are able to operate in the lightly-loaded PDCCH as long as they are constructed according to the currently defined rules for PDCCH blind decoding. The trade-off is however a less efficient PDSCH since the PDCCH is always over-dimensioned.

Although direct PCFICH collision between the macro- and picocells can be avoided by proper cell-ID planning, PCFICH may still suffer from CRS interference. Symbol shifting can help to mitigate this issue, but in some cases system constraints may render this method unviable. If the PCFICH is not demodulated properly, it may result in huge performance loss due to a wrong CFI value. When the PCFICH is not reliable, it may be possible to set the CFI value semi-statically and inform the UE by RRC signalling [41]. Another possibility is to always set the PHICH symbol span value to the maximum, so that PCFICH detection is not necessary.

Unnecessary radio link failure (RLF) may be declared due to a poor average link quality arising from the aggressor's PDSCH interference on the victim's CRS, despite that connection still being possible with enhanced ICIC applied. This RLF issue may be mitigated by nulling REs of the aggressor cell that overlap with the victim cell's CRS location, either by puncturing or rate matching. Puncturing is a backward-compatible technique but it brings performance degradation for the aggressor. Rate matching around the muted locations, on the other hand, is applicable to future releases only. Another solution is for Release-10 UEs to support restricting measurements at certain resources only [8], which is already supported by Release-10 specifications in conjunction with almost blank subframes (ABS, see below).

Subframe level shifting – Though symbol level shifting can be applied to prevent PBCH interference, for TDD part of the PBCH still interferes with the SSCH. To alleviate this, an s subframe shift can be applied on top of the k symbol shift. For FDD, s can take any value. Figure 5.18 shows an FDD system with $s = 2$ in which the PSCH/SSCH/PBCH interference across different layers are avoided.

On the other hand, subframe level shifting is only applicable to TDD configurations 0, 1 and 2 with 5 ms DL/UL switch-point periodicity, and only $s = 5$ is possible due to UL-DL interference [21].

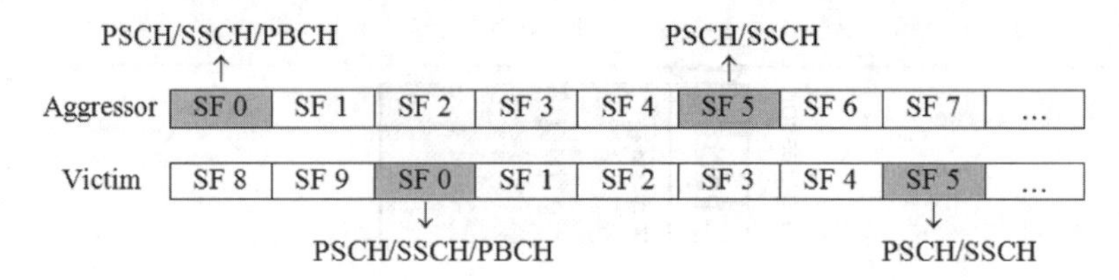

Figure 5.18 Subframe shifting for FDD, $s = 2$ to avoid interference of PSCH, SSCH and PBCH on different layers.

Fake uplink subframe – This is a TDD-specific enhanced ICIC method, in which a UL subframe is configured without any transmission except ACK/NACK. A potential drawback is that the ACK/NACK transmitted at the fake uplink subframe could be heavily interfered, which may result in unnecessary retransmissions [23].

MBSFN subframe – A multi-cast/broadcast single frequency network (MBSFN) subframe, as its name suggests, is designed to carry multi-cast/broadcast information. It is characterized by the presence of CRS, PCFICH and PHICH in its PDCCH region, which carries only uplink grants. As a result, the PDCCH span is at most two OFDM symbols in MBSFN subframes. The rest of the OFDM symbols beyond the PDCCH region are reserved for multi-cell data, and there is no unicast data or CRS. All UEs are aware of the allocation of MBSFN and non-MBSFN subframes regardless of whether multi-cast/broadcast reception is supported by individual UE. With symbol and/or subframe shift, the victim's control channels could still suffer interference from the aggressor's PDSCH. If no multi-cell data is present, MBSFN subframes can be applied on the aggressor to mitigate the interference due to their lack of PDSCH and CRS. The MBSFN subframe can also be applied in conjunction with symbol and/or subframe shift for better interference handling, as shown in Figure 5.19.

However, for both FDD and TDD, certain subframes cannot be configured as MBSFN subframes. These subframes are those carrying PSCH/SSCH, PBCH or paging information, namely subframes $\{0, 4, 5, 9\}$ in FDD and $\{0, 1, 5, 6\}$ in TDD. The possible MBSFN subframe configuration is summarized in Table 5.3 [34].

Due to the configuration restrictions as shown in Table 5.3, it would be important for the ACK/NACK feedback in response to an uplink grant for a cell-edge UE to be reliably received [22]. This can be ensured by sending an uplink grant at a subframe so that the corresponding

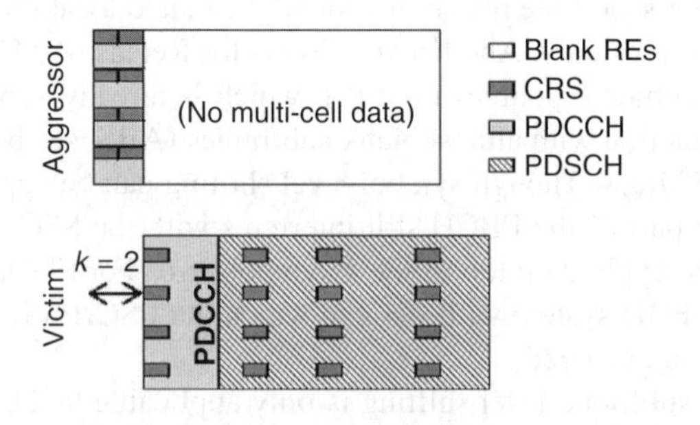

Figure 5.19 MBSFN subframe with $k = 2$ symbol shift.

Table 5.3 Subframes that cannot be configured as MBSFN are marked with ×

		Subframe number									
TDD Configuration (D = downlink S = special U = uplink)	UL-DL switch-point periodicity	0	1	2	3	4	5	6	7	8	9
0 (DSUUU)	5 ms	×	×	×	×	×	×	×	×	×	×
1 (DSUUD)	5 ms	×	×	×	×		×	×	×	×	
2 (DSUDD)	5 ms	×	×	×			×	×	×		
3 (DSUUU DDDDD)	10 ms	×	×	×	×	×	×				
4 (DSUUD DDDDD)	10 ms	×	×	×	×		×				
5 (DSUDD DDDDD)	10 ms	×	×	×			×				
6 (DSUUU DSUUD)	10 ms	×	×	×	×	×	×	×	×	×	†
FDD	NA	×				×	×				×

† See main text below for further explanation on this subframe

ACK/NACK response would be in a subframe that can be configured as an MBSFN subframe at the aggressor side for enhanced ICIC. On the other hand if an uplink grant is scheduled at subframe 9 in TDD configuration 6 (marked with † in Table 5.3), its corresponding ACK/NACK sent on subframe 0 may be heavily interfered because it cannot be configured as an MBSFN subframe.

Almost blank subframe – Almost blank subframe (ABS) is a fully backward compatible non-regular subframe, sharing many similarities with MBSFN subframe such as [15]:

- PSCH/SSCH transmission occurs in subframes 0 and 5 (FDD) and 0, 1, 5 and 6 (TDD),
- PBCH transmission occurs in subframe 0 and
- no unicast transmission.

ABS is characterized by a lack of PDCCH and the presence of CRS throughout the subframe. The presence of CRS is to ensure backward compatibility, but it creates CRS interference to other heterogeneous cells. CRS interference on the PDSCH can be further avoided by applying ABS in conjunction with MBSFN subframe. ABS can also be applied together with symbol and/or subframe shift.

In terms of HARQ, ABS has greater flexibility as it does not have similar configuration constraints for MBSFN subframes as recorded in Table 5.3, though significant interference may be present in subframes with SCH or PBCH transmission [22].

It should be noted that both MBSFN and ABSs require some coordination between the aggressor and victim cells, in which the subframe pattern that employs these non-regular

Aggressor
ABS
ABS
Victim
Serve cell edge UEs
Serve cell edge UEs

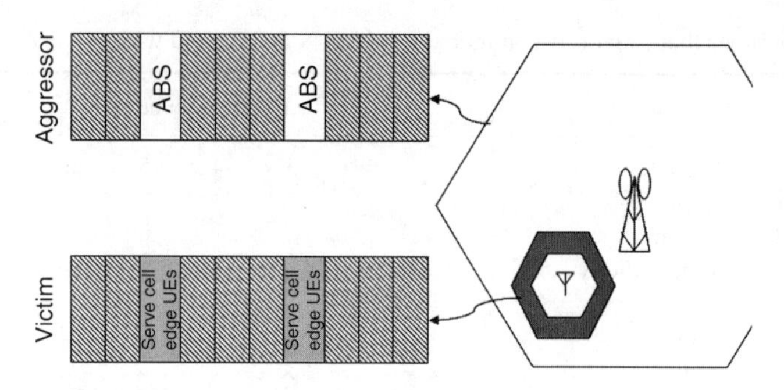

Figure 5.20 ABS subframe assignment and picocell with CRE.

subframes need to be known. In the absence of backhaul, such configuration can be static and communicated over the non-standardized operation and maintenance (OAM) interface, or the configuration is simply predefined. Figure 5.20 shows an example of the ABS subframe assignment used in conjunction with picocell CRE. For UEs located in the shaded cell range expansion zone, the macrocell aggressor assigns ABS to certain subframes so that these UEs do not suffer too much interference from the macrocell.

If backhaul is latency is negligible, for example with remote radio heads (RRHs, see section 5.3.2.4), dynamic ABS assignment is also possible. In such case, blanking coordination may also be extended to frequency domain as shown in Figure 5.21.

Even with ABS configured, the victim cell may still suffer from CRS interference. In [42], it has been shown that the performance of CRS interference cancellation is robust despite the presence of strong CRS interference. However, interference cancellation is a receiver-based implementation and would not be directly specified. Another receiver-based method is by selectively puncturing the REs that are heavily interfered by CRS, taking account of the tradeoff between interference reduction gain and puncturing loss [48].

5.3.2.3 Power Domain eICIC

In the current specifications for Release-8/9/10, only the maximum transmission power is defined for the low-power nodes. The exact power setting scheme is hence implementation dependent, making interference control for LPNs suboptimal. For instance, it is important to control the transmission power of the CSG to decrease the size of the coverage holes seen by the macro UEs.

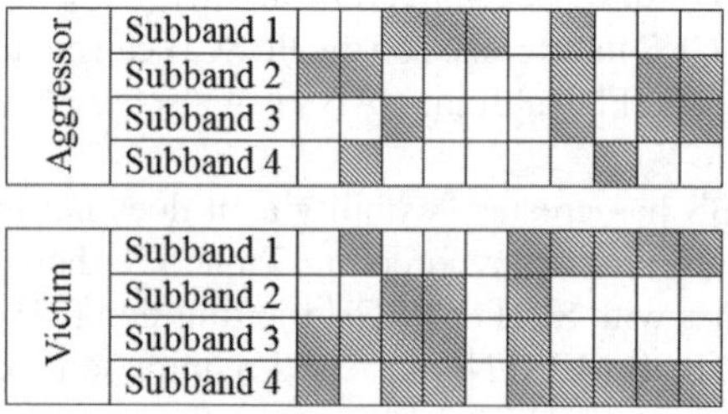

Aggressor
Subband 1
Subband 2
Subband 3
Subband 4
Victim
Subband 1
Subband 2
Subband 3
Subband 4

Figure 5.21 Dynamic blanking.

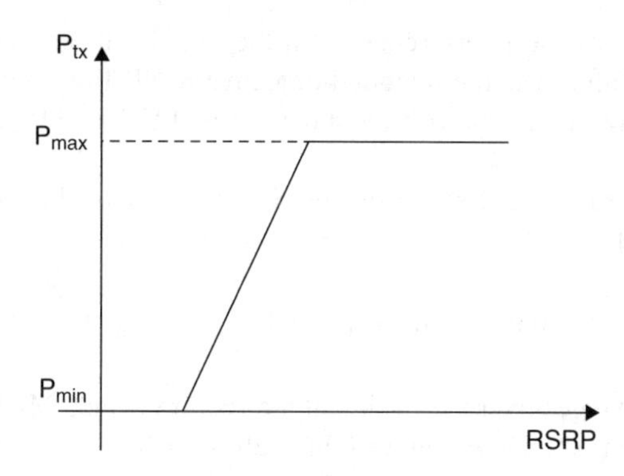

Figure 5.22 Illustration of power control formula based on (5.2).

Power setting based on strongest receiving power of macro at the femto [25] – This relies on the femtocell performing RSRP measurement P_M from the strongest macrocell, so that its transmission power P_{tx} is set according to:

$$P_{tx} = \max(\min(\alpha \cdot P_M + \beta,\ P_{max}), P_{min})\,[\text{dBm}], \tag{5.2}$$

in which P_{max} and P_{min} are the maximum and minimum transmission power of the femtocell as defined in the specifications, α and β are parameters that control the slope and dynamic range of power control respectively. An example of the transfer function according (5.2) is shown in Figure 5.22.

The advantage of this scheme is that it requires no information exchange between base stations or any form of MUE measurement. The transmission power is dependent on the path loss plus shadowing from the macrocell. On the other hand, if control over power settings at individual CSG from the operator side is desired, this is still possible via OAM.

In some cases, the penetration loss due to the wall and the indoor path loss between the MUE and the interfering CSG can be taken into account for more accurate interference control [7]. This can be done by two methods [26]:

- RSRP measurement report from MUE based on the femtocell to the macrocell (i.e., femto to MUE path loss) – the macrocell transmits the RSRP report to the femto via backhaul,
- femtocell overhearing uplink transmission of MUE and using the UL RSRP to approximate the DL RSRP, but the MUE uplink resource block allocation needs to be first communicated to the femtocell over the backhaul.

Power setting based on SINR measured by UEs – By controlling the objective SINR of femto-UE (FUE), the interference suffered by MUE in proximity to a CSG can be reduced, which is achieved by setting the transmission power of the femtocell based on [27]:

$$P_{FUE\_received} = 10\log_{10}(10^{Intf/10} + 10^{N_0/10}) + x[\text{dBm}] \tag{5.3}$$

$$P_{HeNB} = \max(\min(PL_{est} + P_{FUE\_received},\ P_{max}), P_{min})[\text{dBm}] \tag{5.4}$$

In which *Intf* represents the interference from the macrocell detected by the FUE, N_0 is the background noise value, x is the intended objective SINR limit for the FUE, PL_{est} is the estimated path loss between the femtocell and the served FUE, and P_{max} and P_{min} are defined according to (5.2).

Another power control method relies on the SINR report of the MUE according to the following equation [26]:

$$P_{tx} = \max(\min(\alpha \cdot P_{SINR} + ß, P_{max}), P_{min})[dBm] \tag{5.5}$$

In this case, X2 exchanges over the backhaul are required since MUE is not connected to, and hence is not reporting to, the aggressor femtocell.

PBCH power control and muting [15] – The PBCH comprises four independently self-decodable units, transmitted repeatedly every 10 ms. As shown in Figure 5.9, the PBCH occupies the centre 72 subcarriers over four OFDM symbols. Most FUEs are in good channel condition, which implies that a FUE can decode the PBCH without receiving the full set of four repeated PBCHs. The femtocell may transmit PBCH at very low power (e.g., be smaller than the receiving power of MeNB at Femto) when transmitting PBCH, or mute some of the PBCH transmission [50].

5.3.2.4 Spatial Domain Enhanced ICIC

When multiple transmit antennas are available at the heterogeneous cells, spatial domain enhanced ICIC is possible. In fact, most of the spatial domain schemes described in this section would also fall under the scope of coordinated multi-point (CoMP) in 3GPP, since these techniques would also require dynamic coordination among multiple geographically separated transmission points [35]. This dynamic coordination requires a backhaul with negligible latency, so that the UE reports used would not be outdated due to a time varying mobile channel.

With the advancement of fibre optics, remote radio heads (RRHs) may be an important deployment scenario, in which the transmitting antennas are physically separated but fibre connected to the baseband processing unit (BBU). This is shown in Figure 5.23, in which each

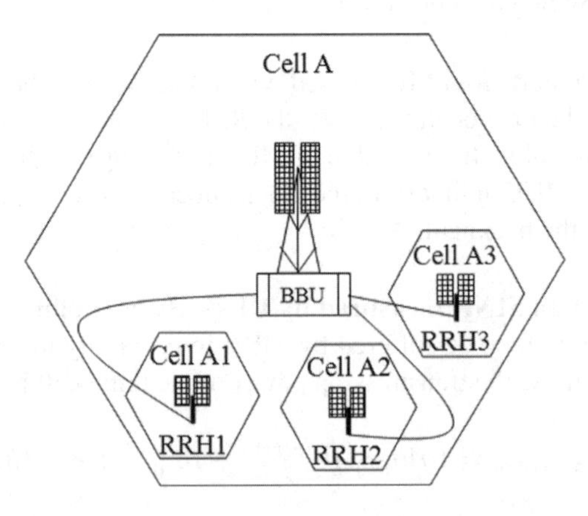

Figure 5.23 RRHs are physically separated but fibre connected to the baseband processing unit.

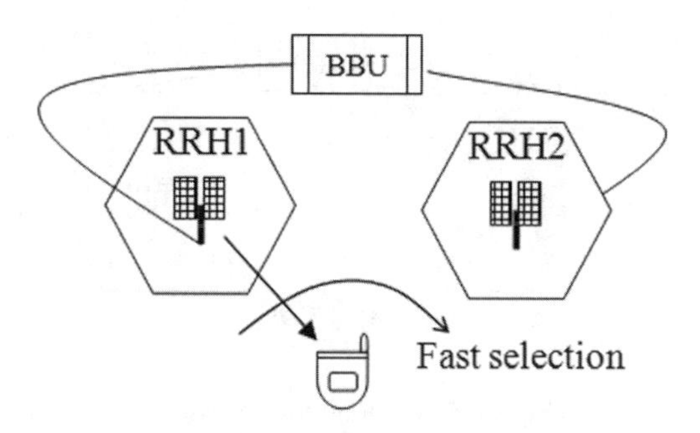

Figure 5.24 Dynamic cell selection.

RRH creates a new cell. RRH enables centralized scheduling and fast inter-cell interference management over RRH-created cells so that better inter-cell orthogonality can be achieved. Synchronous transmission among different cell without using GPS is also possible. In fact low-power RRHs overlaid on a high-power one, as shown in Figure 5.23, can be considered as one possible configuration for heterogeneous networks.

Coordinated scheduling or coordinated beam-forming – As the name suggests, the scheduler of a particular cell takes account of the scheduling decisions of other coordinated cells so that interference can be controlled. In fact, legacy ICIC schemes such as RNTP, already available to Release-8/9, can be considered as semi-static coordinated scheduling. When the backhaul latency is negligibly small – for example fibre connected RRHs – dynamic coordination is possible.

An example of dynamic coordinated scheduling is dynamic point selection (DPS), in which the UE measures the downlink channels from multiple cells and requests to be served by the best cell dynamically. For this to be possible, DPS assumes that UE data is available in all the coordinated cells, but data transmission to the UE takes place from only one cell each time, based on the scheduling decision. Dynamic muting may be applied to other coordinated non-best cells simultaneously to minimize interference, as shown in Figure 5.24 [36]. More advanced scheduling schemes can also be applied to further boost throughput [37]. With DPS, the instantaneous cell serving the UE's DL may be different from its UL.

Figure 5.25 shows the concept of coordinated beam-forming, in which the schedulers of the coordinated cells choose their beamformers to serve UEs connected to their own cell while

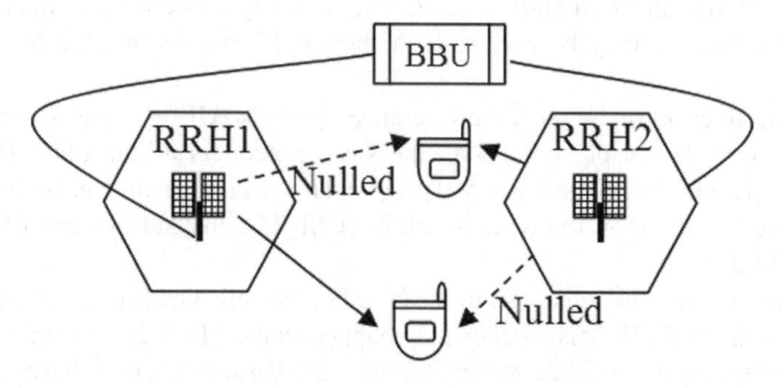

Figure 5.25 Coordinated beam-forming.

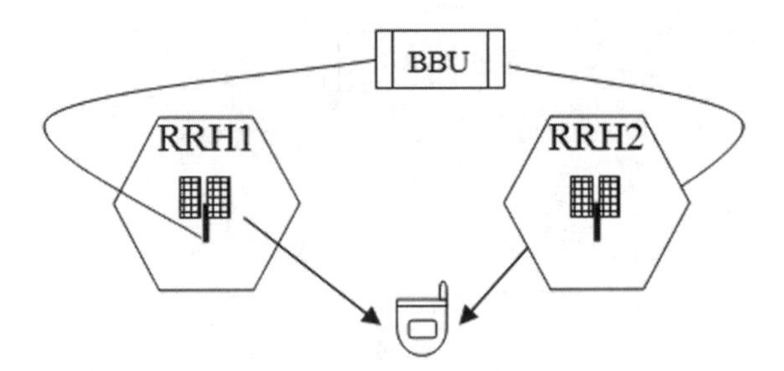

Figure 5.26 Joint transmission.

minimizing interference to other cells. For example, in addition to precoding the matrix index (PMI) that the UE reports to the serving cell for the best channel quality index (CQI) for the observed channel from that cell, the UE may also report best companion PMIs [45] that will create the least interference, based on the observed channel from the other coordinated cells. Once the serving cell has received this UE report, it exchanges the best companion PMIs for the reported resource blocks with the other cells in coordination over the fibre backhaul. The coordinated cells will take this information into account during scheduling, so that the PMI chosen for its served UEs will generate the least inter-cell interference. This coordination is only possible with negligible backhaul latency, otherwise the PMI reports would be outdated due to changes to the channel conditions.

Instead of the best PMI report, a related concept is for the UE to report the worst companion PMIs [46] that will cause the most interference for a particular channel conditions from the other cells, so that this worst companion PMIs will be avoided by the other cells' schedulers for that particular time/frequency resources.

In the context of heterogeneous networks, since the number of transmit antennas for the macrocell may not be the same as the LPNs, the UE may also need to know the antenna configuration so that it can use the correct set of codebooks for best/worst companion PMI report.

Joint transmission (JT) – Joint transmission uses signals from multiple cells to serve a UE's PDSCH, effectively turning interference into useful signals as illustrated by Figure 5.26. JT requires that the UE data is available to all coordinating cells, and the channel state information (CSI) reported from the UE for all of these cells is exchanged. Subsequently, the cells can compute the precoders coherently or non-coherently [38], so that the received signal quality can be improved and/or interference actively cancelled [3]. Inter-transmission point phase information may be needed for coherent JT, but not needed for non-coherent JT [35].

With the introduction of UE-specific reference symbols (UERS) in Release-10 in which the reference symbols are precoded with the same precoders as the PDSCH, JT may be supported transparently from the UE's perspective. However, measurements will need to rely on channel state information reference symbols (CSI-RS), instead of the CRS as has been done in Release-8/9.

The configuration of other channels in the coordinated cells needs to be addressed because JT only deals with PDSCH. Mismatches may happen in the PDCCH size and the CRS shifts between the different cells, as illustrated by Figure 5.27. Various solutions have been proposed, with throughput reduction [39].

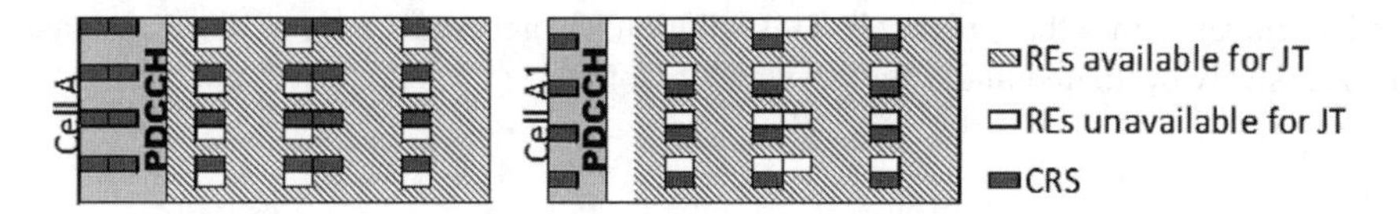

Figure 5.27 Mismatches in JT due to different configurations of PDCCH and CRS.

Noting that the problems shown in Figure 5.27 occur when the cells have different cell IDs, it has been proposed in [40] that the low-power RRHs use the same cell ID as the high-power RRH that they are fibre connected to. In this case, data demodulation would have to rely on UERS, and CSI measurement on transmission point-specific CSI-RS. Not only does this same cell ID approach solve the configuration mismatch issue in JT, but it also solves the collision problem of other control channels, as shown in Figure 5.28. In addition, no handover is needed

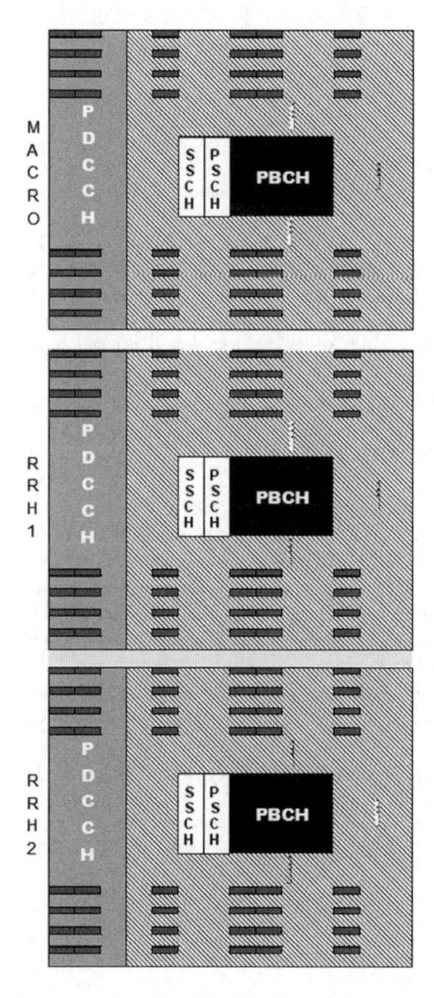

Figure 5.28 Control channels are soft combined in the air with the same cell ID approach, and there is no control-to-control interference.

when UE moves within the macrocell. This solution has generated a lot of discussion and has been extensively evaluated under the purview of CoMP.

5.4 Conclusion

In this chapter we have summarized enhanced ICIC techniques that are suitable for handling interference in heterogeneous networks deployment. These techniques are divided into frequency, time, power and spatial domains, and they can be combined when necessary. Information exchange among different cells is performed over the backhaul, and when its latency is very small, dynamic enhanced ICIC is possible. Otherwise, coordination is semi-static. If no backhaul is available – for example between the macrocell and the femtocell – enhanced ICIC becomes more challenging. When working with a TDD system, time domain schemes need to be designed carefully to prevent excessive UL-DL interference.

References

1. 3GPP, 'TR 36.913 v10.0.0, Requirements for further advancements for E-UTRA (LTE-Advanced)', April 2011.
2. ITU, 'Report ITU-R M.2135-1, Guidelines for evaluation of radio interface technologies for IMT-Advanced', December 2009.
3. 3GPP, 'TR 36.814 v9.0.0, Evolved Universal Terrestrial Radio Access (E-UTRA); Further advancements for E-UTRA physical layer aspects', March 2010.
4. P. Mogensen, Na Wei, I. Z. Kovacs, F. Frederiksen, A. Pokhariyal, K. I. Pedersen, T. Kolding, K. Hugl, M. Kuusela, 'LTE Capacity Compared to the Shannon Bound', VTC Spring 2007, pp. 1234–1238.
5. A. Khandekar, N. Bhushan, Tingfang Ji, V. Vanghi, 'LTE-Advanced: Heterogeneous networks' European Wireless Conference, 2010, pp 978–982.
6. NTT DoCoMo Inc., 'REV-080026 Proposals for LTE-Advanced Technologies', 3GPP TSG TAN IMT Advanced Workshop, April 2008.
7. 3GPP, 'TR 36.921 v10.0.0, FDD Home eNodeB (HeNB) Radio Frequency (RF) requirements analysis (Release 9)', April 2011.
8. CMCC, 'R1-105094 LS on eICIC progress in RAN1', August 2010.
9. Qualcomm Incorporated et al, 'R3-102241 CR: X2 interface for HeNB mobility enhancement', August 2010.
10. 3GPP, 'TR 36.423 v10.3.0, E-UTRAN; X2 application protocol (X2AP) (Release 10)', September 2011.
11. 3GPP, 'TR 36.213 v10.3.0, E-UTRA; Physical layer procedures (Release 10)', September 2011.
12. CATT et al., 'R1-103417 WF on Identification of Co-channel Problem and Needs in Macro-Pico', May 2010.
13. CATT et al., 'R1-103416 WF on Identification of Co-channel Problem and Needs in Macro-Femto', May 2010.
14. Qualcomm, 'R1-104818 Enabling communication in harsh interference scenarios', August 2010.
15. CMCC, 'R1-105081 Summary of the description of candidate eICIC solutions', August 2010.
16. Nokia Siemens Networks, Nokia, 'R1-101924 Macro+HeNB performance with escape carrier or dynamic carrier selection', April 2010.
17. Huawei, 'R1-104247 Concrete Proposal for Frequency Domain Solution', January 2010.
18. Kyocera, 'R1-104356 Downlink Evaluation of Lightly Loaded Control Channel Transmission in Macro-Pico Deployment', August 2010.
19. Texas Instruments, 'R1-102831 Rel-8/9 compatible PDCCH interference mitigation schemes for HetNets', May 2010.
20. LG Electronics, 'R1-104659 Evaluation of control channel coordination in co-channel CSG deployment', August 2010.
21. Motorola, 'R1-103924 Downlink Control Protection in LTE TDD for non-CA Heterogeneous Networks', July 2010.
22. CATT, 'R1-104345 Further Analysis on Time Domain Solutions in HetNet', August 2010.
23. ITRI, 'R1-104367 Discussion on time domain eICIC solutions in TDD system', August 2010.

24. Intel Corporation (UK) Ltd, 'R1-102814 Non-CA based PDCCH Interference Mitigation in LTE-A', May 2010.
25. Nokia Siemens Networks, Nokia, 'R1-104463 Autonomous Power Setting for HeNB Cells', August 2010.
26. Alcatel-Lucent, Alcatel-Lucent Shanghai Bell, 'R1-104414 HeNB Power Setting Specifications', August 2010.
27. CATT, 'R1-103495 DL Power Setting in Macro-Femto', July 2010.
28. NTT DOCOMO, 'R1-102307 Interference Coordination for Non-CA-based Heterogeneous Networks', April 2010.
29. Samsung, 'R1-102223 Performance Evaluation of Femto-based HetNet', April 2010.
30. I2R, 'R1-104732 eICIC for HeNB UL and MUE DL based on HeNB UL Measurement', August 2010.
31. Nokia Siemens Networks, Nokia, 'R1-103823 HeNB power setting performance under different access constraints', July 2010.
32. MediaTek Inc., 'R1-104544 Inter-cell Interference Mitigation for Uplink Channels in Heterogeneous Networks', August 2010.
33. F. Gunnarsson et al., 'Downtilted Base Station Antennas – A Simulation Model Proposal and Impact on HSPA and LTE Performance', VTC-Fall 2008, pp 1–5.
34. CATT, 'R1-101783 Interference coordination for DL CCH considering legacy UE', April 2010.
35. 3GPP TR 36.819 v11.0.0, 'Coordinated Multi-Point Operation for LTE Physical Layer Aspects (Release 11)', September 2011.
36. NTT DOCOMO 'R1-090314 Investigation on Coordinated Multipoint Transmission Schemes in LTE-Advanced Downlink',.
37. Minghai Feng; Xiaoming She; Lan Chen; Kishiyama, Y.; 'Enhanced Dynamic Cell Selection with Muting Scheme for DL CoMP in LTE-A', VTC Spring 2010.
38. NTT DOCOMO 'R1-084252 Views on Coordinated Multipoint Transmission/Reception in LTE-Advanced', November 2008.
39. NTT DOCOMO 'R1-112600 Investigation of Specification Impact for Rel. 11 CoMP', August 2011.
40. Ericsson, ST-Ericsson 'R1-110461 Baseline Schemes and Focus of CoMP Studies', January 2011.
41. NTT DOCOMO 'R1-104942 Views on eICIC Schemes for Rel-10', August 2010.
42. Qualcomm Incorporated 'R1-104818 Enabling communication in harsh interference scenarios', August 2010.
43. Huawei, HiSilicon 'R1-106165 Interference coordination for common channels in HetNet', November 2010.
44. Ericsson, ST-Ericsson 'R1-111323 Remaining details for CA based HetNet in Rel-10', May 2011.
45. Alcatel-Lucent 'R1-090926 Best Companion reporting for improved single-cell MU-MIMO pairing', February 2009.
46. Alcatel-Lucent, Alcatel-Lucent Shanghai Bell 'R1-100944 Performance of coordinated beamforming with multiple PMI feedback', February 2010.
47. LG Electronics 'R1-113272 Considerations on PBCH and PSS/SSS for FeICIC', October 2011.
48. Samsung 'R1-113085 PDSCH performance evaluation for FeICIC', October 2011.
49. Huawei, 'R1-104308 Understanding the Time Domain eICIC Schemes', August 2010.
50. ITRI, 'R1-104368 Considerations on PBCH eICIC for CSG HeNB', August 2010.
51. Alcatel-Lucent, Alcatel-Lucent Shanghai Bell 'R1-112411 Scenarios for Further Enhanced Non CA-based ICIC for LTE', August 2011.

6

Cognitive Radios to Mitigate Interference in Macro/femto Heterogeneous Networks

Shin-Ming Cheng[1] and Kwang-Cheng Chen[2]
[1]*National Taiwan University of Science and Technology, Taiwan*
[2]*National Taiwan University, Taiwan*

6.1 Introduction

Heterogeneous network (HetNet) deployment [1, 2], where low-power and small-coverage local nodes are distributed in the coverage of a macro node, is considered as a promising solution to achieve *universal frequency reuse*. The local nodes such as pico, femto and relay nodes deployed at coverage holes could extend coverage and increase spectral utilization. Moreover, the small coverage area of local node facilitates a large number of concurrent transmissions and improves spatial reuse, thereby potentially yielding enhanced wireless capacity. When all heterogeneous nodes share the same spectrum, two kinds of interference appear:

Cross-tier interference. The aggressor (e.g., a local node) and the victim of interference (e.g., a macro node user) belong to different tiers.

Intra-tier interference. The aggressor (e.g., a local node) and the victim (e.g., a neighbouring local node user) belong to the same tier.

Under the impact of interference, some resource blocks (RBs) can not be utilized simultaneously, which challenges HetNet toward universal frequency reuse. Thus, interference mitigation in HetNet receives much attention in both areas of academic and industry [3–5]. Typically, interference mitigation techniques are classified into following two categories [6]:

Heterogeneous Cellular Networks, First Edition. Edited by Rose Qingyang Hu and Yi Qian.

Interference coordination. This ensures orthogonality between mutual interfering transmitted signals in the following domains:

- *time-frequency*: In the orthogonal frequency division multiple access (OFDMA) system, it simply allocates resource at the different RBs.
- *location/space*: By controlling the transmission power according to the distance to the victim, the aggressor can avoid causing harmful interference to the victim.
- *antenna spatiality*: Transmissions via uncorrelated spatial paths in multiple input multiple output (MIMO) would not disturb each other.

Interference cancellation. If orthogonality can not be achieved, coding techniques such as sphere decoding or dirty paper coding (DPC) can be exploited to allow the victim to actively cancel interfering signals from the desired signal.

In this chapter, one kind of local node deployed at home, known as the femtocell, is investigated due to its attractive benefits for both subscribers and operators. As shown in Figure 6.1, each femtocell is composed of a base station (femto-BS) and multiple mobile stations (femto-MSs). The underlying macrocell is composed of a BS (macro-BS) and multiple MSs (macro-MSs). The advantages for such two-tier networks are described as follows:

- *Coverage extension*: The femtocells can be deployed in a house to avoid dead spots and extend coverage.
- *Capacity gain*: The indoor femto-MSs can enjoy high quality of transmission due to the short distance from the femto-BS, and the number of concurrent transmissions that can be accommodated in the network can be increased, thereby yielding enhanced coverage and

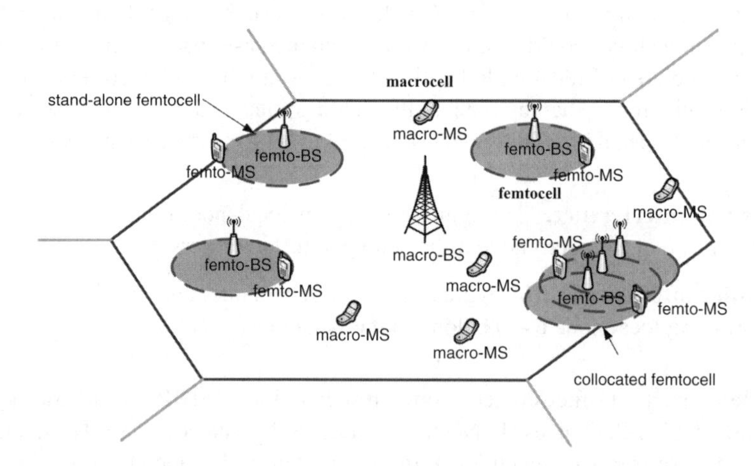

Figure 6.1 Network architecture of macro/femto heterogeneous networks. Adapted from S.-M. Cheng, S.-Y. Lien, F.-S. Chu, and K.-C. Chen, "On exploiting cognitive radio to mitigate interference in macro/femto heterogeneous networks", IEEE Wireless Communications magazine, Vol. 18, Issue 3, pp. 40–47, June 2011.

capacity. Short range communication can bring spatial gain (more terminals can coexist) and energy consumption.

- *Offloading*: By deploying femto-BSs, operators can reduce the cost of deploying macro-BSs since traffic loads transfer to femto-BSs.
- *Additional information*: The macro can provide useful control information to the femto-BS to facilitate operations such as interference mitigation.
- *Power consumption*: The short transmission range is detrimental to the power consumption.
- *Diversity*: Terminals can choose from femtocell and macrocell by channel condition or other information. The downlink can perform similar selection.

According to the deployed location, there are two types of femtocells, namely, stand alone and collocated. A stand alone femtocell locates far from other femtocells. According to signal attenuation models defined by LTE-Advanced [1, 7], a stand alone femtocell only needs to mitigate cross-tier interference from the overlying macrocell. On the other hand, the coverage of a collocated femtocell overlaps with that of other collocated femtocells. Therefore, a collocated femtocell needs to mitigate both intra-tier and cross-tier interferences.

Typically, the spectrum sharing among nodes is typically achieved via *cooperation* or *coexistence* [8]. Interference mitigation in such two-tier heterogeneous networks faces practical challenges from the following aspects [9]:

- *Random deployment*: Since femto-BSs are installed and deployed by users without network planning, they can appear anywhere and act as aggressors.
- *Restricted/closed access*: Since femto-BSs are paid for by the customers, it is reasonable that only users defined by the owners are allowed to access femto-BSs. In this case, the unauthorized users could only connect to the macro-BS even if a femto-BS exists in vicinity and thus the users suffer heavy cross-tier interference.
- *No coordination between macro-BS and femto-BS*: The delay of connection via wired back-haul is too long to admit any cooperation [7], which means that centralized interference mitigation approaches are not feasible.
- *Backward compatibility*: The legacy macro-MS and macro-BS must be supported for market penetration. Thus, any additional operation on the MS side or the current macrocell protocol is not suggested.

Obviously, femto-BSs distributively mitigating interference in a coexistent fashion by using local information without explicit signalling might be more appropriated. As can be seen throughout this chapter, we enable cognitive radio (CR) technology [10] on femto-BSs as a solution for coexistence to tackle the above challenges. A CR-enabled femto-BS could automatically sense the environment, interpret the received signalling from the macro-BS and surrounding femto-BSs, and intelligently allocate resource [11–17]. Consequently, macro-BSs and CR-enabled femto-BSs are respectively analogous to primary and secondary users in the CR model. With CR capability, femto-BS could actively acquire knowledge about the environment without the aid of a macrocell in a decentralized fashion and automatically prevent it from disturbing the macro or surrounding femto transmissions.

This chapter provides an overview of how CR facilitates interference mitigation in two-tier heterogeneous networks from both theoretical analysis and method design perspectives.

A comprehensive comparison of the interference mitigation approaches with respect to the requirement of leveraged information is introduced in Section 6.2. This section also carefully examines the existing information acquisition mechanisms and identifies the effects of CR on interference mitigation. Section 6.3 presents the system model of the considered two-tier networks. In Section 6.4, we study possible CR-enabled cross-tier interference mitigation approaches, including orthogonal radio resource assignment in time-frequency, geometry spatial and antenna spatial domains, as well as cross-tier interference cancellation via novel decoding techniques. The simulation results are also investigated in this section. By exploiting information acquired by CR, recent innovations such as game theory and the Gibbs sampler are explored to mitigate intra-tier interference among femto-BSs in Section 6.5. In Section 6.6, we consider the interference mitigation when applying macro/femto as the service architecture for machine-to-machine (M2M) communications. Finally, Section 6.7 draws some conclusions to this chapter.

6.2 Information Requirement and Acquisition for Interference Mitigation

A multitude of studies have been recently proposed to study the downlink spectrum-sharing problems between macrocell and femtocells, considering cross-tier interference control [13, 18–25], among femtocells considering intra-tier interference mitigation [26–29], or both [11,12,14,15,17,30]. In these control schemes, femto-BSs acquire *side information* about the macrocell and surrounding femtocells to achieve interference mitigation. The problems that arise are 'how much information is required' and 'how to acquire information'. By answering these questions, the role of information in interference mitigation is identified conceptually. The information of other network nodes leveraged by femto-BS for interference mitigation can be classified as follows [10]:

- *Activity information*: A femto-BS knows which RBs are unoccupied by other BSs. For example, if the received interference power of an RB, for example reference signal received power (RSRP) in an LTE network, exceeds a certain threshold, femto-BS identifies that the RB is allocated for macro-MS [24].
- *Channel information*: A femto-BS knows the channel statistics or gains between some BSs and some MSs, which can be obtained by recent innovations such as cognitive radio network tomography [31] without heavy overheads.
- *Location information*: A femto-BS knows the distance between itself to an MS/BS, and thus the channel strengths to the MS/BS is known. For macro-MS, this implies that knowledge related to its behaviour is acquired, such as the specific macro-MS allocated to an RB or the location in which the assigned macro-MS will use an RB. Typically, such scheduling or zone allocation information [20] is encapsulated in PDCCH (LTE) or DLMAP (WiMAX) by encoding with identities of served MSs. We can simply assign femto-BSs a special user identity for decoding such information broadcast channel. Moreover, femto-BS could overhear the feedback information from the MS to determine its location and thus the femto-BS could adjust its own access parameters accordingly [32].
- *Codebook/message information*: A femto-BS knows the codebooks of the other BSs. The codebook information could be obtained from the periodical broadcasting information and

the message information might be obtained after decoding. While this is impractical for an initial transmission [10], the assumption holds for a message retransmission where the femto-BS hears the first transmission and decodes it [33].

If femto-BS is aware of spectrum activity, it could simply avoid allocating the same RB occupied by a macro-BS, which achieves orthogonality in the time-frequency domain. While the channel condition is known, the femto-BS could adjust its transmission power to ensure that the receiving signals at other nodes' receivers remain below the constraint. Thus, the concurrent femto-BS and other node transmissions may occur, which implies that they are orthogonal in the space domain. Alternatively, we could utilize multiple antennas to guide femto-BS signals away from the other nodes' receivers. Knowledge of other nodes' messages and/or codebooks can be exploited to cancel interference seen at the femto-MS by using coding techniques.

In current distributed interference coordination solutions, the method used to retrieve information for interference mitigation plays a key role in obtaining the performance gains. Typically, following mechanisms are applied:

- *Exchanging information among BSs.* In this case, femto-BSs and the macro-BS could directly exchange information about their allocation usages (e.g., activity), connection behaviour (e.g., channel conditions), as well as resource demands (e.g., codebooks or messages) [20,23,28–30,34]. By being aware of the present actions and future intentions of the macro-BS and surrounding femto-BSs, perfect interference mitigation can be achieved. Obviously, a common control channel and message exchanging procedure must exist among BSs, and it is typically performed via wired backhaul, but this is not feasible due to the constraint of no macro–femto coordination [7]. Moreover, the heavy communication overheads make this mechanism inefficient in dense femtocell deployments.
- *Receiving measurement reports from femto-MSs.* In this case, femto-MS periodically performs measurement and feeds back reports to its serving femto-BS [12, 26, 35]. By analyzing the report, the femto-BSs can acquire activity and channel condition about the immediate environment of each femto-MS, which facilitates interference mitigation. However, performing measurement may consume quite a bit of power, and this may not be feasible for femto-MSs that are typically power limited. Moreover, imposing new operation on MS side may incur backward incompatibility problem.
- *Building CR into femto-BS itself.* By adopting traditional spectrum sensing techniques, the detection of macro-BS and surrounding femto-BS signals can be achieved without any coordination [11–13, 15–17]. For example, if the received interference power of an RB exceeds a certain threshold, femto-BS identifies that the RB is occupied and retrieves the activity [15]. Thus, CR-enabled femto-BSs could automatically configure itself and mitigate both tier interferences.

Orthogonality in time-frequency and antenna spatial domains can be achieved with activity and channel condition information acquired by CR. When codebooks and message information are further retrieved, the cancellation approach can be exploited. The following sections discuss possible CR-enabled interference mitigation approaches in detail.

6.3 Descriptions of System Models

6.3.1 *Two-tier Network Architecture*

As shown in Figure 6.2, the downlink of an orthogonal frequency division multiple access (OFDMA) system with two-tier macro- and femtocells is considered. Typically, OFDMA-based systems divide system bandwidth into $|\mathcal{R}| = 100$ basic time-frequency units of resource blocks (RBs). Denote the RBs allocated by macro-MS by macro RB. The *co-channel* deployment and the *closed access* policy are assumed. macro- and femto-BS transmit to only one user at any given RB at full power P_m and P_f respectively, which implies that transmission power is maintained constant across the RBs. Perfect synchronization in time and frequency is assumed.

Denote $\mathcal{H} \in \mathbb{R}^2$ as the interior of a reference macrocell of serving area radius R_m, which consists of multiple macro-MSs and one macro-BS located in the centre of $\mathcal{H}$. The spatial distribution of macro-MSs is assumed to follow a homogeneous Poisson point process (PPP) with density μ_m and the locations of macro-MSs are denoted as $\Phi_m = \{X_i\}$. The macrocell is overlaid with the femto-BSs of radius R_f, which are randomly distributed on $\mathbb{R}^2$ according to a homogeneous PPP with intensity λ_f. We let $\Phi_f = \{Y_i\}$ denote the locations of the femto-BSs. Each femto-BS is coupled with relevant CR capabilities, such as spectrum sensing, interference management and efficient spectrum allocation and sharing. Please note that we concentrate on the case where the intra-macrocell interference is dominant and thus the interference originated by other macrocells is not considered.

6.3.2 *Channel Model*

We consider path loss attenuation effects, Rayleigh fading with unit average power $\mathcal{G}$, penetration loss due to walls β and background noise power per RB N_0 in our channel model.

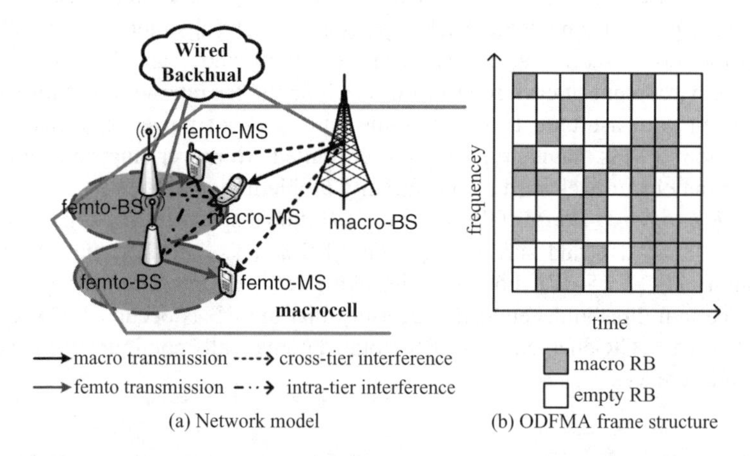

Figure 6.2 Network model and frame structure of macro/femto heterogeneous networks. Adapted from S.-M. Cheng, S.-Y. Lien, F.-S. Chu and K.-C. Chen, 'On exploiting cognitive radio to mitigate interference in macro/femto heterogeneous networks', IEEE Wireless Communications magazine, Vol. 18, Issue 3, pp. 40–47, June 2011. Reproduced with permission © IEEE 2011.

The path loss exponent of transmission is denoted by γ. The success reception of a transmission at MS depends on whether the signal-to-interference-plus-noise ratio (SINR) observed by the MS is larger than a SINR threshold (denoted by η). To guarantee the transmission quality, there is an outage constraint at MS with maximum outage probability ϵ.

6.3.3 Traffic Model

The reception success of a primary transmission at PR depends on whether the channel can support the information rate R. To guarantee the transmission quality, there is an outage constraint at PR with maximum outage probability (approximating the decoding error probability) ϵ. Once decoding errors occur, stop-and-wait (SAW) automatic repeat-reQuest (ARQ) is adopted. We assume no constraint on the number of retransmissions, that is, the packet is repeatedly retransmitted until it is successfully received at PR. On the other hand, no ARQ for secondary transmission is assumed due to the opportunistic nature.

6.3.4 CR-enabled Operations

After powering on, femto-BS is registered, authenticated and associated into the network via the backhaul. The operator provides the necessary information such as carrier frequency and location to femto-BS in the pre-operational state for self-configuration on coverage-related parameters and the neighbouring list [7]. As CR is introduced into femto-BS, extra information for the sensing operation has to be provided by the network operator.

6.4 Cross-tier Interference Mitigation

This section elaborates the various cross-tier interference mitigation approaches based on the information acquired by CR.

6.4.1 Interference Coordination: Orthogonality in the Time/Frequency Domain

With activity of the macrocell that indicates which RB is occupied in a frame, each femto-BS prevents the allocation of these occupied RBs to its femto-MSs. An instinct algorithm (i.e., Algorithm 1) is proposed at femto-BS in our previous work [24]:

Algorithm 1 Cognitive resource block management (CRBM)

1: Femto-BS periodically senses the channel to identify which RB is occupied by the macrocell. The sensing period is T_s frames and each channel sensing persists for one frame. Among the T_s frames, one frame is defined as the 'sensing frame' where femto-BS performs channel sensing. The remaining $T_s - 1$ frames are regarded as the 'data frames' for data transmission and reception. Note that femto-BS can not perform data transmission and reception within the sensing frame. We assume that all femtocells are synchronized and have the same frame as the sensing frame.

2: Femto-BS senses the received interference power on each RB within the sensing frame.
3: **if** the received interference power on an RB exceeds a certain threshold **then**
4: The RB is identified as being occupied by the macrocell since all femtocells senses at the same time.
5: **else**
6: The RB is unoccupied by the macrocell.
7: In subsequent data frames, femto-BS only allocates unoccupied RBs sensed in the sensing frame to its femto-MSs. Please note that the unoccupied RBs may be allocated by the collocated femto-BSs at the same time, which incurs heavy intra-tier interference. We could apply strategic game or Gibbs sampler in the following section to mitigate intra-tier interference.

As shown in Figure 6.3, CRBM suggests that femto-MSs and macro-MSs operate on orthogonal RBs, and no cross-tier interference is introduced, which is similar to the *interweave* paradigm in the CR model [10]. The received interference power has been adopted by 3GPP LTE-Advanced as a mandatory sensing quantity in base stations [36]. Therefore, CRBM can be applied to LTE-Advanced without any hardware modifications. In addition, the received interference power can be in the downlink or the uplink, for the femtocell to mitigate interference from macro-BS to femto-MS in the downlink case and interference from macro-MS located within the coverage of the femtocell to femto-BS in the uplink case.

6.4.2 *Interference Coordination: Orthogonality in the Antenna Spatiality Domain*

If MIMO is supported in the heterogeneous networks, we could uncorrelate the transmitter–receiver pairs and arrange the macro and femto transmissions on different spatial paths. As

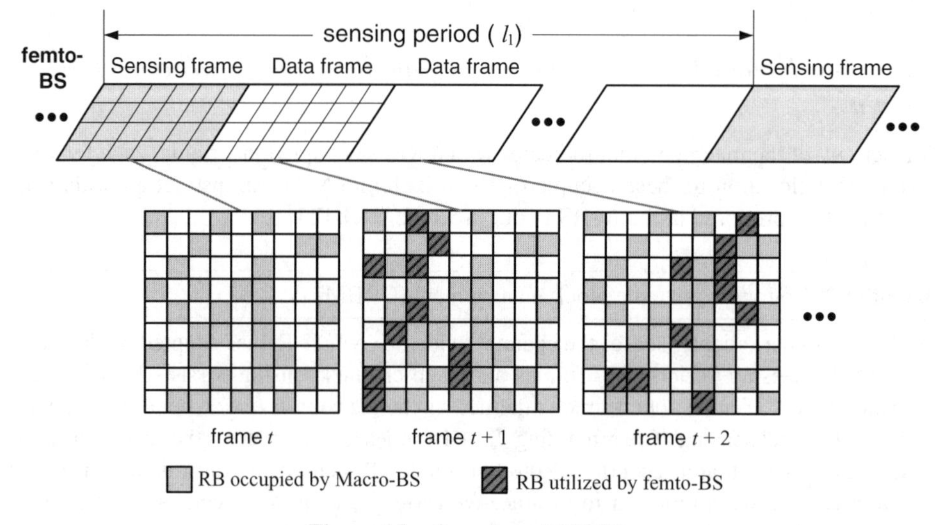

Figure 6.3 Operation of CRBM.

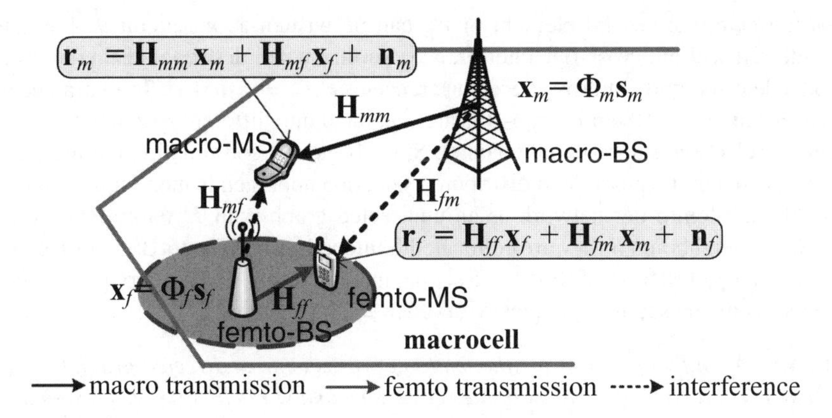

Figure 6.4 Signal representation of macro-BS, femto-BS and MS. Adapted from S.-M. Cheng, S.-Y. Lien, F.-S. Chu, and K.-C. Chen, "On exploiting cognitive radio to mitigate interference in macro/femto heterogeneous networks", IEEE Wireless Communications magazine, Vol. 18, Issue 3, pp. 40–47, June 2011.

shown in Figure 6.4, the transmitted signal vectors of femto-BS and macro-BS are respectively denoted as $\boldsymbol{x}_f \in R^i$ and $\boldsymbol{x}_m \in R^n$, where i and j are antenna numbers of femto-BS and macro-BS respectively. Please note that $\boldsymbol{x}_f$ ($\boldsymbol{x}_m$) is generated by multiplying the data vector $\boldsymbol{s}_f \in R^p$ ($\boldsymbol{s}_m \in R^q$) with the pre-coding matrix $\boldsymbol{\Phi}_f \in R^{i\times p}$ ($\boldsymbol{\Phi}_m \in R^{j\times q}$). Moreover, $\mathbf{H}_{ff} \in R^{k\times i}$, $\mathbf{H}_{mf} \in R^{k\times i}$, $\mathbf{H}_{fm} \in R^{k\times j}$ and $\mathbf{H}_{mm} \in R^{k\times j}$ are respectively channel matrixes from femto-BS to femto-MS, from femto-BS to macro-MS, from macro-BS to femto-MS and from macro-BS to macro-MS, where k is the antenna number of MS. The received signal vectors at femto-MS and macro-MS can be respectively represented as $\boldsymbol{r}_f = \mathbf{H}_{ff}\boldsymbol{x}_f + \mathbf{H}_{fm}\boldsymbol{x}_m + \boldsymbol{n}_f$ and $\boldsymbol{r}_m = \mathbf{H}_{mm}\boldsymbol{x}_m + \mathbf{H}_{mf}\boldsymbol{x}_f + \boldsymbol{n}_m$, where $\boldsymbol{n}_f \in R^k$ and $\boldsymbol{n}_m \in R^k$ are noise vectors. Thus, for the receive vector $\boldsymbol{r}_f$ at femto-MS ($\boldsymbol{r}_m$ at macro-MS), $\mathbf{H}_{fm}\boldsymbol{x}_m$ ($\mathbf{H}_{mf}\boldsymbol{x}_f$) is regarded as the cross-interference from macro-BS (femto-MS).

From the following example, we demonstrate that both macro-MS and femto-MS can detect the desired signal without effects of cross-tier interference by separating spatial paths. Assuming that $i = j = k = 2$, the data vectors of femto-BS and macro-BS are respectively set as $\boldsymbol{s}_f = \begin{bmatrix} a \\ 0 \end{bmatrix}$ and $\boldsymbol{s}_m = \begin{bmatrix} 0 \\ b \end{bmatrix}$, which means that femto-BS and macro-BS load their data symbols on uncorrelated spatial paths. In this way, the received signal vector at macro-MS can be represented as:

$$\boldsymbol{r}_m = \mathbf{H}_{mm}\Phi_m\begin{bmatrix} 0 \\ b \end{bmatrix} + \mathbf{H}_{mf}\Phi_f\begin{bmatrix} a \\ 0 \end{bmatrix} + \boldsymbol{n}_m \triangleq \widehat{\mathbf{H}}_{mm}\begin{bmatrix} 0 \\ b \end{bmatrix} + \widehat{\mathbf{H}}_{mf}\begin{bmatrix} a \\ 0 \end{bmatrix} + \boldsymbol{n}_m. \tag{6.1}$$

If both $\widehat{\mathbf{H}}_{mm}$ and $\widehat{\mathbf{H}}_{mf}$ can be estimated at macro-MS, the received vector $\boldsymbol{r}_m$ can be transformed to $\boldsymbol{r}'_m$ as:

$$\boldsymbol{r}'_m = \widehat{\mathbf{H}}^{\dagger}_{mf}\boldsymbol{r}_m = \widehat{\mathbf{H}}^{\dagger}_{mf}\widehat{\mathbf{H}}_{mm}\begin{bmatrix} 0 \\ b \end{bmatrix} + \begin{bmatrix} a \\ 0 \end{bmatrix} + \boldsymbol{n}'_m. \tag{6.2}$$

As we observe, the second element of $\boldsymbol{r}'_m$ can be written as $\boldsymbol{r}'_m|_m = h_m b + n$, and thus can be detected without cross-tier interference. Similarly, the desired signal α at femto-MS can also be detected without cross-tier interference as $\boldsymbol{r}'_f|_f = h_f a + n$. To guarantee that the transmissions of macro-BS and femto-BSs are separated into different spatial paths, all femto-BSs need to select the same spatial channel. Since backhaul coordination among femtocells might not practically be possible, a distributed selection approach is more appropriate.

We model the femto-tier network as an undirected graph with K femto-BSs, where two femto-BSs are neighbours if they are collocated femtocells. Each femto-BS k utilizes a spatial path s_k. A clique of order n is defined as a set having n nodes in which every pair of nodes are neighbours, while the set of all cliques with order n is represented as $\mathcal{C}(n)$.

Definition 1 *A conflict graph $\mathcal{G}$ for the femto networks is an undirected graph $\mathcal{G} = (\mathcal{N}, \mathcal{E})$, where $\mathcal{N}$ and $\mathcal{E}$ are sets of vertices and edges respectively. Each vertex corresponds to one femto-BS and two vertices m and n are connected with an edge if $\frac{P_f D_f^{-\gamma}}{N_0 + P_f d_{mn}^{-\gamma}} < \eta$, where $d_{kk'}$ is the distance between femto-BSs k and k'. In addition, the conflicting set $\mathcal{C}(k)$ for femto-BS k is defined to be the neighbour set of corresponding vertices in the graph $\mathcal{G}$, i.e., $\mathcal{C}(k) = \{q|(q, k) \in \mathcal{E}\}$.*

A global energy function $E(\boldsymbol{s})$ is defined as $E(\boldsymbol{s}) = \sum_k \sum_{\mathcal{B} \in \mathcal{C}(k)} V(\mathcal{B})$, where $V(\mathcal{B})$ is a potential function which associates non-negative real numbers to all subsets of nodes in $\mathcal{B}$. A local energy function of node k is defined as $E_k(s_k, s_{i \neq k}) = \sum_k \sum_{k \in \mathcal{B}, \mathcal{B} \in \mathcal{C}(k)} V(\mathcal{B})$. In our case, the global energy is defined as minus of the total intra-tier interference (i.e., all femto-BSs experienced from all other femto-BSs). Since the Gibbs sampler can minimize global energy (and thus maximize total interference), all femto-BSs select the same spatial channel. The potential function is consequently defined as follows:

$$\begin{cases} V(\mathcal{B}) = 0, & \text{for } \mathcal{B} \in \{k\}, \\ V(\mathcal{B}) = -I(k, k'), & \text{for } \mathcal{B} \in \{k, k'\}, \\ V(\mathcal{B}) = 0, & \text{for } |\mathcal{B}| \geq 3. \end{cases} \tag{6.3}$$

Please note that $I(k, k')$ is the intra-tier interference from femto-BS k' to femto-BS k if both of them are using the same spatial channel. The local energy function can then be specified as:

$$E_k = -\sum_{k' \neq k} I(k, k'). \tag{6.4}$$

Given the global and local energy functions, the Gibbs sampler, which originated from statistical physics [37], provides a procedure to minimize global energy (and thus maximize total interference), where each node updates its state by sampling a random variable over the state set S according to a distribution $p(s)$:

$$p(s) = \frac{e^{-\frac{E_k(s, s_{i \neq k})}{T}}}{\sum_{s' \in S} e^{-\frac{E_k(s', s_{i \neq k})}{T}}}. \tag{6.5}$$

Please note that this distribution only relates to local energy, which implies that the femto-BS could only exploit local information sensed from surrounding femto-BSs to achieve global optimization. The temperature parameter T is a function of run time t and formed as $T = \frac{T_0}{\log_2(2+t)}$, where T_0 is a constant. With a logarithmically decreasing T, the state with less energy will be chosen with high probability. By [37], the global state will converge to the state with minimum global energy, consequently, all femto-BSs are selecting the same spatial channel. In Algorithm 2, we summarize several critical steps in our approach.

Algorithm 2 Gibbs Sampler based spatial path selection

1: Compute the temperature parameter $T = \frac{T_0}{\log_2(2+t)}$.
2: Compute local energy for each spatial path.
3: For each spatial path, compute probability of interference by local energy according to Gibbs distribution with temperature T.
4: Sample a spatial path randomly by the interference probability.

6.4.3 Interference Cancellation: Coding Techniques

If femto-BS could acquire more information such as message and codebooks at macro-BS, interference at femto transmission from macro-BS could be mitigated by applying some precoding or decoding techniques even though orthogonality is not achieved. Specifically, the transmitter-side of femto transmission could exploit such information to completely cancel the interference from the macro-BS by performing a precoding technique such as DPC. Here we focus on the alternative approach where interference is subtracted at the receiver side of the femto transmission by using decoding techniques and leveraging opportunity that arise during the macro retransmission [38]. Specifically, the femto-BS listens and successfully decodes a macro packet in the initial transmission so that during the retransmissions it can eliminate the interference caused by the macro-BS. The process of cross-tier interference cancellation includes two states:

$$\begin{cases} S_0 \text{ macro-BS transmits a new packet; femto-MS keeps silence;} \\ \quad \text{femto-BS overhears primary signal,} \\ S_1 \text{ macro-BS retransmits the old packet; femto-MS transmits a new packet;} \\ \quad \text{femto-BS mitigates interference.} \end{cases} \tag{6.6}$$

The successful reception of a macro transmission at macro-MS depends on whether the receiving SINR is higher than a predefined SINR threshold. In the infrastructure HetNet, the (re)transmissions or (re)receptions of macro and femto packets are centralized-scheduled by macro-BS and femto-BS respectively and then are notified to each MS. With CR, the femto-BS could actively acquire scheduling results and ACK/NAK feedback information from macrocell and determine the retransmission periods of the macro packet for concurrent transmission. If the overheard primary packet could be successfully decoded, interference could be perfectly subtracted [38]. If not, femto-BS combines signals received in states S_0 and S_1 in a maximizing SINR fashion [39] to retrieve capacity gains.

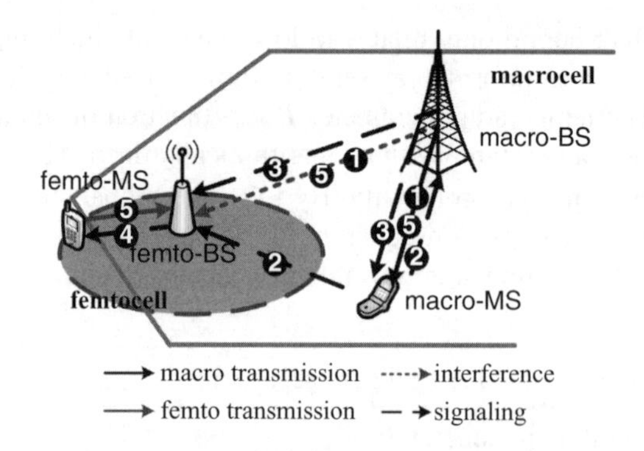

Figure 6.5 Procedure of interference cancellation. Adapted from S.-M. Cheng, S.-Y. Lien, F.-S. Chu, and K.-C. Chen, "On exploiting cognitive radio to mitigate interference in macro/femto heterogeneous networks", IEEE Wireless Communications magazine, Vol. 18, Issue 3, pp. 40–47, June 2011.

By summarizing the above results, we propose a novel interference cancellation mechanism in two-tier networks, where the corresponding procedures are illustrated in Figure 6.5 and are described in Algorithm 3.

Algorithm 3 Cognitive interference cancellation for retransmission

1: The process starts from state S_0 where femto-BS actively senses when macro-BS transmits packets to macro-MS to obtain the prior knowledge about the interference, that is, $x_m^{(t)}$.
2: CR-enabled femto-BS overhears the feedback ACK/NAK sent by macro-MS.
3: **if** an ACK message is received **then**
4: Femto-BS discards the knowledge $x_m^{(t)}$ and jumps to Step 1.
5: **else**
6: Femto-BS keeps $x_m^{(t)}$.
7: The macro-BS then schedules the failed-decoded packet for retransmission, determines the transmission slot $t + \Delta t$, and notifies the scheduling results to the macro-MS, which is also sensed by the femto-BS.
8: The femto-BS utilizes overheard knowledge to determine retransmission slot of the macro packet (i.e., $t + \Delta t$), schedules the same slot for femto packet reception and notifies the femto-MS as the transmitter at $t + \Delta t$.
9: During the retransmission slot of the macro packet, the process transients to state S_1 and the femto-MS uploads the packet to the femto-BS. The femto-BS utilizes the prior information $x_m^{(t)}$ to cancel the interference from the macro-BS $x_m^{(t+\Delta t)} = x_m^{(t)}$ and thus extracts $x_f^{(t+\Delta t)}$ destined to itself.

6.5 Intra-tier Interference Mitigation

After identifying a set of available radio resources orthogonal to that of the macrocell (in time-frequency or antenna spatial domain), the subsequent challenge is that multiple femto-BSs may identify the same set of available RBs, especially when these femto-BSs locate closely with each other as collocated femto-BSs. In this case, without an effective scheme to share these available radio resources, collocated femto-BSs may suffer heavy intra-tier interference. In this section, we consequently propose two schemes enabling an autonomous coordination of available RBs among collocated femto-BSs.

6.5.1 Strategic Game for Collocated Femtocells

Since no coordination (no available interface) among femto-BSs for information exchange exists in 3GPP LTE-Advanced [7], an instinctive solution for each collocated femto-BS is to randomize the RBs' utilization among the set of available RBs [19]. That is, if there are K_c collocated femtocells and there are M_a available RBs that have been identified, each collocated femto-BS utilizes $\lfloor M_a/K_c \rfloor$ available RBs, then each collocated femto-BS randomizes the utilization of these $\lfloor M_a/K_c \rfloor$ available RBs. This solution is referred as 'equal division'. However, equal division may not be effective when each collocated femtocell has diverse demands.

To determine the maximum number of available RBs that can be utilized by each collocated femto-BS in a distributed way for practical operations, game theory is well suited to be the foundation to facilitate our ultimate goal. Since it is not feasible to distinguish priority among collocated femto-BSs, no collocated femto-BS can make the decision prior to other collocated femto-BS, which well motivates the adoption of the strategic game (that is, the one-shot game [40]). Consequently, a strategic game can be formed to find the maximum number of RBs that can be utilized by each collocated femto-BS such that the total number of RBs utilized by all collocated femto-BSs without interference is optimized. In Algorithm 4, we consequently propose strategic game based resource block management (SGRBM) for autonomous intra-tier interference mitigation among collocated femtocells, which replaces Step 3 of the CRBM, while other steps in the CRBM still need to be performed by collocated femto-BSs.

Algorithm 4 Strategic game based resource block management

1: Each collocated femto-BS finds the corresponding optimum strategy profile according to the developed strategic game, by which each collocated femto-BS can obtain the optimum number of unoccupied RBs (denoted by L) available to be allocated to its femto-MSs.
2: In each data frame, each collocated femto-BS can allocate l unoccupied RBs ($l \leq L$) to its femto-MSs, according to the actual demand.
3: These l unoccupied RBs are allocated in the randomized manner.

This scheme achieves a better performance than that of equal division, since the optimum decision is autonomously made in each collocated femto-BS by considering all possible decisions that may be taken by other collocated femto-BSs. In addition, rather than being calculated on-line in each collocated femto-BS, the optimal solution can be calculated off-line by the table lookup as long as the total number of available RBs and the number of collocated femto-BSs are known. With the aid of CR, these two parameters can be acquired at femto-BSs.

6.5.2 Gibbs Sampler for Collocated Femtocells

Unlike choosing the same spatial path, as in Section 6.4.2, here we adopt the Gibbs sampler to achieve orthogonal RB allocation among collocated femto-BSs. When two collocated femto-BSs perform sensing at the same time, they may choose the same unoccupied RB for data transmission. This results in decoding failure at the femto-MS due to heavy mutual interference. Thus, we propose a random initial sensing mechanism along with sensing period to achieve exclusive sensing among femto-BSs. By applying slotted Aloha, initially each femto-BS tosses a coin independently in each frame with probability $\widehat{p}$ and performs sensing if it gets heads (i.e., if $B_n(\widehat{p}) = 1$, where $B_n(\widehat{p})$ are i.i.d. Bernoulli random variables with parameter $\widehat{p}$). Once femto-BS k performs sensing at its initial sensing frame, it senses again after the sensing period with l_k frames, as depicted in Figure 6.6.

With the global and time-invariant information of femto network, the operator could estimate the probability $\widehat{p}$ for initial random sensing. Under the assumption that the spatial distribution of femto-BSs follows a homogeneous PPP, the set of sensing femto-BSs $\Phi_f^{\widehat{p}} = \{X_k : B_k(\widehat{p}) = 1\}$ is modelled as a subset with density $\widehat{\lambda}_f = \widehat{p}\lambda_f$. For each femto-BS in $\Phi_f^{\widehat{p}}$, the transmission qualities of all its serving femto-MSs will be guaranteed even when other sensing femto-BSs select the same RBs for transmission. In particular, the received SINR at femto-MS located at cell boundary will guarantee:

$$\mathbb{P}\left(\frac{G_f P_f D_f^{-\gamma}}{N_0 + I_{f,f}} \geq \eta\right) = 1 - \epsilon. \tag{6.7}$$

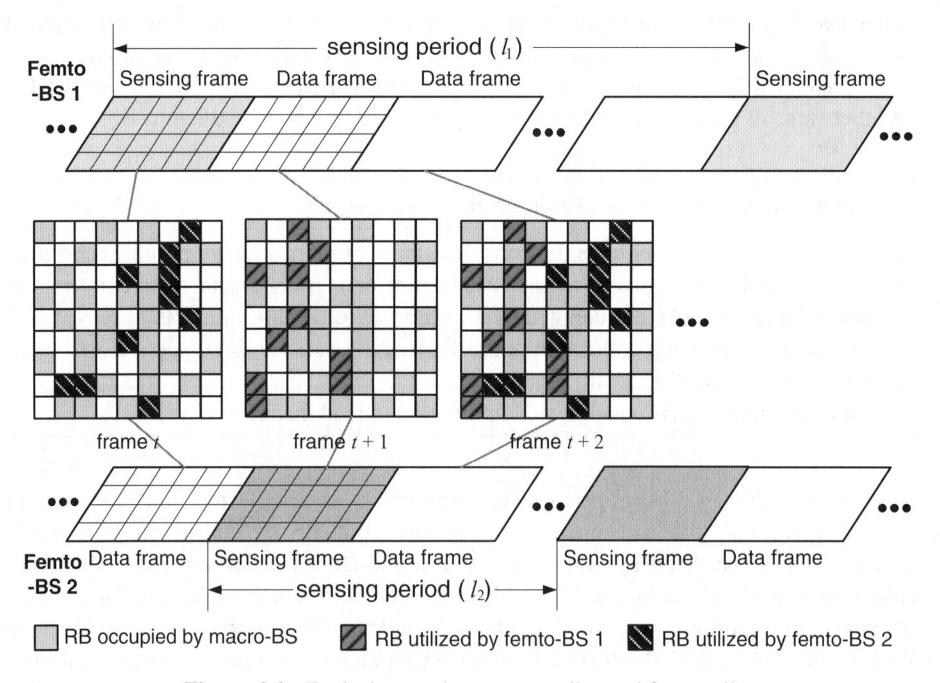

Figure 6.6 Exclusive sensing among collocated femtocells.

By the stationary characteristic of homogeneous PPP [41], the interference measured by the typical femto-MS is representative of the interference seen by all other femto-MSs. Thus $I_{f,f} = \sum_{X_k \in \Phi_f^{\widehat{p}} \setminus \{X_0\}}$.

$G_{X_k} P_f \|X_k\|^{-\gamma}$ is the intra-tier interference from surrounding femto-BSs to a reference femto-MS located at the origin, where G_{X_k} and $\|X_k\|$ are respectively the channel gain and the distance between femto-BS at X_k and the typical femto-MS. (6.7) can be evaluated as:

$$\begin{aligned} 1-\epsilon &= \mathbb{P}\left[G_f \geq \frac{\eta}{P_f D_f^{-\gamma}}(N_0 + I_{f,f})\right] \\ &= \mathbb{E}\left[\exp\left(-\frac{\eta}{P_f D_f^{-\gamma}} N_0\right)\right]\mathbb{E}\left[\exp\left(-\frac{\eta}{P_f D_f^{-\gamma}} I_{f,f}\right)\right] \\ &\overset{(a)}{=} \exp\left(-\frac{\eta}{P_f D_f^{-\gamma}} N_0\right)\exp(-\widehat{\lambda}_f D_f^2 \eta^{\frac{2}{\gamma}} K_\gamma), \end{aligned} \tag{6.8}$$

where (a) follows by:

$$\mathbb{E}[\exp(-sI_{f,f})] = \exp\left(-2\pi\widehat{\lambda}_f \int_0^\infty \frac{u}{1+\frac{u^\gamma}{sP_f}} du\right) = \exp\left(-\widehat{\lambda}_f P_f^{2/\gamma} s^{2/\gamma} K_\gamma\right), \tag{6.9}$$

where $K_\gamma = \frac{2\pi^2}{\gamma \sin(2\pi/\gamma)}$ [41]. Then we have the maximum allowable sensing density as $\widehat{\lambda}_f = \frac{-\ln(1-\epsilon) - \frac{\eta}{P_f D_f^{-\gamma}} N_0}{D_f^2 \eta^{\frac{2}{\gamma}} K_\gamma}$. The operator calculates $\widehat{p} = \widehat{\lambda}_f / \lambda_f$ as the probability for initial sensing and distributes it to all femto-BSs for self-configuration. Even in the worse case that all sensing femto-BSs select the same RBs for transmission, the resulting intra-tier interference is still constrained. In the realistic environment, femto-BSs may not be uniformly distributed, and thus a decoding failure at femto-MS may occur.

By adopting the proposed exclusive sensing, a conflict graph defined in definition 1 is built to model the interfering relation among femto-BSs and acts as the basis for the Gibbs sampler to achieve intra-tier interference-free allocation. In these cases, the global energy can be defined as the of total interference that all femto-BSs experienced from all other femto-BSs. We use the following definitions:

$$\begin{cases} V(\mathcal{B}) = 0, & \text{for } \mathcal{B} \in \{k\}, \\ V(\mathcal{B}) = I(k,k'), & \text{for } \mathcal{B} \in \{k,k'\}, \\ V(\mathcal{B}) = 0, & \text{for } |\mathcal{B}| \geq 3. \end{cases} \tag{6.10}$$

The local energy function can then be specified as follows:

$$E_k = \sum_{k' \neq k} I(k,k') \tag{6.11}$$

From the mathematical foundation of the Gibbs sampler, we can achieve the minimum of global energy function by exploiting the local energy function, that is, each femto-BS would finally allocate the different RBs [42] in a distributed way. In Algorithm 5, we consequently propose the Gibbs sampler based resource block management for intra-tier interference mitigation among collocated femtocells, which replaces Step 3 of the CRBM, while other steps in the CRBM still need to be executed by collocated femto-BSs.

Algorithm 5 Gibbs sampler based resource block management (GSRBM)

1: Each collocated femto-BS computes the temperature parameter: $T = \frac{C}{\log_2(2+t)}$
2: Each femto-BS computes local energy for each unoccupied RB.
3: For every RB, compute $p(s)$ according to Gibbs distribution with temperature T.
4: Sample a random variable over all RBs by the interference probability.

Obviously, the random initial sensing can not completely prevent conflicting femto-BSs from simultaneously sensing and observing the same unoccupied RB, especially when femto-BSs are not uniformly distributed. An RB may be simultaneously selected by conflicting femto-BSs for data transmission, causing a collision, which results in decoding failure at the femto-MS. Moreover, the critical issues encountered in wireless media such as fast fading, shadowing and noise power uncertainty may lead to decoding failure at femto-MS. When femto-BS n receives an indication of decoding failure at RB r, self-healing will be executed, where the position of RB r allocated by femto-BS n is adjusted to avoid the cascading occurrences of collisions in the following frames. In this way, femto-BS n uses the interference observed in its last sensing frame to calculate the Gibbs distribution with local energy function. Note that although it is possible for mutually conflicting femto-BSs to receive indications of decoding failure and rearrange the RBs simultaneously, the probability that they rearrange to the same RB at the same time is negligible.

6.5.2.1 Numerical Results

This subsection conducts simulation experiments to evaluate the performance of proposed GSRBM approach in terms of intra-tier interference mitigation. The simulation experiments are based on Matlab software according to 3GPP dual-strip model [1]. The transmission powers of macro-BS and femto-BS are respectively set as 46 dBm and 20 dBm, and the number of antennas at macro-BS, femto-BS and users are 2. The setups related to path loss and shadowing are according to 3GPP dual strip model in Table 6.1. 2.1.1.2-8 of [1]. Please note that it is critical to practically simulate the penetration loss since the signal strength degradation due to the inner wall (the wall inside the apartment block; 20 dB) and outer wall (the wall of the apartment block; 5 dB) can effectively reduce the cross-tier and intra-tier interference. We consider total 144 femto-BSs uniformly and non-uniformly distributed in a macrocell. Except the GSRBM, we investigate two variants as follows.

- Gibbs without healing. The self-healing operation for collision recovery is disabled.
- Greedy allocation [43]. RBs with minimum intra-tier interference are selected.
- Randomized allocation [19]. RBs are randomly selected.

Table 6.1 Simulation parameters

Parameter	Value
carrier frequency	2 GHz
transmission bandwidth	10 MHz
number of antennas	2
macro transmission power	46 dBm
femto transmission power	20 dBm
active femto-BS ratio	$\alpha = 25\%$ or 50%
indoor macro-MS ratio	$\beta = 35\%$ or 80%
path loss and shadowing	Table A.2.1.1.2-8 in [1]
penetration loss	Table A.2.1.1.2-8 in [1]
channel model	Rayleigh fading
minimum macro-BS to MS distance	35 m
minimum femto-BS to MS distance	3 m
minimum macro-BS to femto-BS distance	75 m
antenna pattern of femto-BS	omni-directional with 5 dB gain
antenna pattern of macro-BS	Table A.2.1.1-2 in [1]

Figure 6.7 investigates the aggregated intra-tier interference occurring at a reference femto-MS at each time slot for all approaches among the uniformly distributed femto-BSs. Obviously, the Gibbs sampler in RB allocation highly outperforms greedy and randomized allocations in terms of interference power and convergence speed when power control is disabled. Another intuitive observation is that the existence of self-healing can quickly adjust RB allocation to the

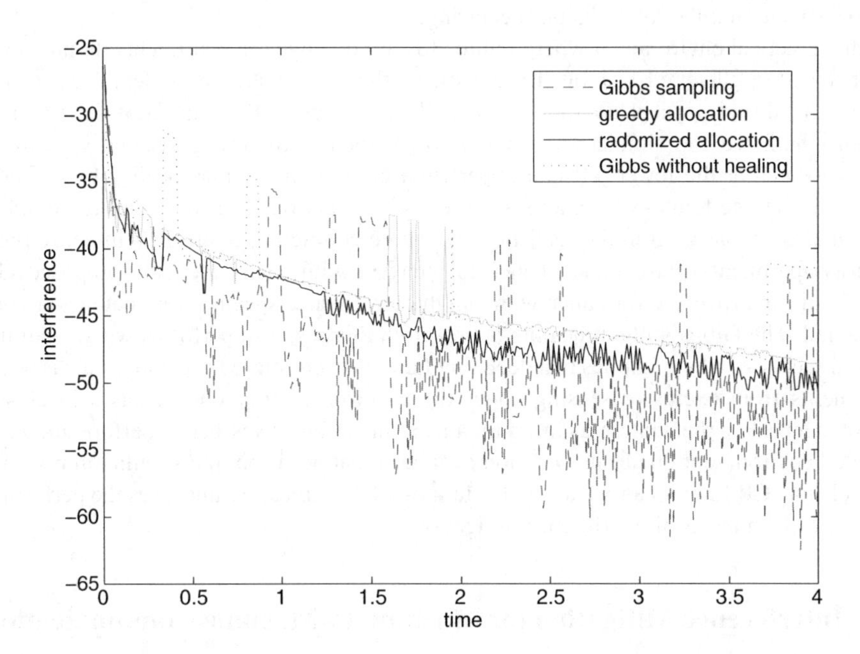

Figure 6.7 Aggregated intra-tier interference received at one femto-MS among the uniform-distributed femto-BSs.

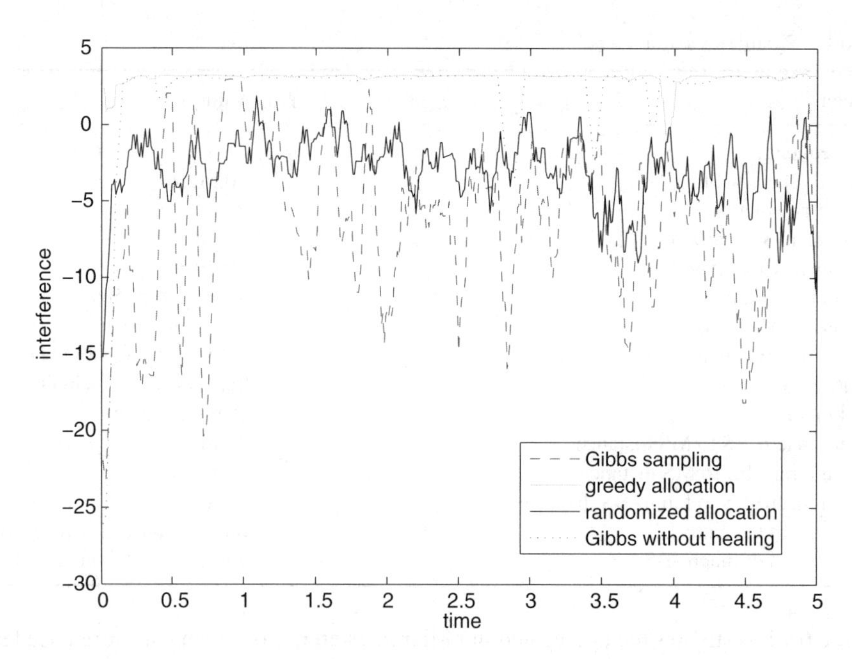

Figure 6.8 Aggregated intra-tier interference received at one femto-MS among the beta-distributed femto-BSs.

optimal distribution. Moreover, randomized allocation performs better than greedy allocation due to a smaller number of collisions occurring.

In the practical environment where femto-BSs are deployed in apartments, femto-BSs are distributed in a clustered fashion. By exploiting the Beta distribution Beta$(2, 5)$ to model the horizontal and vertical coordinates of half of the femto-BSs and Beta$(5, 2)$ to model the other half, we can simulate a scenario where femto-BSs are aggregated in two dense regions located at the lower left and upper right corners of the macrocell. The aggregated interference at one femto-MS under such a realistic scenario is illustrated in Figure 6.8. The figure indicates that even in the realistic case where conflicting femto-BSs in dense regions (such as apartments) have more chance to interfere with each other, the proposed Gibbs sampler still improves performance with a high convergence speed, given that power control is disabled. The Gibbs without healing approach in this scenario performs worse than in the uniform scenario, which implies that decoding failures occur more frequently. This is because the benefits from exclusive sensing under the uniform assumption decreases in clustered environments. For the same reason, randomized allocation gives better performance in the clustered environment due to its collision avoidance nature. From the simulation results, we conclude that CR functionality executed at femto-BSs significantly improves the performance at femto-MSs in terms of interference mitigation.

6.6 Interference Mitigation for Machine-to-Machine Communications

There is an increasing concern for integrating real-world physical activities and the cyber-world of computations and communications to design efficient cyber-physical systems (CPS)

[44]. These systems consist of networks of embedded computations and communication machines which monitor and control the physical entities via sensors and actuators on electrical power grids, transportation vehicles and traffic roads, robotic systems, healthcare and medical machines, environmental control and smart buildings. To provide ubiquitous communications among such a large number of machines without additional cyber-infrastructure deployment costs, connecting all these machines by leveraging cellular communication systems turns out to be an effective and efficient solution. As a result, a promising realization of CPS can be the machine-type communication (MTC) in LTE-Advanced [45, 46]. To ensure the effective operation of CPS, the reliable information delivery of physical events is essential, especially in the most challenging and variant environments of wireless links.

Furthermore, to preserve the precious spectrum resources, communications/connections among machines ideally should exploit the spectrum of the existing radio networks (such as LTE-Advanced [47, 48]), and thus the appearance of numerous machines induces catastrophic impacts on the existing radio network. It is desirable to separate the management of machine-to-machine (M2M) communications and human-to-human (H2H) communications onto different BSs to significantly reduce the burden on each BS. One possible solution is applying femto-BSs to deal with M2M communications while the macro-BS still deals with H2H communications. Under such a solution, how to distributively share the limited radio resources according to the QoS demand of machines without inducing harmful interference so that numerous machines could coexist with the existing radio network emerges as a primary challenge.

Under the design constraint of simplicity without imposing additional complexities on current radio network protocols and operations, CR technology is a well-suited solution for M2M communications. Under this framework, if each machine performs periodic channel sensing to identify the radio resource usage of existing macrocells and only utilizing radio resources identified as unoccupied by existing macrocells, interference can thus be mitigated. However, under this framework, all machines in the femtocell have to perform channel sensing and report sensing results to the femto-BS. This leads to an enormous amount of feedback data, which makes it infeasible for CR to be directly applied to femtocells.

To achieve autonomous interference mitigation by leveraging the CR technology in the femtocell, the key is to reduce the amount of feedback. One effective solution is to significantly reduce the number of machines required to perform channel sensing. In addition, when numerous machines directly execute monitoring or control subprocesses in CPS simultaneously, a considerable amount of information is funnelled toward the same destination and has to utilize common resources. Even with the aid of CR, the available radio resource might not be sufficient, and regional hotspots might be spawned. The resulting large delays and wasteful packet drops violate QoS requirements [49].

We therefore adopt a powerful technology known as *compressive sensing* [50], widely applied to signal compression and restoration, by which the ordinary sparse signal is sampled with a sampling rate far lower than the Shannon/Nyquist sampling rate, while the signal can still be recovered with a high probability. By using compressive sensing, a proportion of the machines (instead of all machines) are required to perform channel sensing and report sensing results to the femto-BSs, and so the interference levels at any given locations within the coverage of the femtocells can be constructed, which is referred as the *spectrum map*. As a result, the amount of channel information feedback is significantly reduced and the CR technology can be applied to femtocells for autonomous interference mitigation. We consequently propose the spectrum map based resource management (SMRM) for femtocells to construct the spectrum map [51], which enables a successful realization of CPS and M2M communications in LTE-Advanced.

6.6.1 Background of Compressive Sensing

Based on the realisation that a small sample of a sparse signal contains enough information for its reconstruction, compressive sensing (CS) [50, 52, 53] shows high promise for distributed schemes. CS is a new paradigm of performing sampling and compression simultaneously thus significantly reducing the sampling rate. Traditional approaches acquire the entire signal and process it to extract the information. CS acquires only a small number of linear measurements that preserve the structure of the signal, and the signal is then reconstructed from these measurements using an optimization process. Specifically, CS uses the measuring matrix $\mathbf{A}$ to measure the signal $\mathbf{x}$ and acquire the measurements $\mathbf{y}$: $\mathbf{y}_{R\times 1} = \mathbf{A}_{R\times N}\mathbf{x}_{N\times 1}$, where the dimension $R \ll N$. To reconstruct $\mathbf{x}$ from $\mathbf{y}$, there are two necessary conditions:

1. Sparsity of signal $\mathbf{x}$: The N-dimensional vector signal $\mathbf{x}$ is K-sparse, that is, there are only $K \ll N$ elements in $\mathbf{x}$ that are non-zero. When $\mathbf{x}$ is binary, $\mathbf{x}$ is K-sparse if there are only $K \ll N$ elements in $\mathbf{x}$ that are non-zero, or there are only $K \ll N$ elements in $\mathbf{x}$ that are zero.
2. Restricted isometry property (RIP) of measurement matrix $\mathbf{A}$: for any K-sparse signal $\mathbf{x}$, the $R \times N$ matrix $\mathbf{A}$ has the property that:

$$\|\mathbf{x}\|_2(1-\delta) \leq \|\mathbf{A}\mathbf{x}\|_2 \leq \|\mathbf{x}\|_2(1+\delta). \tag{6.12}$$

where $0 \leq \delta \leq 1$.

Then, given $\mathbf{A}$ and $\mathbf{y}$, l_1 minimization problem with $R \geqslant cKlog(N/K)$ is able to recover $\mathbf{x}$ by:

$$\mathbf{x}^* = \text{argmin}\|\mathbf{x}\|_1 \text{ subject to } \mathbf{y} = \mathbf{A}\mathbf{x} \tag{6.13}$$

6.6.2 SMRM for Femtocells

To mitigate interference, the femto-BS has to avoid allocating occupied RBs (by macrocells). As a result, the key idea of interference mitigation is that all machines should autonomously estimate the RBs usage of the macrocell and report the sensing result (that is, identifying which RB is occupied by the macrocell in a subframe at any given location, i.e., the spectrum map). However, this approach is infeasible due to the potentially unacceptable amount of uplink reporting data. As a result, the number of machines required to perform channel sensing has to be significantly reduced. To resolve this issue, we have to adopt the compressive sensing for the spectrum map construction. For this goal, the coverage of the femtocell is divided into N isotropic 'grids' indexed by $n = 1, \ldots, N$, as shown in Figure 6.9. With the aid of the spectrum map, interference can be mitigated by simply allocating unoccupied RBs to machines. Such an SMRM is proposed in Algorithm 6.

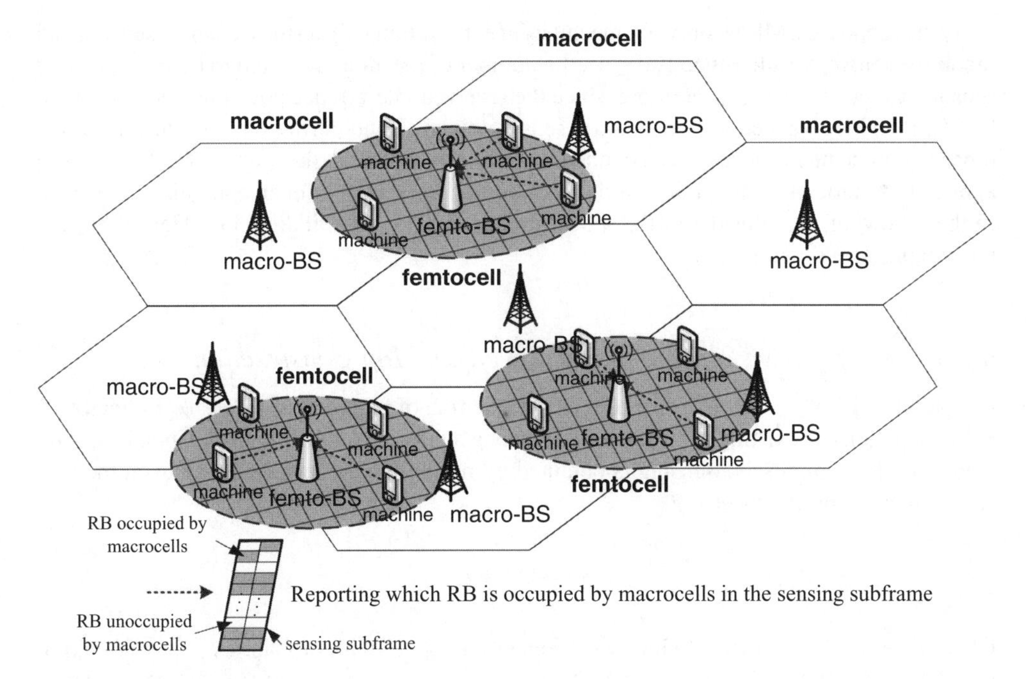

Figure 6.9 Compressive sensing is applied to construct the spectrum map of the femtocell by channel sensing results of a proportion of the machines. The coverage of the femtocell is divided into N isotropic 'grids'.

Algorithm 6 Spectrum map based resource management (SMRM)

1: The femto-BS periodically allocates subframes for machines to perform channel sensing to identify which RB is occupied. The sensing period is T subframes and each channel sensing persists for one subframe (Figure 6.3). A subframe is referred to as a 'sensing subframe' if the subframe is allocated for performing channel sensing; otherwise, it is referred as a 'data subframe'. All machines can not perform data transmissions/receptions within a sensing subframe.
2: Within a sensing subframe, each machine performs channel sensing with probability q to measure the received interference power on each RB.
3: **if** the received interference power on an RB exceeds a certain threshold **then**
4: This RB is identified as being occupied; otherwise, the RB is unoccupied. The machine reports the sensing result to the femto-BS.
5: The femto-BS constructs the spectrum map based on the channel sensing report (the method of constructing the spectrum map will be elaborated later).
6: In subsequent data subframes, femto-BS only allocates unoccupied RBs sensed in the sensing subframe to machines.
7: The femto-BS also extracts the following parameters from the result of reports: (i) the traffic load of existing macrocells, (ii) the RBs' allocation correlation probability of existing macrocells and (iii) the fraction of correlated RBs allocation of existing macrocells, which are detailed later.

By the proposed SMRM, only Vq machines in expectation will perform channel sensing and report the sensing result. Since $Vq \leq V$, the number of machines required to perform channel sensing can be reduced. Furthermore, since the received interference power had been adopted by 3GPP LTE-Advanced as a mandatory sensing quantity and corresponding sensing results reporting procedures had also been defined by LTE-Advanced [7], the proposed SMRM can be applied to femtocells without any hardware modifications or additional signalling overheads. In the following, we consider to the spectrum map construction in Step 3 of Algorithm 6 by the compressive sensing.

6.6.3 *Compressive Sensing for the Spectrum Map Construction*

Denote $\Psi = [\psi_1 \psi_2 \ldots \psi_N]^{\mathrm{T}}$ as the true RB occupation of macrocells, where ψ_n indicates M RB occupations of existing macrocells in a sensing subframe on the nth grid. Upon receiving feedback, the compressive sensing is obtained by multiplying a sampling matrix on the true RB occupation of macrocells Ψ,

$$\mathbf{y} = \mathbb{A}\Psi + \varepsilon \tag{6.14}$$

where $\mathbb{A}$ is an $R \times N$ matrix with each element taking '1' with probability $q\frac{V_n}{V}$, and taking '0' with probability $1 - q\frac{V_n}{V}$, where V_n is the number of machines within the nth grid. The spectrum map can be constructed by searching the minimum l_1 norm of Ψ,

$$\Psi^* = \arg\min \|\Psi\|_1 \ s.t. \ \|\mathbb{A}\Psi - \mathbf{y}\|_2 \leq \epsilon \tag{6.15}$$

by applying *second order corn programming* [54] if $R = O(K \log \frac{NM}{K})$, where l_p norm of Ψ is $\|\Psi\|_p := (\sum_{n=1}^{N} |\psi_n|^p)^{1/p}$, K is the sparsity of $\mathbb{A}$, $\epsilon = \|\Phi\varepsilon\|_2$ and Φ is a random basis.

To successfully construct the spectrum map by the compressive sensing with an acceptable quality, it is suggested that two conditions must be considered [50]: (i) Ψ shall be sparse. (ii) The selection of $\mathbb{A}$ shall satisfy the RIP. Since, in a typical situation in the urban environment, most RBs in a subframe could be occupied by macrocells, Ψ is typically sparse. From [55], although it is suggested that the RIP can be achieved if each machine senses the channel the i.i.d. and symmetry Bernoulli random variable, the spectrum map can be constructed with an acceptable quality when each machine senses the channel with probability q of the Bernoulli random variable.

6.6.4 *Performance Evaluations*

In this section, we evaluate the performance of the proposed SMRM by adopting the system parameters of LTE-Advanced in Table 6.1. The network deployment for the simulation is shown in Figure 6.9, where three femtocells coexist with six macrocells. The femtocells are capable of the proposed SMRM and thus CR technology, while the SMRM is not applied to macrocells. We evaluate the performance of the spectrum map construction by compressive

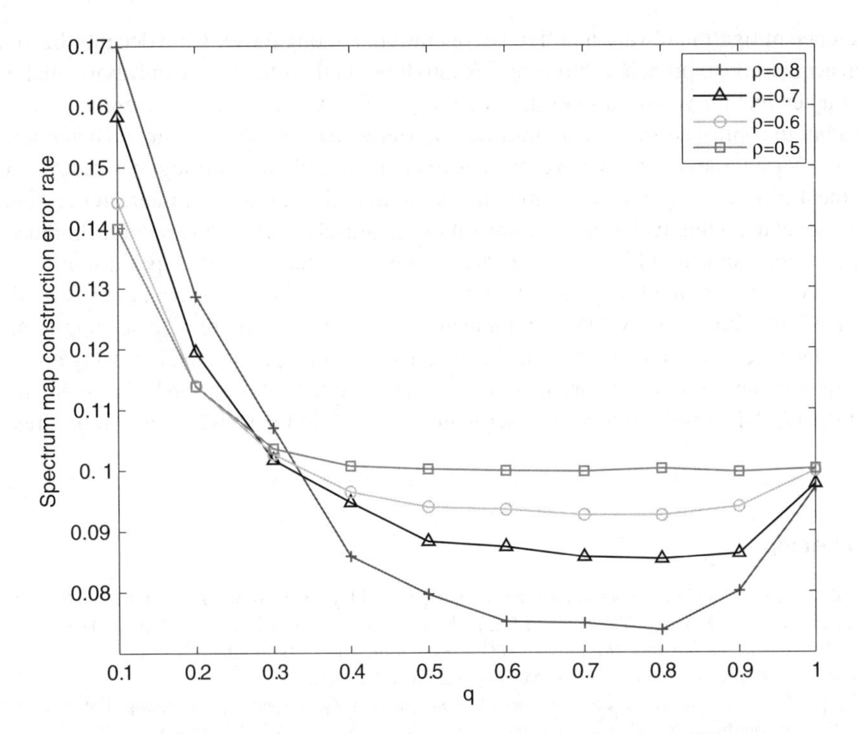

Figure 6.10 Error rate of the spectrum map construction under different q.

sensing. Figure 6.10 shows the error rate of the spectrum map construction. The error rate is measured by:

$$\xi = \frac{\sum_{n=1}^{N} \sum_{m=1}^{M} \Upsilon(m, n)}{NM} \tag{6.16}$$

where $\Upsilon(m, n) = 1$ if there is an error on the spectrum map construction on the mth RB in a subframe on the nth grid; otherwise, $\Upsilon(m, n) = 0$. We can observe from Figure 6.10 that, when macrocells have a typically high traffic load ($\rho = 0.8$), there is only a 7.5% error (7.5 RBs are occupied by macrocells but they are estimated as unoccupied, or 7.5 RBs are not occupied by macrocells but they are estimated as occupied, among 100 RBs in a frame) if $q \in [0.6, 0.8]$. Please note that only 60% of machines are adopted for channel sensing. This result suggests the effectiveness of the SMRM on the spectrum map construction. We may particularly note that, in the compressive sensing, a large q may lead to the oversampling issue to degrade the performance.

6.7 Conclusion

In current two-tier heterogeneous networks, the unique features of random deployment, restricted access, no macro-femto coordination and backward compatibility challenge interference mitigation toward an effective operation. Among the existing decentralized information acquisition approaches, building CR into femtocells could tackle the above challenges. This chapter has provided an overview of the possible CR-enabled interference mitigation approaches to control cross-tier and intra-tier interference in OFDMA femtocell heterogeneous networks. Various approaches have been investigated, including orthogonal radio resource assignment in time-frequency and antenna spatial domains, as well as interference cancellation via novel decoding techniques. Based on the acquired information, novel techniques such as game theory and the Gibbs sampler are exploited to achieve better performance, which shows the potential of applying CR. Moreover, we consider the scenario where CR is adopted to mitigate interference of M2M communications and apply compressive sensing to significantly reduce the signalling overheads. Yielding a limited complexity and imposing no impacts to the state-of-the-art femtocell architecture, the CR-enabled solution can be smoothly applied to 3GPP LTE-Advanced femtocells to serve urgent needs in the standardization progress.

References

1. 3GPP. E-UTRA: Further Advancements for E-UTRA Physical layer aspects 3GPP TR 36.814 v9.0.0, 2010b.
2. A. Damnjanovic, J. Montojo, Y. Wei, T. Ji, T. Luo, M. Vajapeyam, T. Yoo, O. Song and D. Malladi. A survey on 3gpp heterogeneous networks. *IEEE Wireless Commun. Mag.*, 18(3), 10–21, 2011.
3. 3GPP. 3GPP Report of TSG RAN WG1 Meeting #62, v0.1.0, 2010a.
4. D. Lopez-Perez, I. Guvenc, G. de la Roche, M. Kountouris, T. Q. S. Quek and J. Zhang. Enhanced intercell interference coordination challenges in heterogeneous networks. *IEEE Wireless Commun. Mag.*, 18(3), 22–30, 2011.
5. A. Barbieri, A. Damnjanovic, T. Ji, J. Montojo, Y. Wei, D. Malladi, O. Song and G. Horn. LTE femtocells: System design and performance analysis. *IEEE J. Sel. Areas Commun.*, 30(3), 586–594, 2012.
6. G. Boudreau, J. Panicker, N. Guo, R. Chang, N. Wang and S. Vrzic. Interference coordination and cancellation for 4G networks. *IEEE Commun. Mag.*, 47(4), 74–81, 2009.
7. 3GPP. Evolved Universal Terrestrial Radio Access (E-UTRA) and Evolved Universal Terrestrial Radio Access (E-UTRAN); Overall description; Stage 2 3GPP TS 36.300 v10.0.0, 2010d.
8. J. M. Peha. Sharing spectrum through spectrum policy reform and cognitive radio. *Proc. IEEE*, 97(4), 708–719, 2009.
9. M. Yavuz, F. Meshkati, S. Nanda, A. Pokhariyal, N. Johnson, B. Raghothaman and A. Richardson. Interference management and performance analysis of UMTS/HSPA+ femtocells. *IEEE Commun. Mag.*, 47(9), 102–109, 2009.
10. A. Goldsmith, S. A. Jafar, Marić I. and S. Srinivasa. Breaking spectrum gridlock with cognitive radios: An information theoretic perspective. *Proc. IEEE*, 97(5), 894–914, 2009.
11. G. Gur, S. Bayhan and F. Alagoz. Cognitive femtocell networks: an overlay architecture for localized dynamic spectrum access. *IEEE Wireless Commun. Mag.*, 17(4), 62–70, 2010.
12. J. Xiang, Y. Zhang, T. Skeie and L. Xie. Downlink spectrum sharing for cognitive radio femtocell networks. *IEEE Syst. J.*, 4(4), 524–534, 2010.
13. S. M. Almalfouh and G. L. Stuber. Interference-aware radio resource allocation in OFDMA-based cognitive radio networks. *IEEE Trans. Veh. Technol.*, 60(4), 1699–1713, 2011.
14. S. M. Cheng, S. Y. Lien, F. S. Chu and K. C. Chen. On exploiting cognitive radio to mitigate interference in macro/femto heterogeneous networks. *IEEE Wireless Commun. Mag.*, 18(3), 40–47, 2011b.
15. S. Y. Lien, Y. Y. Lin and K. C. Chen. Cognitive and game-theoretical radio resource management for autonomous femtocells with QoS guarantees. *IEEE Trans. Wireless Commun.*, 10(7), 2196–2206, 2011b.
16. A. Attar, V. Krishnamurthy and O. N. Gharehshiran. Interference management using cognitive base-stations for UMTS LTE. *IEEE Wireless Commun. Mag.*, 49(8), 152–159, 2011.
17. S. M. Cheng, W. C. Ao, F. M. Tseng and K. C. Chen. Design and analysis of downlink spectrum sharing in two-tier cognitive femto networks. to appear in *IEEE Trans. Veh. Technol.*, 2012.

18. D. Lopez-Perez, A. Valcarce, G. de la Roche and J. Zhang. OFDMA femtocells: A roadmap on interference avoidance. *IEEE Commun. Mag.*, 47(9), 41–48, 2009b.
19. V. Chandrasekhar and J. G. Andrews. Spectrum allocation in tiered cellular networks. *IEEE Trans. Commun.*, 57(10), 3059–3068, 2009.
20. K. Sundaresan and S. Rangarajan. Efficient resource management in OFDMA femto cells. *Proc. ACM MobiHoc 2009*, 33–42, 2009.
21. Z. Bharucha, H. Haas, G. Auer and I. Cosovic. Femto-cell resource partitioning. *Proc. IEEE GLOBECOM 2009 workshops*, 2009.
22. Y. Y. Li, M. Macuha, E. S. Sousa, T. Sato and M. Nanri. Cognitive interference management in 3G femtocells *Proc. IEEE PIMRC 2009*, 1118–1122, 2009.
23. R. Madan, J. Borran, A. Sampath, N. Bhushan, A. Khandekar and T. Ji. Cell association and interference coordination in heterogeneous LTE-A cellular networks. *IEEE J. Sel. Areas Commun.*, 28(9), 1479–1489, 2010.
24. S. Y. Lien, C. C. Tseng, K. C. Chen and C. W. Su. Cognitive radio resource management for QoS guarantees in autonomous femtocell networks. *Proc. IEEE ICC 2010*, pp. 1–6, 2010.
25. F. S. Chu and K. C. Chen. Mitigation of macro-femto co-channel interference by spatial channel separation. *Proc. IEEE VTC 2011 Spring*, 2011.
26. D. Lopez-Perez, A. Ladanyi, A. Juttner and J. Zhang. OFDMA femtocells: A self-organizing approach for frequency assignment *Proc. IEEE PIRMC 2009*, 2202–2207, 2009a.
27. S. Brahma and M. Chatterjee. Mitigating self-interference among IEEE 802.22 networks: A game theoretic perspective. *Proc. IEEE GLOBECOM 2009*, 2009.
28. J. H. Yun and K. G. Shin. CTRL: A self-organizing femtocell management architecture for co-channel deployment. *Proc. ACM MobiCom 2010*, 2010.
29. Y. S. Liang, W. H. Chung, G. K. Ni, I. Y. Chen, H. Zhang and S. Y. Kuo. Resource allocation with interference avoidance in OFDMA femtocell networks. to appear in *IEEE Trans. Veh. Technol.*, 2012.
30. C. H. Ko and H. Y. Wei. On-demand resource-sharing mechanism design in two-tier OFDMA femtocell networks. *IEEE Trans. Veh. Technol.*, 60(3), 1059–1071, 2011.
31. C. K. Yu, K. C. Chen and S. M. Cheng. Cognitive radio network tomography. *IEEE Trans. Veh. Technol.*, 59(4), 1980–1997, 2010.
32. S. Huang, X. Liu and Z. Ding. Distributed power control for cognitive user access based on primary link control feedback. *Proc. IEEE INFOCOM 2010*, 2010.
33. S. M. Cheng, W. C. Ao and K. C. Chen. Efficiency of a cognitive radio link with opportunistic interference mitigation. *IEEE Trans. Wireless Commun.*, 10(6), 1715–1720, 2011a.
34. X. Chu, Y. Wu, D. Lopez-Perez and X. Tao. On providing downlink services in collocated spectrum-sharing macro and femto networks. *IEEE Trans. Wireless Commun.*, 10(12), 4306–4315, 2011.
35. R. Radaydeh and M. Alouini. Switched-based interference reduction scheme for open-access overlaid cellular networks. to appear in *IEEE Trans. Wireless Commun.*, 2012.
36. 3GPP. E-UTRA physical layer measurements 3GPP TS 36.214 V9.2.0, 2010c.
37. S. Gemana and D. Geman. Stochastic relaxation, Gibbs distributions and the Bayesian restoration of images. *Journal of Applied Statistics*, 20(5), 25–62, 1993.
38. R. A. Tannious and A. Nosratinia. Cognitive radio protocols based on exploiting hybrid ARQ retransmissions. *IEEE Trans. Wireless Commun.*, 9(9), 2833–2841, 2010.
39. W. Chen, K. B. Letaief and Z. Cao. Network interference cancellation. *IEEE Trans. Wireless Commun.*, 8(12), 5982–5995, 2009.
40. D. Fudenberg and J. Tirole. *Game Theory*. The MIT Press, 1991.
41. A. M. Hunter, J. G. Andrews and S. P. Weber. Transmission capacity of ad hoc networks with spatial diversity. *IEEE Trans. Wireless Commun.*, 7(12), 5058–5071, 2008.
42. B. Kauffmann, F. Baccelli, A. Chaintreau, V. Mhatre, K. Papagiannaki and C. Diot. Measurement-based self organization of interfering 802.11 wireless access network. *Proc. IEEE INFOCOM 2007*, 1451–1459, 2007.
43. L. G. U. Garcia, K. I. Pedersen and P. E. Mogensen. Autonomous component carrier selection: interference management in local area environments for LTE-advanced. *IEEE Commun. Mag.*, 47(9), 110–116, 2009.
44. E. A. Lee. Cyber physical systems: Design challenges *Proc. IEEE ISORC 2008*, 363–369, 2008.
45. 3GPP. Service requirements for machine-type communications 3GPP TS 22.368 v11.0.0, 2010e.
46. 3GPP. System improvement for machine-type communications 3GPP TR 23.888 V1.0.0, 2010f.
47. K. Doppler, M. Rinne, C. Wijting, C. B. Ribeiro and K. Hugl. Device-to-device communication as an underlay to LTE-advanced networks. *IEEE Commun. Mag.*, 47(12), 42–49, 2009.

48. S. Y. Lien, K. C. Chen and Y. Lin. Toward ubiquitous massive accesses in 3GPP machine-to-machine communications. *IEEE Commun. Mag.*, 49(4), 66–74, 2011a.
49. K. Karenos and V. Kalogeraki. Traffic management in sensor networks with a mobile sink. *IEEE Trans. Parallel Distrib. Syst.*, 21(10), 1515–1530, 2010.
50. Donoho D. Compressed sensing. *IEEE Trans. Inf. Theory*, 52(4), 1289–1306, 2006.
51. S. Y. Lien, S. M. Cheng, S. Y. Shih and K. C. Chen. Radio resource management for QoS guarantees in Cyber-Physical Systems. to appear in *IEEE Trans. Parallel Distrib. Syst.*, 2012.
52. E. Candes, J. Romberg and T. Tao. Stable signal recovery from incomplete and inaccurate measurements. *Comm. Pure Appl. Math.*, 59(8), 1207–1223, 2006.
53. J. Haupt, W. U. Bajwa, M. Rabbat and R. Nowak. Compressed sensing for networked data. *IEEE Signal Process. Mag.*, 25(2), 92–101, 2008.
54. M. A. T. Figueiredo, R. D. Nowak and S. J. Wright. Gradient projection for sparse reconstruction: Application to compressed sensing and other inverse problems. *IEEE J. Sel. Areas Commun.*, 1(4), 586–597, 2007.
55. S. Mendelson, A. Pajor and N. Tomczak-Jaegermann. Uniform uncertainty principle for Bernoulli and subgaussian ensembles. *Constructive Approximation*, 28(3), 277–289, 2008.

7

Game Theoretic Approach to Distributed Bandwidth Allocation in OFDMA-based Self-organizing Femtocell Networks

Chenxi Zhu and Wei-Peng Chen
Fujitsu Laboratories of America, USA

7.1 Introduction

The femtocell is a new concept rapidly being developed in the wireless communication industry [1,2]. A femtocell (called a home eNB or HeNB in the 3GPP standard) is a low-cost miniature base station placed in a user's home or small office to provide private (or semi-public), high throughput cellular access to its subscribers. It uses the homeowner's wired broadband access (DSL, cable modem or fibre) as a backhaul to connect to the wireless operator's network, and operates in the spectrum licensed to the wireless operator. Besides improving the user's indoor access quality of the service (QoS), it leads to significant savings for the operator. It reduces the traffic load from the operator's macrocellular network, and increases the overall capacity in a given area through aggressive spatial spectrum reuse. By using the user's internet access as the backhaul, the wireless operator reduces its capital expenditure and operation cost significantly. Architectures and protocols of femtocells are still being standardized in the 3GPP standard body [3,4].

A major difference between the traditional macrocellular network and a femtocell network is in their deployments. A macrocell network is carefully designed and optimized by the operator. A femtocell network, on the other hand, is essentially unplanned because the femtocells are placed at random locations by the users within a deployment area, with little control of the

Heterogeneous Cellular Networks, First Edition. Edited by Rose Qingyang Hu and Yi Qian.

operator. Because the available channels for femtocell deployment (or for co-deployment of femto and macrocells) is very limited, a large number of femtocells in an area have to share the same channel. As the density of femtocell increases, mutual interference between adjacent femtocells becomes a significant issue [5, 6]. One way to reduce the interference between femtocells is to control the bandwidth allocated to each femtocell. The traditional approach of configuring macrocells using network planning tools is no longer applicable. This is because of the high cost of planning and optimizing a network with tens of thousands or even millions of nodes, and the difficulty of obtaining detailed geometry and propagation information for the femtocells. Besides, a user may turn on, turn off or move the femtocell to a different location at any time, making the information collected by the operator frequently out of date. In order for an operator to run a large femtocell network successfully, the network needs to constantly (re)configure and optimize itself in real time. Algorithms allowing a femtocell network to self-organize and self-optimize without human intervention (a.k.a. SON algorithms) is key to successful large-scale deployment of femtocell networks [7, 8]. Interference management in macrocell-femtocell overlay network has been studied in [9, 10], and interference control and resource allocation in femtocell networks has been studied in [11–13].

In this chapter we develop a distributed algorithm for allocating the channel bandwidth for femtocell networks autonomously, using the theory of non-cooperative games [14]. Distributed algorithms with low complexity are particularly attractive because of the large size of the network. In this non-cooperative game, every femtocell is a player trying to maximize its own benefit. Femtocells do not communicate with each other either directly or indirectly. Different femtocells interact through the interference they generate to each other, and no explicit communication is necessary between the femtocells. Each femtocell relies on the SINR reported by its own users (UEs) as a basis for the dynamic adjustment of its used bandwidth. We prove that under a wide range of conditions, there exists a unique Nash equilibrium in the network, and the convergence to the Nash equilibrium is guaranteed. The operation performed at each femtocell is simple, robust and efficient. It can be extended to a wide range of network control architectures, ranging from fully distributed to centralized and hybrid architecture. It can also be used to manage the interference in a heterogeneous network of mixed macro, pico and femtocells.

The rest of the chapter is organized as follows. In Section 7.2, we develop a framework for the distributed bandwidth allocation scheme using game theory. We then discuss the convergence of the distributed bandwidth allocation scheme in Section 7.3, and the choice of utility function and the specific bandwidth allocation scheme in Section 7.4. Simulation results in a wide range of scenarios are presented in Section 7.5. Section 7.6 provides further discussion and future directions. Finally, a summary is given in Section 7.7.

7.2 Distributed Bandwidth Allocation

Assume that in the network there are N femtocells operating in the same channel. We assume that the average DL transmission power density (average transmission power per subcarrier) for femtocell i (HeNB_i) is fixed at p_i in an occupied subcarrier, while the occupied bandwidth can be adjusted with the scheme developed here. For an OFDMA-based system such as LTE [15] or WiMAX [16], this can be realized by changing the number of physical resource blocks (PRBs) used in LTE or PUSC subchannels in WiMAX. Let FW be the bandwidth of the entire channel (typically 5 or 10 MHz) and $w_i \in [0, 1]$ be the normalized bandwidth used by HeNB$_i$ with respect to *FW*. $w_i = 1$ means that HeNB$_i$ is using all the resource blocks in *FW*. Different HeNBs do not negotiate with each other on what PRBs they use. In order to achieve diversity in the entire channel bandwidth *FW*, HeNB$_i$ chooses its w_i PRBs with equal likeliness across the entire channel bandwidth, and this choice of PRBs can vary from frame to frame. When combined with pseudo-randomized resource block hopping in the DL resource allocation, the transmission from an HeNB is spread across the entire *FW*. The interference in the network is effectively whitened. This allows us to treat w_i as a continuous variable as all PRBs are now equivalent. The details of the transmission in HeNB$_i$ using w_i is omitted for simplicity, as this does not affect the scheme developed here. The transmission in femtocell *i* is modelled as transmission from HeNB$_i$ to a single UE$_i$, where UE$_i$ can be considered either as a representative UE, or a (hypothetical) composite UE representing all the real UEs in the cell. A femtocell installed in a residence or office usually has a rich multi-path scattering environment. The channels between the HeNB and its multiple UEs in the femtocell will likely exhibit different and independent frequency selective fadings. Assuming that either a frequency-selective-, or a frequency-diversity-based proportional fair scheduling algorithm is used by the HeNB, it is reasonable to assume that, as an approximation, an HeNB does not have any preference of a portion of the bandwidth *FW* over another portion. As a consequence, we omit the frequency selectivity between an HeNB and *all* the UEs it serves for the study of bandwidth allocation across all the HeNBs.

The $SINR_i$ of the transmission from HeNB$_i$ to UE$_i$ on those used subcarriers depends on transmission power and bandwidth usage in all the other femtocells:

$$SINR_i = SINR_i(W) = \frac{G_{ii}p_i}{\sum_{j\neq i} G_{ij}p_j w_j + \eta_i} = \frac{h_{ii}}{\sum_{j\neq i} h_{ij} w_j + \eta_i}, \tag{7.1}$$

where G_{ij} is the *effective flat channel* power gain from HeNB$_j$ to UE$_i$, $p_j > 0$ is the average transmission power per occupied subcarrier of HeNB$_j$, and η_i is the noise level per subcarrier at UE$_i$ including both receiver thermal noise and background interference, such as interference from macro eNBs. The G_{ij} can be computed as $G_{ij} = \frac{1}{K}E[\sum_{k=1}^{K} |g_{ij}^k|^2]$, where g_{ij}^k is the channel coefficient of subchannel *k* from HeNB$_j$ to UE$_i$, and the expectation is taken with respect to channel fast fading. The normalized bandwidth usage w_j is the same as the probability that HeNB$_j$ uses a subchannel that is occupied by HeNB$_i$. It can be considered as a 'thinning factor' per bandwidth after the interference from HeNB$_j$ is spread over the entire *FW* by the random permutation of its used resource blocks. Let

$$h_{ij} = G_{ij}p_j,$$

and $W = [w_0, \ldots, w_{N-1}]^T$. Because HeNB$_i$ only cares about the SINR of those resources (subcarriers) that it occupies, its own bandwidth w_i does not appear in the equation. This makes it different than the SINR definition in the classic CDMA power control problem (for example, [17]), where the control variables (transmission power levels) appear in both the nominator and the denominator. The problem and the solution in the current work, although resembling the power control work in the literature to some degree, are different from the previous work on power control.

We assume that an HeNB_i keeps its average transmission power density p_j per used subcarrier constant. The transmission rate R_i can be approximated as a function of $SINR_i$ using the Shannon channel capacity:

$$R_i = w_i \text{In}\,(1 + \beta \cdot SINR_i), \tag{7.2}$$

where $0 < \beta < 1$ represents the gap between the realized modulation and coding scheme (MCS) and the Shannon capacity. We use natural logarithm instead of base 2 logarithm here for simplicity. The QoS provided by HeNB_i is determined by throughput R_i in its femtocell. In general the utility function $U_i(R_i)$ is an increasing and cave function of cell rate R_i. If its own QoS is the only objective of every HeNB_i when adjusting its bandwidth usage w_i, every femtocell becomes most greedy and tends to use all the available bandwidth ($w_i = 1$), because its rate $R_i(w_i)$ is an increasing function of w_i. In order to make every femtocell conscious of its bandwidth usage (thus regulating the interference that it causes to the other cells), we introduce a fictional price for its bandwidth usage. The net utility in femtocell i is defined as:

$$NU_i(w_i, R_i) = U_i(R_i(w_i)) - c_i w_i, \tag{7.3}$$

where $c_i > 0$ is the (fictional) price paid per unit bandwidth. Including the bandwidth price term makes a femtocell aware of the network resources it uses and enables it to be self-regulating. The price c_i is for network resource management purpose here and may not be directly related to the monetary price paid by the femtocell user. By using the distributed bandwidth allocation scheme, every HeNB_i tries to maximize a utility function $U_i(R_i)$ representing the QoS that it provides in its femtocell, while paying a price for using the bandwidth w_i. This way a balance between individual cell QoS and inter-cell interference can be achieved. With an abuse of notation, we rewrite NU_i as:

$$NU_i(w_i, R_i) = NU_i(W) = NU_i(w_i, W_{-i}), \tag{7.3a}$$

where W_{-i} represents the entries of W except w_i. Assume that there is no communication, direct or indirect, between different HeNBs. The only interaction between them is through the interference that they cause to each other. Autonomously, HeNB_i adjusts its used bandwidth w_i to maximize its own net utility NU_i. Checking the first-order optimality condition of (7.3a) by taking derivative with respect to w_i:

$$\left.\frac{dNU(w_i, R_i)}{dw_i}\right|_{w_i^*} = \left.\frac{dU(R_i(w_i))}{dR_i}\frac{dR_i(w_i)}{dw_i}\right|_{w_i^*} - c_i = 0. \tag{7.4}$$

Let $f_i(x) = \frac{dU_i(x)}{dx}$. Using (7.2), (7.4) can be rewritten as:

$$f_i(R_i^*)\,\text{In}(1 + \beta \cdot SINR_i(W^*)) = c_i, \tag{7.5}$$

and the globally optimal bandwidth for HeNB_i is:[1]

$$w_i^* = \frac{1}{\text{In}(1+\beta \cdot SINR_i(W^*))} f^{-1}\left(\frac{c_i}{\text{In}(1+\beta \cdot SINR_i(W^*))}\right). \tag{7.6}$$

Because of the constraint $0 \leq w_i \leq 1$, we modify (7.6) as:

$$w_i^* = \min\left(\frac{1}{\text{In}(1+\beta \cdot SINR_i(W^*))} f^{-1}\left(\frac{c_i}{\text{In}(1+\beta \cdot SINR_i(W^*))}\right), 1\right). \tag{7.6a}$$

How the requirement that $w_i \geq 0$ is taken care of will be clear later. Because $SINR_i$ is a function of W, (7.6a) does not provide a direct solution to W^*. It is a set of N fixed point equations. However, it provides the basis for an iterative algorithm of updating bandwidth usage w_i. Iterative algorithms are well suited for distributed optimization in wireless networks, such as distributed power control [17–19]. We assume that all the HeNBs update their bandwidths periodically in a synchronous manner,[2] with the superscript t representing the t-th update interval. During the t-th iteration, HeNB_i updates its bandwidth w_i^t based on its throughput R_i^t in the t-th interval:

$$\begin{aligned} w_i^{t+1} &= \min\left(\frac{1}{\text{In}(1+\beta \cdot SINR_\text{i}^\text{t})} f_i^{-1}\left(\frac{c_i}{\text{In}(1+\beta \cdot SINR_i^t)}\right), 1\right) \\ &= \min\left(\frac{w_i^t}{R_i^t} f_i^{-1}\left(\frac{c_i w_i^t}{R_i^t}\right), 1\right) = \omega_i(W^t). \end{aligned} \tag{7.7}$$

Note that $R_i^t = R_i^t(W^t) = w_i^t \text{In}(1+\beta \cdot SINR_\text{i}^\text{t}(W^t))$. In vector form, the bandwidth usage of the entire network is updated as:

$$W^{t+1} = \Omega(W^t) \tag{7.8}$$

Note that throughput R_i^t in cell i depends on the bandwidth usage W^t of all the femtocells in the t-th interval. The length of the update interval should be long enough for an HeNB to measure the throughput in its femtocell with per slot fluctuation smoothed out, while short enough to accommodate changes in the network dynamics. These changes include events such as change of load or number of UEs in an HeNB, an HeNB being turned on or turned off, or changes in the effective channel gain (G) or noise (N_i) due to the changing radio environment. The function $f^{-1}(\cdot)$ depends on the utility function $U(\cdot)$ and will be discussed in detail in the following sections. The process by which each HeNB updates its bandwidth usage, based on the bandwidth usage of other HeNBs in the network, can be viewed as a non-cooperative game where each player (HeNB_i) tries to maximize its own benefit (net utility NU_i) by adjusting its strategy w_i. Within the framework of non-cooperative game theory, the most fundamental question is the existence of Nash equilibrium $W^* = [w_0^*, \cdots, w_{N-1}^*]^T$, where W^* satisfies:

$$NU_i\left(w_i^*, W_{-i}^*\right) \geq NU_i\left(w_i', W_{-i}^*\right) \tag{7.9}$$

[1]We will verify that w_i^* is indeed the global maximizer of NU_i by checking the second-order derivative later.
[2]The synchronous requirement is made for notational simplicity and will be relaxed here.

for all $0 \le w_i' \le 1, 0 \le i \le N-1$. Under what condition a Nash equilibrium exists, its uniqueness and the convergence of the iterative procedure in (7.8), will be investigated in the next section.

7.3 Convergence Analysis

Yates established in [17] that an iterative algorithm $P(t+1) = I(P(t))$ converges to a unique fixed point $P^* = I(P^*)$ when the function $I(P)$ is a *standard* function. *I(P)* is standard if it satisfies the following three conditions:

- *Positivity:* $I(P) > 0$,
- *Monotonicity:* for $P' \ge P, I(P') \ge I(P)$,
- *Scalability:* for $\mu > 1, \mu I(P) > I(\mu P)$.

For vector inequalities, these conditions need to hold element-wise. Standard functions are widely used in distributed transmission power controls in wireless networks [11–13]. We now investigate whether, and under what condition, the bandwidth updating function $\Omega(W)$ is a standard function. It is sufficient to show that the individual bandwidth updating function $\omega_i(W)$ is standard, as this proves the inequalities element-wise.

- *Positivity*

This requires:

$$\omega_i(W) = \min\left(\frac{1}{\text{In}(1+\beta \cdot SINR_i(W))} f_i^{-1}\left(\frac{c_i}{\text{In}(1+\beta \cdot SINR_i(W))}\right), 1\right) > 0, \tag{7.10}$$

for all $0 < W \le 1$. Because $SINR_i(W) > 0$, positivity requires that:

$$f^{-1}(x) > 0, \quad \text{for} \quad x > 0. \tag{7.11}$$

- *Monotonicity*

For $W' \ge W, SINR_i(W') \le SINR_i(W)$:

$$\frac{1}{\text{In}(1+\beta \cdot SINR_i(W'))} \ge \frac{1}{\text{In}(1+\beta \cdot SINR_i(W))}. \tag{7.12}$$

A necessary and sufficient condition for monotonicity is that the function:

$$g_i(x) = x f_i^{-1}(c_i x)$$

is a non-decreasing function for $x > 0$.

• *Scalability*

Scalability requires that for $\mu > 1$:

$$\frac{\mu}{\text{In}(1+\beta\cdot SINR_i(W))}f_i^{-1}\left(\frac{c_i}{\text{In}(1+\beta\cdot SINR_i(W))}\right)$$
$$> \frac{1}{\text{In}(1+\beta\cdot SINR_i(\mu W))}f_i^{-1}\left(\frac{c_i}{\text{In}(1+\beta\cdot SINR_i(\mu W))}\right) \tag{7.13}$$

or equivalently:

$$\mu g_i\left(\frac{1}{\text{In}(1+\beta\cdot SINR_i(W))}\right) > g_i\left(\frac{1}{\text{In}(1+\beta\cdot SINR_i(\mu W))}\right). \tag{7.13a}$$

Because for $\mu > 1$:

$$SINR_i(\mu W) = \frac{h_{ii}}{\sum_{j\neq i} h_{ij}\mu w_j + \eta_i} > \frac{h_{ii}}{\sum_{j\neq i} h_{ij}\mu w_j + \mu\eta_i} = \frac{1}{\mu}SINR_i(W), \tag{7.14}$$

hence

$$\frac{1}{\text{In}(1+\frac{\beta}{\mu}SINR_i(W))} > \frac{1}{\text{In}\,(1+\beta\cdot SINR_i(\mu W))}. \tag{7.15}$$

When $g_i(x)$ is a non-decreasing function of x:

$$g_i\left(\frac{1}{\text{In}\left(1+\frac{\beta}{\mu}SINR_i(W)\right)}\right) \geq g_i\left(\frac{1}{\text{In}(1+\beta\cdot SINR_i(\mu W))}\right) \tag{7.16}$$

$$\frac{1}{1+\frac{\beta}{\mu}SINR_i(W)}f_i^{-1}\left(\frac{c_i}{\text{In}\left(1+\frac{\beta}{\mu}SINR_i(W)\right)}\right)$$
$$\geq \frac{1}{\text{In}(1+\beta\cdot SINR_i(\mu W))}f_i^{-1}\left(\frac{c_i}{\text{In}(1+\beta\, SINR_i(\mu W))}\right) \tag{7.16a}$$

A sufficient condition for inequality (7.16*a*) is:

$$\mu g_i\left(\frac{1}{\text{In}(1+\beta\cdot SINR_i(W))}\right) \geq g_i\left(\frac{1}{\text{In}\left(1+\frac{\beta}{\mu}\cdot SINR_i(W)\right)}\right), \tag{7.17}$$

or

$$\frac{\mu}{\text{In}(1+\beta\cdot SINR_i(W))}f_i^{-1}\left(\frac{c_i}{\text{In}(1+\beta\cdot SINR_i(W))}\right)$$
$$\geq \frac{1}{\text{In}\left(1+\frac{\beta}{\mu}SINR_i(W)\right)}f_i^{-1}\left(\frac{c_i}{\text{In}\left(1+\frac{\beta}{\mu}SINR_i(W)\right)}\right) \tag{7.17a}$$

The inequality (7.17a) holds when the following two inequalities hold:[3]

$$\frac{\mu}{\ln(1+\beta\cdot SINR_i(W))} > \frac{1}{\ln\left(1+\frac{\beta}{\mu}SINR_i(W)\right)} \tag{7.18}$$

and

$$f_i^{-1}\left(\frac{c_i}{\ln(1+\beta\cdot SINR_i(W))}\right) > f_i^{-1}\left(\frac{c_i}{\ln\left(1+\frac{\beta}{\mu}SINR_i(W)\right)}\right) \tag{7.19}$$

Condition (7.18) is true because for $x > 0$:

$$\begin{aligned}\mu\ln\left(1+\frac{x}{\mu}\right)-\ln(1+x) &= \ln\left(\left(1+\frac{x}{\mu}\right)^{\mu}\right)-\ln(1+x)\\ &= \ln(1+x+O(x^2))-\ln(1+x)>0.\end{aligned} \tag{7.20}$$

Inequality (7.19) requires that $f_i^{-1}(x)$ is a decreasing function for $x > 0$.

Combining (7.11), (7.12) and (7.19), we get a set of three sufficient conditions to guarantee the convergence of the bandwidth $\omega_i(W)$ through (7.7):

For $x > 0$

$$f^{-1}(x) > 0; \tag{7.21a}$$

$$xf_i^{-1}(x)\text{ does not decrease with } x; \tag{7.21b}$$

$$f_i^{-1}(x)\text{ decreases with } x. \tag{7.21c}$$

When $\Omega(W)$ is a standard function, the network converges to a unique Nash equilibrium W^* with (7.8). At W^*, $w_i^* = \omega_i(w_i^*, W_{-i}^*)$, every femtocell HeNB$_i$ has maximized its own net utility $NU_i^* = NU_i(w_i^*, W_{-i}^*)$ at w_i^*, given the bandwidth usage W_{-i}^* in the rest of the network. As the net utility $NU_j(W)$ of every other femtocell $j \neq i$ decreases with w_i, it is not feasible to find another W' that Pareto dominates W^*. As a consequence, while the femtocells approach the Nash equilibrium through iterative updates of W, the system achieves Pareto optimality at the same time.

7.4 Choice of Utility Function and its Parameters

The convergence of w_i^t depends on the shape of $f_i^{-1}(x)$, therefore it is dependent on the utility $U_i(x)$, where the variable x has a meaning of aggregated throughput in the femtocell. We now investigate what type of utility functions satisfies the convergence requirement. The most widely used utility function is the proportional fairness function with the following form [20]:

$$U_i(x) = \alpha_i\cdot\ln(x), \tag{7.22}$$

[3] Although the inequality (7.17a) holds if we loosen the condition 19 by replacing '>' with '≥', the stronger condition ('>') is required because $f(x)$ will be ill-defined in the case of equality.

where $\alpha_i > 0$. It can be verified that it satisfies all the three conditions in (7.21). However, under this definition the bandwidth update (7.7) becomes:

$$w_i^{t+1} = \min\left(\frac{\alpha_i}{c_i}, 1\right), \tag{7.23}$$

and it converges to a Nash equilibrium where an HeNB$_i$ gets $w_i^* = \min(\frac{\alpha_i}{c_i}, 1)$. Because it does not reflect the interference between the femtocells, this fixed assignment is clearly not desirable. Consequently the regular notion of proportional fairness function (7.22) cannot be applied here. In general, the utility function $U_i(x)$ representing the QoS in a femtocell should be an increasing and concave function of the cell throughput. The concavity of $U_i(x)$ also ensures that the second-order derivative $\frac{\partial^2 NU_i(W)}{\partial w_i^2}$ is negative, therefore guarantees that the Nash equilibrium W^* indeed maximizes the net utility for a femtocell. A class of functions meeting this requirement is the isoelastic utility function with the form:

$$U_i(x) = -\alpha_i x^{-k_i}, \alpha_i > 0, k_i > 0. \tag{7.24}$$

It is straightforward to verify that this utility function satisfies the conditions in (7.21). It follows that:

$$f_i^{-1}(x) = (\alpha_i k_i)^{\frac{1}{k_i+1}} \cdot x^{\frac{-1}{k_i+1}}, \tag{7.25}$$

$$w_i^{t+1} = \min\left(\frac{w_i^t}{R_i^t} f_i^{-1}\left(\frac{c_i w_i^t}{R_i^t}\right), 1\right)$$

$$= \min\left(\left(\frac{\alpha_i k_i}{c_i}\right)^{\frac{1}{k_i+1}} \left(SE_i^t\right)^{\frac{-k_i}{1+k_i}}, 1\right) = \min\left(\gamma_i \left(SE_i^t\right)^{-l_i}, 1\right), \tag{7.26}$$

where $SE_i^t = \frac{R_i^t}{w_i^t}$ is the spectrum efficiency in femtocell i at time t, and:

$$0 < l_i = \frac{k_i}{k_i + 1} < 1, \gamma_i = \left(\frac{\alpha_i k_i}{c_i}\right)^{\frac{1}{k_i+1}} > 0. \tag{7.27}$$

Without loss of generality, we can choose $\alpha_i = 1$ for $\forall i, 0 \le i \le N-1$. Equation (7.26) provides a simple algorithm for distributed bandwidth allocation, and (l_i, γ_i) are two control parameters that can be tuned to adjust the Nash equilibrium of the system. A HeNB$_i$ only needs to calculate the average spectrum efficiency SE_i^t in its femtocell, and updates its own bandwidth usage w_i^{t+1} following (7.26). Because the only intensive computation in (7.26) is x^{-l_i}, a table of x^{-l_i} can be pre-computed and stored. This way this operation is reduced to a simple table lookup. The requirement that all the femtocells update their bandwidth simultaneously is actually not necessary, because the standard condition guarantees convergence in both synchronous and asynchronous cases [17]. According to [17], the convergence to the unique Nash equilibrium is not dependent on the initial bandwidth W^0 either, while $W^0 > 0$ is necessary in reality. Different HeNB$_i$ can be assigned different (l_i, γ_i) to provide desirable individual service levels in different femtocells. Convergence is guaranteed as long as (l_i, γ_i) meet the requirements in (7.27) in all the femtocells.

It is helpful to investigate how the choice of parameters (l_i, γ_i) affects the bandwidth adjustment. Given (l_i, γ_i), the bandwidth of HeNB$_i$ is adjusted based on the spectrum efficiency SE_i^t. Because it is difficult to analyse the system equilibrium directly, we focus the following qualitative analysis on the core of the bandwidth update scheme in (7.26):

$$\psi\left(SE_i^t\right) \triangleq \gamma_i\left(SE_i^t\right)^{-l_i}. \tag{7.28}$$

This is a decreasing function of the spectrum efficiency SE_i^t. As the transmission quality improves in a femtocell, the HeNB reduces the bandwidth it occupies. This agrees with the notion that a femtocell should not take radio resources excessively and cause unnecessary interference to the other cells, and is a desirable property from the network point of view. A femtocell with poor SINR is given more bandwidth to compensate for its poor transmission quality. This way a certain degree of fairness is achieved across the network. Because

$$SE_i^t \cdot \psi\left(SE_i^t\right) = \gamma_i\left(SE_i^t\right)^{1-l_i} \tag{7.29}$$

increases with spectrum efficiency, the throughput (and the QoS) of a femtocell increases with its SINR, even as its bandwidth allocation decreases. This agrees with the expectation that the QoS improves with SINR and is desirable both for the femtocell users and the network operator.

For different l_i values, the most significant change lies in the region $SE_i^t < 1$. In this region, $\psi(SE_i^t)$ increases more dramatically with decreasing SE_i^t as l_i increases. This shows that, as l_i increases, a femtocell with poor SINR is allocated more spectrum resource more generously. As the spectrum efficiency increases ($SE_i^t \geq 1$), the decrease of $\psi(SE_i^t)$ with SE_i^t becomes much gentler and is less affected by choice of l_i. Consequently choosing parameter l_i provides a way to tune the overall fairness across the network, with larger l_i treats femtocells with poor SINR more favourably. From (7.29), we can see that as l_i approaches 1, the cell throughput will become less sensitive to the spectrum efficiency (or SINR) in the cell, so the network operates in a rate-fair sense. From (7.29), we can see that, as l_i approaches 0, the allocated bandwidth becomes less sensitive to the spectrum efficiency (or SINR) in the cell, so the network operates in a resource-fair sense. Therefore, the parameter l_i (equivalently k_i) can be selected based on the fairness consideration in the network.

After l_i (k_i) is determined, c_i needs to be selected. For a given k_i:

$$\psi\left(SE_i^t\right) \propto c_i^{\frac{-1}{1+k_i}} \tag{7.30}$$

decreases with c_i. So c_i determines how much bandwidth a femtocell i occupies. This agrees with the fact that c_i is the unit price for bandwidth, and the higher the price, the less bandwidth a femtocell uses. When the bandwidth prices for all the femtocells go up, every femtocell will reduce its transmission. In the case where macrocells and femtocells are deployed in the same carrier, this effectively reduces the interference caused to the macrocells by the femtocells. For an individual femtocell HeNB$_i$, adjusting its bandwidth price c_i provides a way to customize its QoS. This is useful when the operator wants to classify femtocell customers into different categories (e.g., by grouping them into different classes and charging different rates) and provides a means for service differentiation. These control parameters can be assigned to the femtocells by a SON server and may be changed in real time to adjust the bandwidth and QoS of a femtocell. Numerical examples to tune the network by varying these parameters will be given in Section 7.5.

It is interesting to compare the current work on distributed bandwidth allocation with that of [19], where a similar problem formulation (equation 4 in [19]) is used for distributed power allocation in a CDMA network. As stated previously, the definition of SINR in these two works involves the control variables (bandwidth W in this work and transmission power P in [19]) in different ways. In current work Wonly appears in the denominators, while in [19] P appears in both the nominator and the denominator. This difference makes the two problems fundamentally different, despite some similarities at a first glance. In [19] the cell utility is defined directly as a function of the SINR, and a sigmoid function is adopted. When we translate this utility function to a function of cell throughput, the resulting function does not correspond easily to any usual fairness notion used in the network literature. In our current work, we define utility directly as an increasing and concave function of the cell throughput. This is more natural. Although proportional fairness cannot be used, our choice of isoelastic utility function is in line with the notion of (p, α)-proportionally fair developed in [21]. This utility function allows the network operator to define and implement different policies easily. In our case this requires a simple change of (l_i, γ_i). This gives the network operator more flexibility.

7.5 Simulation Results

We now verify the bandwidth allocation scheme through numerical simulations. We simulate a LTE femtocell network with parameters given in Table 7.1. All the femtocells operate in the same carrier and are located at users' homes, and the users are at home when accessing their private femtocells. Table 7.2 summarizes the default choice of parameters in the

Table 7.1 LTE femtocell simulation parameters

Simulation parameters	Values
Carrier frequency	2 GHz
System bandwidth	10 MHz
Number of subcarriers	600
Number of PRBs	50
Shadowing	
Indoor	5 dB
outdoor	10 dB
Penetration	
Indoor-to-indoor (L_{iw})	5 dB
Outdoor-to-indoor (L_{ow})	10 dB
Femtocell separation	40 m
Femtocell radius	20 m
Average HeNB-UE distance	10 m
UE noise figure	7 dB
MCS-Shannon capacity gap (β)	8 dB

Table 7.2 Default control parameters

Control parameters	Values
α_i	1
k_i	1
c_i	1
l_i	0.5
γ_i	1

control algorithm. The path loss is calculated for indoor femtocells using the following model from [22]:

$$PL\,(dB) = \max\left(15.3 + 37.6log_{10}\,(d)\,, 38.46 + 20log_{10}\,(d)\right) + 0.7d_{2D,indoor} + L_{iw} + L_{ow} \tag{7.31}$$

where d is the transmitter to receiver distance and $d_{2D,indoor}$ is the indoor distance between the femtocell and UE. L_{iw} and L_{ow} are the indoor and outdoor penetration loss respectively.

7.5.1 Convergence Studies

We first verify that the algorithm converges to the unique Nash equilibrium under arbitrary initial conditions. Figure 7.1 shows the convergence of the bandwidth and cell throughput in a network of 20 femtocells. Every femtocell randomly picks its initial bandwidth w_i^0 uniformly in (0, 1). The bandwidth converges to the equilibrium within the first four iterations. From different initial bandwidth W^0, the network always converges to the same point. This confirms that the Nash equilibrium is the unique solution to the fixed point given by equation (7.8). From the numerous simulations we conducted, it appears that the speed of the convergence is insensitive to the initial bandwidth W^0, the network size or the choice of individual control parameter (l_i, γ_i). In a larger network of 100 cells, the system also converges within four frames (Figure 7.2). This implies that the iterative algorithm is extremely robust and works well under a wide variety of situations. The scenario that a femtocell is turned on or turned off in the network is naturally accommodated by the scheme, because the system simply converges to the next equilibrium point quickly when the network changes. It also means that when the network operator (the SON server) tunes the network by adjusting (l_i, γ_i) of individual femtocells, the operation point of the system converges to the next equilibrium quickly and smoothly.

7.5.2 Bandwidth Allocation and Network Tuning

We now look at how the bandwidth is allocated to the femtocells by the distributed algorithm. In order to see the relationship between the interference environment of the femtocells and the bandwidth allocation clearly, we construct a fictional network of seven femtocells arranged in

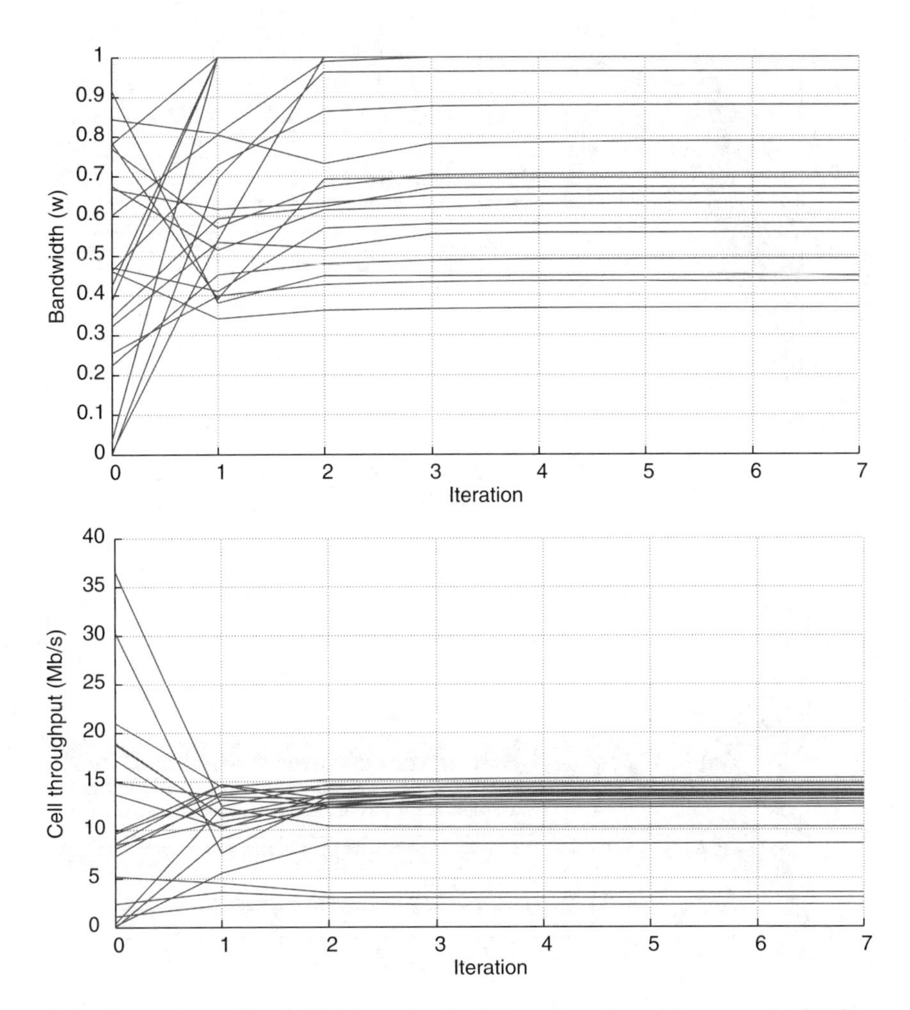

Figure 7.1 Convergence of bandwidth (upper) and cell throughput (lower) in a network of 20 femtocells.

a linear array 20 m apart. With shadow fading disabled in the simulation, the path loss matrix is as follows (dB):

$$\begin{bmatrix} 58.66 & 84.73 & 95.26 & 101.43 & 105.80 & 109.19 & 111.96 \\ 84.73 & 58.66 & 84.73 & 95.26 & 101.43 & 105.80 & 109.19 \\ 95.26 & 84.73 & 58.66 & 84.73 & 95.26 & 101.43 & 105.80 \\ 101.43 & 95.26 & 84.73 & 58.66 & 84.73 & 95.26 & 101.43 \\ 105.80 & 101.43 & 95.26 & 84.73 & 58.66 & 84.73 & 95.26 \\ 109.19 & 105.80 & 101.43 & 95.26 & 84.73 & 58.66 & 84.73 \\ 111.96 & 109.19 & 105.80 & 101.43 & 95.26 & 84.73 & 58.66 \end{bmatrix}$$

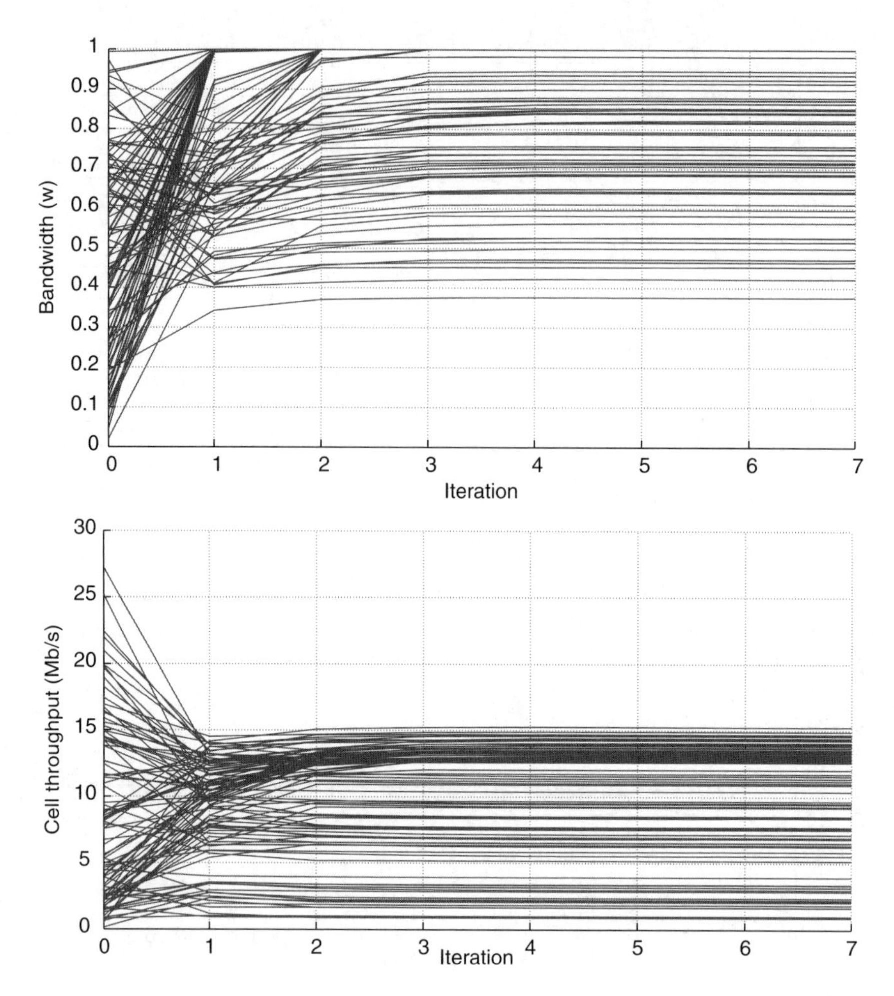

Figure 7.2 Convergence of bandwidth (upper) and cell throughput (lower) in a network of 100 femtocells.

The femtocells are numbered from 0 to 6; Cell 0 and cell 6 are at the two ends of the array and experience the least interference in the network. The interference gets stronger on moving towards the centre of the array (cell 3). From network operator's point of view, one reasonably expects that the QoS provided by the femtocells decrease as one moves towards the centre, and more bandwidth should be given to the centre femtocells to compensate for their higher interference. The bandwidth allocation and cell throughput made by the distributed algorithm with parameter $(l, \gamma) = (0.5, 1)$ are shown in Table 7.3 (case A). It is clear that both expectations are being met by our scheme. We can tune the network towards different fairness considerations by varying the parameter l_i. Decreasing l_i makes the bandwidth allocated more uniformly across the network, and the system operates in a resource-fair manner. The bandwidth allocation and cell throughput obtained with $(l, \gamma) = (0.0909, 0.5325)$ are shown

Table 7.3 Normalized bandwidth (BW) and cell throughput (Th in Mbps) in a linear network of seven cells with different parameter (l, γ) – Case A, default case; Case B, resource-fair; Case C, rate-fair; Case D, service differentiation

Case	Cell #	0	1	2	3	4	5	6
A	BW	0.446	0.476	0.479	0.480	0.479	0.476	0.446
	Th	22.37	19.11	18.97	18.94	18.97	19.11	22.37
B	BW	0.460	0.465	0.466	0.466	0.466	0.465	0.460
	Th	21.08	18.70	18.53	18.50	18.53	18.70	21.08
C	BW	0.425	0.479	0.488	0.490	0.488	0.479	0.425
	Th	19.36	19.28	19.27	19.26	19.27	19.28	19.36
D	BW	0.314	0.459	0.337	0.463	0.337	0.459	0.314
	Th	14.49	19.81	13.50	19.63	13.50	19.81	14.49

in case B. Increasing l_i tunes the network away from resource-fair and towards rate-fair (uniform service quality). This can be seen in case C, where the control parameters $(l, \gamma) = (0.9677, 2.0217)$ are selected to achieve rate-fair across the seven cells. For a given parameter l_i, the bandwidth allocated to a femtocell can be varied by setting the bandwidth price c_i (or equivalently adjusting γ_i). This can be useful when the network operator finds that the QoS in a particular femtocell needs improvement. It can decrease c_i to give HeNB$_i$ more bandwidth. It also enables the operator to provide different service qualities based on user service contracts (and possibly charge different rates). To show service differentiation with bandwidth pricing, in the seven-cell network, we set $(l, \gamma) = (0.5, 0.7071)$ for even numbered cells and $(l, \gamma) = (0.5, 1)$ for odd number of cells. The resulting bandwidth and cell throughput are given in Table 7.3 (case D).

We conclude this section with simulations in a large network of 100 cells. The femtocells are placed in residential houses of a 10×10 grid and shadow fading are included. The histograms of bandwidth allocation and cell throughput when the network operates in resource-fair mode $(l, \gamma) = (0.0909, 0.5325)$ are shown in Figure 7.3. When the network operate in rate-fair mode $(l, \gamma) = (0.9677, 1.116)$, histograms of bandwidth allocation and cell throughput are shown in Figure 7.4. Because realistic path loss and shadowing models are used, different femtocells experience very different interferences and the QoS in these cells become more disparate, even when the network is tuned for the rate-fair case.

7.6 Extensions and Discussions

Although the bandwidth allocation algorithm is developed as a fully distributed scheme, it readily lends itself to a centralized control architecture, where each HeNB$_i$ sends to the SON server its current cell throughput R_i, and the SON server updates w_i with (7.26) and sends it back to the HeNB$_i$. The computation load of the SON server is $O(N)$. With the bandwidth assignment and cell throughput for all the femtocells, the SON server can use this global information to tune the network, by making adjustments to (l_i, γ_i). The process of updating the control parameters is transparent to the HeNB$_i$, which only receives updated w_i. In the distributed control architecture, the HeNB$_i$ can also send to the SON server its current (w_i, R_i). The SON server may update the control parameters (l_i, γ_i) in the same way as in centralized

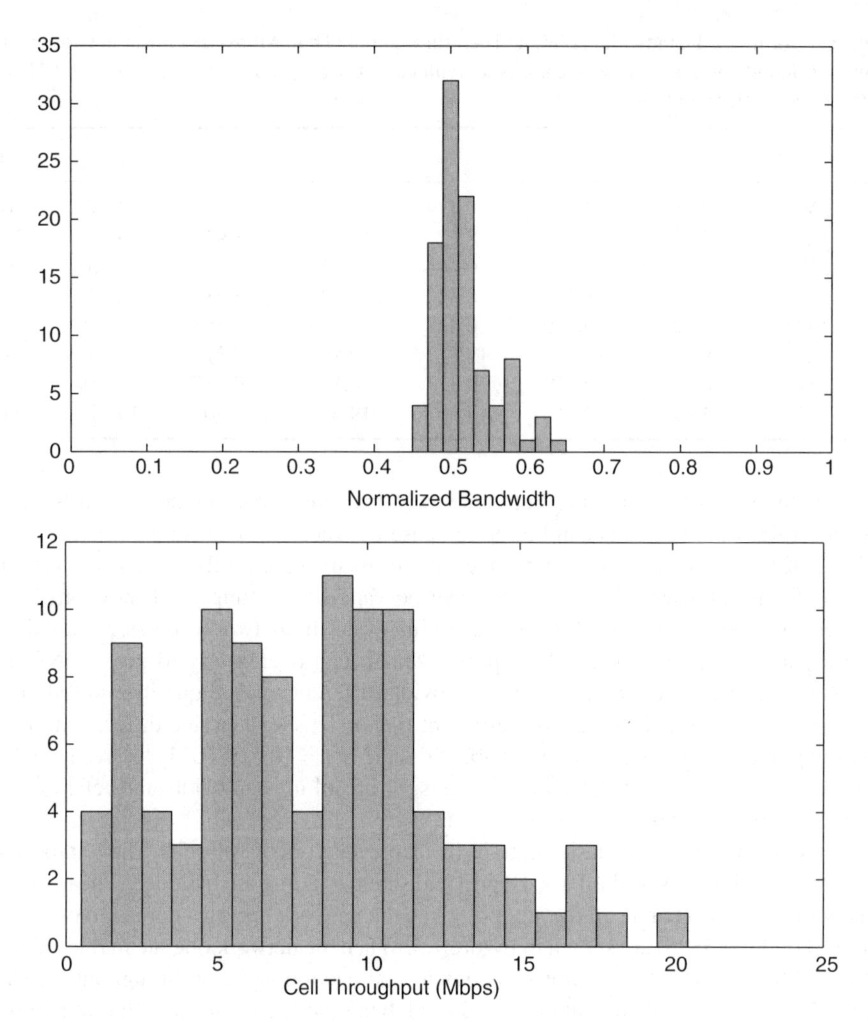

Figure 7.3 Histogram of bandwidth allocation (upper) and cell throughput (lower) in a 100-cell network (resource-fair).

control, but it sends back to the HeNB$_i$ (l_i, γ_i) instead of updated w_i. It is left for the HeNB$_i$ to update its w_i with its updated (l_i, γ_i). This becomes a hybrid control architecture because (l_i, γ_i) and w_i are computed by the SON server and the HeNB$_i$ respectively. In either the centralized or the hybrid control architecture, an HeNB and the SON server only exchange minimal amount of information, so the communication overhead is low for both cases. The distributed/hybrid control architecture is more scalable. The attraction of the centralized control architecture is that it does not require the HeNB to implement the bandwidth calculation algorithm. The parameters w_i and R_i can be transferred through the standard network management interface between the HeNB and the SON server. This is attractive when the HeNBs are from different vendors and do not have the bandwidth update algorithm as a standard feature. The centralized control algorithm relies solely on the SON server and can be upgraded more easily.

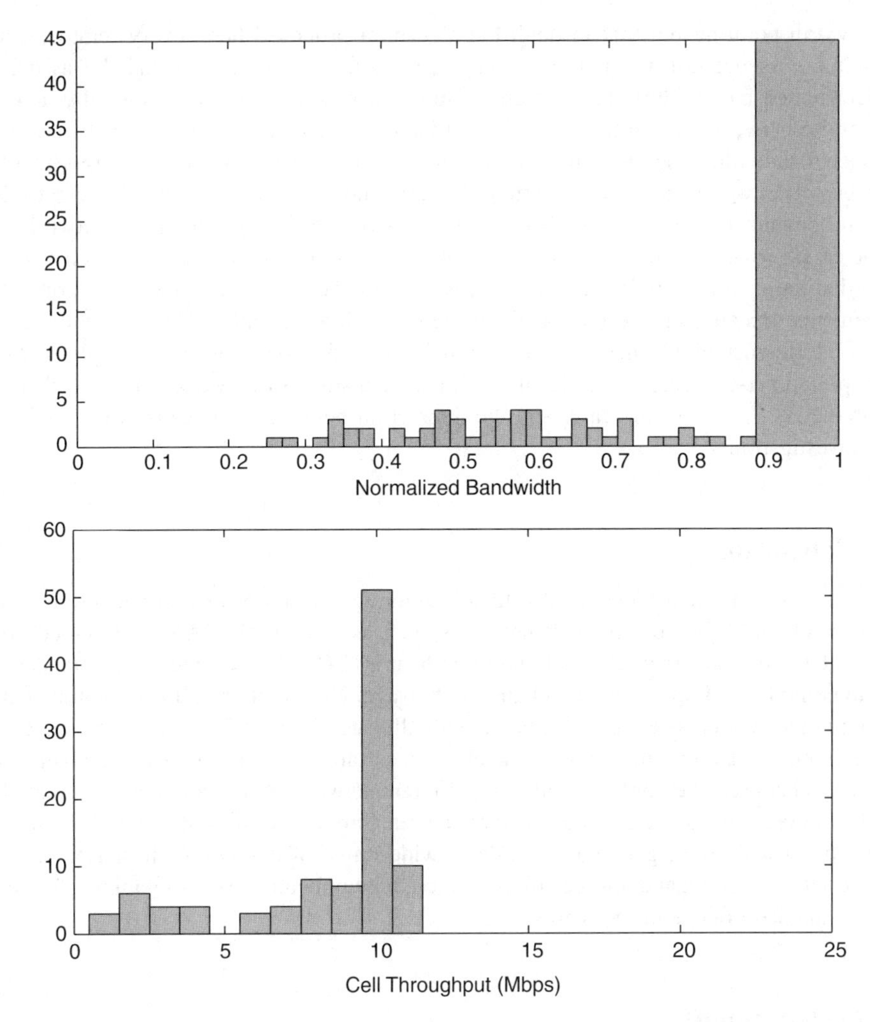

Figure 7.4 Histogram of bandwidth allocation (upper) and cell throughput (lower) in a 100-cell network (rate-fair).

Heterogeneous networks with macro/pico/femtocells deployed in the same channel can be accommodated within the same framework as follows. Macrocells and picocells provide public access to all the users of the wireless operator, and their deployment is usually carefully planned and optimized by the operator for network coverage and capacity. In case the operator chooses to keep the macro and picocells 'as is' (optimized by careful planning and tuning), the transmission from the macro and picocells can be treated as constant background interference to the femtocells. The femtocells can execute the algorithm as previously described. If the operator chooses to keep the macrocells as is and allow the picocells to dynamically adjust the transmission bandwidth, the femtocells and picocells can be treated alike by the algorithm. The operator may want to choose a different set of control parameters (for example, lower bandwidth price c_i) for a picocell than a femtocell to allow it to better service the public.

It is worth pointing out that through the algorithm proposed here, every femtocell tries to maximize its net utility function which incorporates both the QoS and the associated cost (consumed bandwidth). The net utility function was constructed empirically and is a compromise between two mutually conflicting factors. It discourages a femtocell from using too much bandwidth, therefore reduces the interference that a femtocell causes to other femtocells. This way it maintains a certain fairness among the cells. By itself, the net utility function does not represent the QoS in a femtocell *or* the QoS in the entire network. The parameters (l_i, γ_i) need to be chosen carefully. A wide range of choice of (l_i, γ_i) given by (7.27) means that the operating point can be tuned within a wide range to suite different operation requirements. Because all the femtocells in the network are coupled through the direct and indirect interference they generate, the coupling between femtocells is more complicated than the simplified per-cell analysis in Section 7.4. Detailed study of network tuning by changing the parameters (l_i, γ_i) dynamically and the interaction between femtocells is the subject of future investigation.

7.7 Conclusion

We have developed a distributed bandwidth allocation scheme based on non-cooperative game theory for OFDMA-based femtocell networks such as LTE or WiMAX. A femtocell only needs its UEs to report their channel quality indicator (CQI), as required by the standard, in order to compute and update its own bandwidth usage. No direct or indirect communication between femtocells is necessary. The bandwidth allocated to each femtocell is dependent on the interference between femtocells and achieves a certain degree of fairness. The bandwidth allocation scheme can be implemented using different network control architectures, including fully distributed, hybrid and centralized architectures. Simulations have shown that the scheme is very robust and converges quickly under a wide range of conditions. It can be used for dynamic network tuning and service differentiation. It is applicable to self-organizing femtocell networks and heterogeneous networks.

Acknowledgement

The authors wish to thank Prof. R. J. La of University of Maryland for helpful discussions during the preparation of this chapter.

References

1. V. Chandrasekhar and J. G. Andrews, 'Femtocell: A Survey', *IEEE Communication Magazine*, vol. 46, September 2008, pp. 59–67.
2. H. Clausse, L. Ho and L. Samuel, 'An overview of the femtocell concept', *Bell Labs Technical Journal*, vol. 13, issue 1, pp. 221–245, May 2008.
3. G. Horn, '*3GPP Femtocells: Architecture and Protocols*', September 2010, Qualcomm Inc.
4. D. Knisely, T. Yoshizawa and F. Favichia, 'Standardization of femtocells in 3GPP', *IEEE Comm. Magazine*, vol. 47(9), pp. 68–75, September 2009.
5. 3rd Generation Partnership Project, 3GPP RP-100372, 'Enhanced ICIC for non-CA based deployments of heterogeneous networks for LTE'.

6. D. López-Pérez, A. Valcarce, G. de la Roche and J. Zhang, 'OFDMA femtocells: a roadmap on interference avoidance', *IEEE Communication Magazine*, September 2009, pp. 41–48.
7. C. Prehofer and C. Bettstetter, 'Self-Organization in Communication Networks: Principle and Design Paradigms', *IEEE Comm. Magazine*, vol. 43-7, pp. 78–85, July 2005.
8. H. Zeng, C. Zhu and W. Chen, 'System Performance of self-organizing network algorithm in WiMAX Femtocells', Wireless Communication Conference (WICON), 2008.
9. X. Li, L. Qian and D. Kataria, Downlink Power Control in Co-Channel Macrocell Femtocell Overlay, *CISS* 2010.
10. G. Cao, D. Yang, X. Ye and X. Zhang, 'A Downlink Joint Power Control and Resource Allocation Scheme for Co-Channel Macrocell-Femtocell Networks', *IEEE WCNC*, 2011.
11. L. Wang, C. Lee and J. Huang, 'Distributed Channel Selection Principle for Femtocell with Two-tier Interference', *IEEE VTC Spring*, 2010.
12. F. Bernardo, R. Agust, J. Cordero and C. Crespo, 'Self-optimization of Spectrum Assignment and Transmission Power in OFDMA Femtocells', Sixth Advanced International Conference on Telecommunications, 2010.
13. H.-C. Lee, D.-C. Oh and Y.-H. Lee, Mitigation of Inter-Femtocell interference with Adaptive Fractional Frequency Reuse, *IEEE ICC* 2010.
14. D. Fudenberg and J. Tirole, '*Game Theory*', MIT Press, 1991.
15. 3rd Generation Partnership Project, 'Technical Specification Group Radio Access Network: Evolved Universal Terrestrial Radio Access (E-UTRA); LTE Physical Layer – General Description (Release 10)', 2011.
16. IEEE Standard for Local and Metropolitan Area Networks, Part 16: 'Air Interface for Fixed and Mobile Broadband Wireless Access Systems' with Amendment 2: 'Physical and Medium Access Control Layers for Combined Fixed and Mobile Operation in Licensed Bands'.
17. R. D. Yates, 'A Frameworkd for Uplink Power Control in Cellular Radio Systems', *IEEE Journal on Selected Areas in Communications*, vol.13, no.7, pp. 1341–1347, September 1995.
18. C. Saradar, N. B. Mandayam and D. J. Goodman, 'Efficient Power Control via Pricing in Wireless Data Networks', *IEEE Trans. On Communications*, vol. 5, no. 2, pp. 291–303, February 2002.
19. M. Xiao, N. B. Shroff and E.K.P. Chong, 'A Utility-Based Power Control Scheme in Wireless Cellular Systems', *IEEE/ACM Trans. On Networking*, vol.11, no.2, pp. 210–221, April 2003.
20. F. P. Kelly, A. K. Maulloo and D.K.H. Tan, 'Rate control for communication networks: shadow price, proportional fairness and stability', *Journals of Operations Research Society*, 49(3), pp. 237–252, March 1998.
21. J. Mo and J. Walrand, 'Fair End-to-End Window-Based Congestion Control', *IEEE/ACM Trans. On Networking*, vol. 8, no. 5, pp. 556–567, October 2000.
22. 3rd Generation Partnership Project, 3GPP TR 36.814 v 9.0.0, 'Further advancements for E-UTRA physical layer aspects', March 2010.

Part II

Mobility and Handover Management

8

Mobility Management and Performance Optimization in Next Generation Heterogeneous Mobile Networks

Xiaoying Zheng,[1,2] Jiantao Yu,[1,3] Zhenzhen Wei,[4] Honglin Hu,[1,3] Yang Yang,[1,3] and Hsiao-Hwa Chen[5]
[1] *Shanghai Research Center for Wireless Communications, China*
[2] *Shanghai Advanced Research Institute, Chinese Academy of Sciences, China*
[3] *Shanghai Institute of Microsystem and Information Technology, Chinese Academy of Sciences China*
[4] *State Grid Electric Power Research Institute, China*
[5] *National Cheng Kung University, Taiwan*

8.1 Introduction

The next generation mobile networks are on the way to accommodate various applications with different QoS and rate requirements, to support higher mobility at a speed of up to 350km/h, and to deliver services at much higher data rates compared to previous generations. The 3rd Generation Partnership Project (3GPP) is developing and promoting its 3GPP Long Term Evolution (LTE) and LTE-Advanced system specifications, which comprise the existing 3G network and adopt 4G technologies. Though both LTE and LTE-Advanced currently do not fulfil the requirements for 4G standards specified by the ITU-R organization, they have been recognized as a 4G candidate system. In the LTE/LTE-Advanced systems, mobility management is a key component in performance optimization where we are facing a heterogeneous

Heterogeneous Cellular Networks, First Edition. Edited by Rose Qingyang Hu and Yi Qian.

network, which is composed of different types of cells and relays. Also the high mobility requirement makes the mobility management even more challenging. Poor mobility management will result in unnecessary handovers (HO), handover failures, and radio link failures (RLF), and hence system resources are wasted and user experiences deteriorate. In this chapter, we will study various issues arising in mobility management for LTE systems.

Mobility robustness optimization (MRO) addresses the improper handover parameter settings, which can negatively affect user experiences and waste network resources by causing HO ping-pongs, handover failures and radio link failures. In this chapter, based on the mobility state of user equipment (UE), a robustness management algorithm is proposed, which assigns different handover hysteresis parameters for UEs at different speeds. The simulation results show that the algorithm greatly increases the handover success rate and improves the user experience.

Mobility load balancing optimization (MLB) arises when the random arrivals of traffic make the system resource utilization inefficient. There might be a huge number of users in some cells which are overloaded, while there are for fewer users in other cells where resources are not fully utilized. In this chapter, several existing load balancing algorithms will be introduced and a novel algorithm with penalized handovers is designed to achieve the balanced system load.

When the mobility management and load balancing functionalities operate together, there might be some conflicts between their respective decisions. These conflicts can result in dropped calls and damaged system performance. The chapter, reports on simulations and studies the cause of the conflicts, and the impact of the conflicts on the system performance. We will further propose a coordinated solution to avoid these conflicts and make the mobility management and load balancing functionalities collaborated.

Finally, the next generation mobile networks are expected to make extensive use of user-installed femtocells, in order to provide wireless voice and broadband services to customers in the home or office environment. While the deployment of femtocells helps to achieve the goals of spectral efficiency and high speed for a greater number of users, it makes the networks more heterogeneous and the mobility management more complicated. In the chapter, we will discuss the challenges brought by the instalment of femtocells and investigate the solutions.

8.2 Overview of Mobility Management in RRC-connected State

In LTE systems, when are in RRC-connected state, the mobility management is implemented by the handover procedure.[1] A typical procedure to initiate a handover is described as follows:

1. The E-UTRAN NodeB (eNB) sets the measurement configurations for UEs.
2. The measurement reports are sent from UEs to eNB when the reporting event is triggered.
3. A handover decision is made at the eNB based on the measurement reports received from the UE.
4. A handover request message is sent to the selected target eNB from the source eNB.

[1] RRC-connected state refers to a state that an RRC connection is established between the eNB and the UE according to the radio resource control (RRC) protocol specification [5].

Table 8.1 Parameters used in measurement report triggering

Parameter	Description
M_n	the measurement result of the neighbouring cell, not taking into account any offsets
Ofn	the frequency specific offset of the frequency of the neighbour cell
Oc_n	the cell specific offset of the neighbour cell
M_p	the measurement result of the primary cell, not taking into account any offsets
Of_p	the frequency specific offset of the primary frequency
Oc_p	the cell specific offset of the primary cell
Hys	the hysteresis parameter for this event
Off	the offset parameter for this event
TTT	time to trigger
$offset$	$offset = Of_p + Oc_p + off - Of_n - Oc_n$

*Mn, Mp are expressed in dBm in case of reference signal received power (RSRP), or in dB in case of reference signal received quality (RSRQ). *Ofn*, *Ocn*, *Ofp*, *Ocp*, *Hys* are expressed in dB.

5. If the handover request is admitted by the target eNB, then a handover request ACK message is sent back to the source eNB; otherwise, a handover request failure is sent back.
6. Upon receiving the handover request ACK message, the source eNB prepares to handover UE to the target eNB.

In 3GPP LTE standards, the measurement can be reported periodically or triggered by the predefined events. In technical specification 36.331 [2], eight events have been defined to trigger measurement reporting. The widely used event is Event A3, which happens when the signal received from the neighbouring cell becomes offset better than that of the primary cell. The entering condition and leaving condition of Event A3 are shown in (8.1) and (8.2), respectively. Table 8.1 defines the variables used in equations (8.1) and (8.2).

$$M_n + Of_n + Oc_n - Hys > M_p + Of_p + Oc_p + off, \tag{8.1}$$

$$M_n + Of_n + Oc_n + Hys < M_p + Of_p + Oc_p + off. \tag{8.2}$$

The measurement reporting is not triggered immediately after the entering condition (8.1) is satisfied. As shown in Figure 8.1, the entering condition (8.1) needs to be held for a time interval of TTT seconds, then a measurement report is sent to the BS.

Next, we briefly introduce the radio link failure detection mechanism in LTE, which will be used in our simulations. As shown in Figure 8.2 [25], every 10 ms, the UE evaluates the average received signal strength computed by a sliding window algorithm, and reports an in-sync or out-of-sync event if the average signal strength is above or below the thresholds Q_{in} and Q_{out}, respectively. If a number of $N310$ consecutive out-of-sync events are reported, then a timer of $T310$ is started. If a number of $N311$ consecutive in-sync events are reported before the timer $T310$ expires, then the timer $T310$ stops. If the timer $T310$ expires, then a radio link failure is detected. A detailed procedure of RLF detection is shown in Figure 8.3.

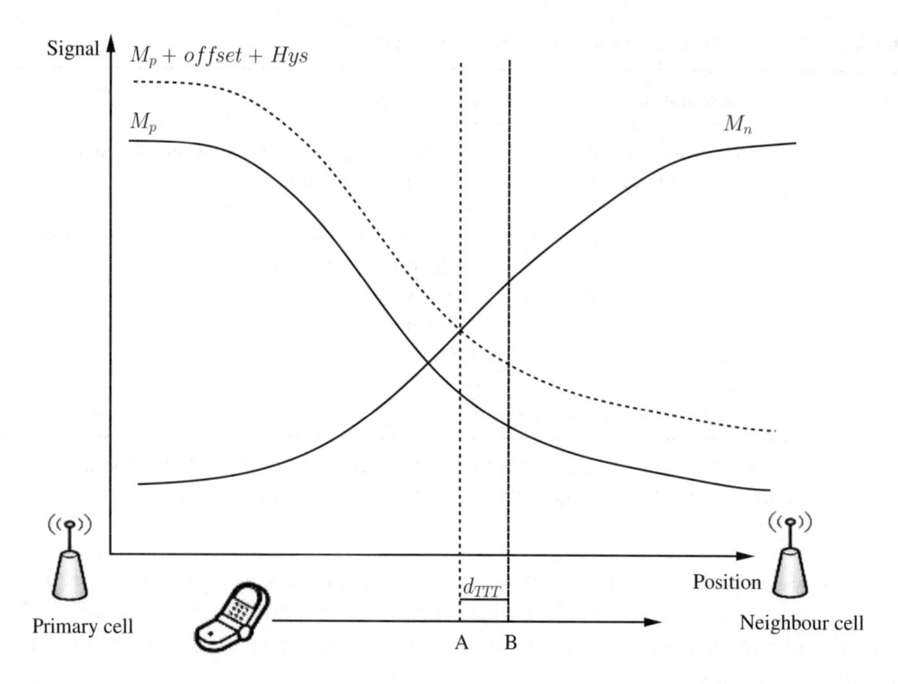

Figure 8.1 Event A3 is triggered after condition (8.1) has been held for the interval of TTT. d_{TTT} is the distance the UE travelled during TTT. Position A: the entering condition (8.1) is satisfied. Position B: Event A3 happens.

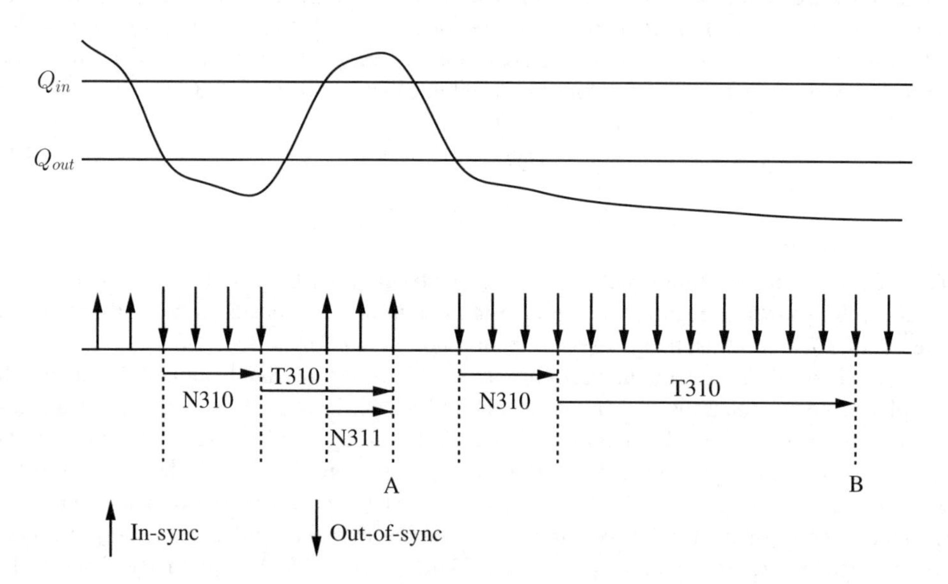

Figure 8.2 RLF detection. Position A: T310 stops. Position B: T310 expires. RLF is detected [25].

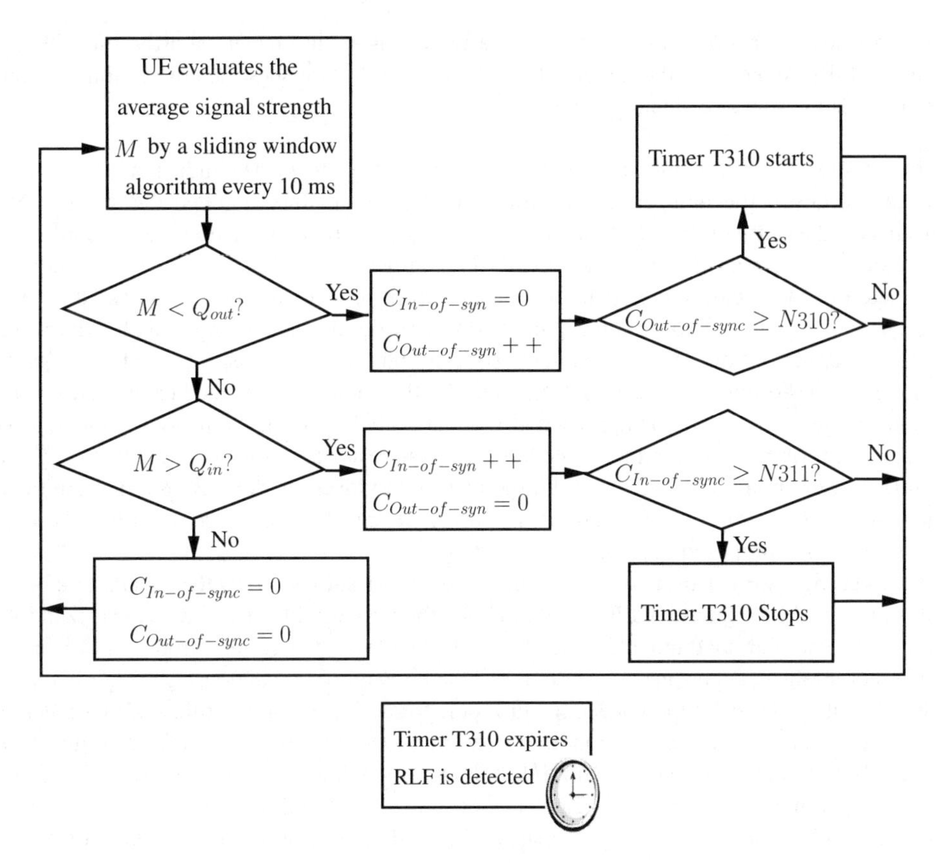

Figure 8.3 The procedure of RLF detection.

8.3 Mobility Robustness Optimization

Mobility robustness optimization has been recognized as an important study item by the standardization body 3GPP [2]. In mobility management, the handover parameters should be properly set to avoid radio link failures when a UE moves from the source cell towards the target cell. Once a radio link failure happens, the eNB which the UE finally reconnects to by RRC reestablishment procedure will report the RLF event to the source eNB by sending a radio failure indication message. The source eNB will differentiate the RLF events based on the cause of the failure and count the respective number. In technical report 36.902 [2], three types of handover related radio link failures are categorized as the following:

- failures due to too-early handover: a radio link failure occurs shortly after the UE successfully connects to the target cell and the UE re-establishes the connection in the source cell,
- failures due to too-late handover: a radio link failure occurs in the source cell before a handover can be initiated or during the handover procedure, and the UE re-establishes the connection in a cell different from the source cell, and

- failures due to handover to a wrong cell: a radio link failure occurs shortly after the UE successfully connects to the target cell, and the UE re-establishes the connection in a cell other than the source cell and the target cell.

The number of radio link failures is the important performance indicator of mobility robustness. Given the numbers of the three types of radio link failures, the source eNB can analyse the cause of the robustness issue, and adjust the handover parameters accordingly.

We discuss the related work. The SOCRATES project studied how to tune the handover parameters of a LTE base station to improve the network performance in terms of call dropping ratio, handover failure ratio and ping-pong handover ratio. In this investigation, by observing the handover performance indicators, the hysteresis and time-to-trigger are tuned accordingly if any of the performance indicators overshoots the threshold. Finally, the best hysteresis and time-to-trigger combination is approached gradually [16]. In [35], the handover parameters (i.e., time-to-trigger, measurement interval and hysteresis) are adjusted by comparing the number of ping-pong handovers and the number of handovers performed in a measurement interval. To further improve the handover performance, an adaptive layer-3 filter based on the user's velocity is proposed in [35].

In this section, we will discuss the mobility robustness issue caused by the different mobility states of UEs. According to 3GPP TS 36.304 [1], the UE mobility states are categorized into three scales: normal, medium and high. Recall the handover procedure introduced in Section 8.1, a measurement report needs to be sent to the eNB before a handover decision is made. When Event A3 is set as the measurement report triggering event, the threshold parameters used in the entering and exiting conditions (8.1) and (8.2) should be carefully set to make the number of handover related radio link failures as small as possible. Assume that all the threshold parameters except the parameter Hys in (8.1) and (8.2) are fixed, and we are only allowed to adjust Hys. Consider a scenario where UEs move at different speeds across the boundary between cells. In Figure 8.4, we show an example where a radio link failure happens when a uniform Hys applies to all the UEs with different mobility states. As we have discussed, when the entering condition (8.1) is satisfied, the measurement report is not sent immediately and should wait a time interval of TTT seconds to be sent. During the interval, the UEs at different speeds have travelled towards the target cell by the distance d_{normal}, d_{medium} and d_{high}, respectively. As shown in Figure 8.4, if a uniform Hys is used to trigger Event A3, the entering condition (8.1) is satisfied when UEs pass the position A. For the UE at a high speed, in the following TTT interval, it will travel the distance of d_{high} and has already passed the position D before a handover procedure to the neighbour cell is initialized. At the position D, the signal received from the primary cell starts to fall below the threshold $M_{failure}$, which makes the UE lose connection with the primary cell and a radio link failure happens. We can use a smaller Hys to make the entering condition (8.1) satisfied at an earlier time and hence avoid the radio link failure for the UE at a high speed. However, with a smaller Hys, a UE at a normal speed may start to handover to the neighbour cell even before the signal received from the neighbour becomes strong enough and later experience a radio link failure due to a too-early handover.

In this section, we propose a solution to tune the parameter Hys based on UE mobility states[32]. From the example shown in Figure 8.4, in order to reduce the number of radio link failures, different Hys values should be used in the Event A3 entering and exiting conditions

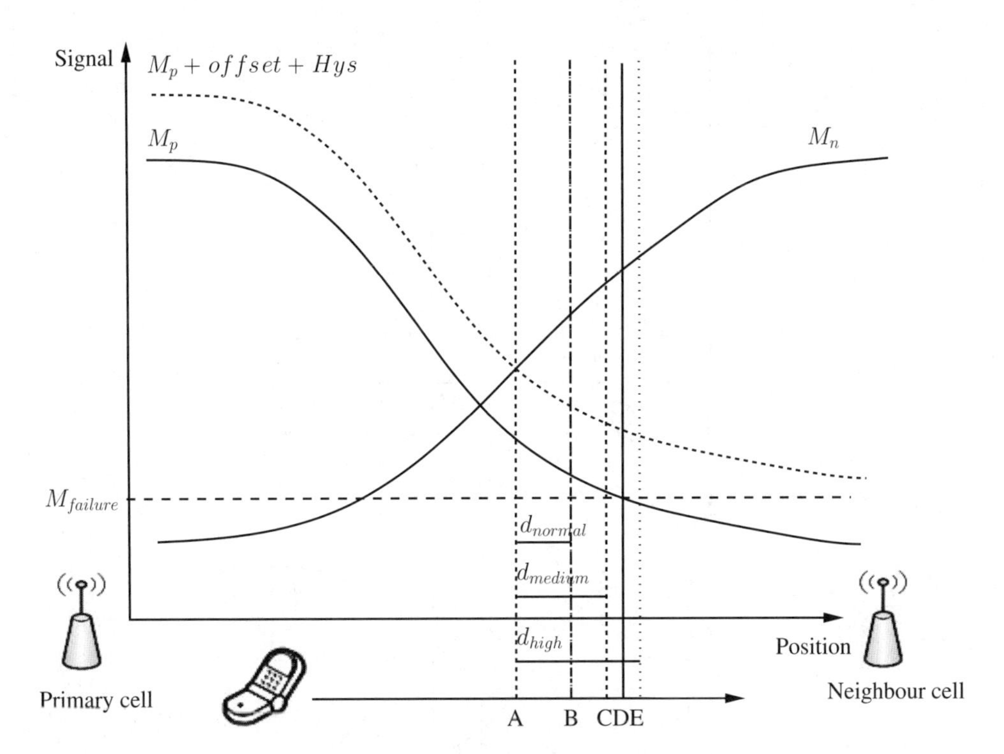

Figure 8.4 Position A: entering condition (8.1) is satisfied. Position B: Event A3 is triggered for a UE at a normal speed. Position C: Event A3 is triggered for a UE at a medium speed. Position D: the signal received from the primary cell starts to fall below $M_{failure}$. A radio link failure happens for a UE at a high speed. Position E: Event A3 is supposed to be triggered for a UE at a high speed.

for UEs with different mobility states. As shown in Figure 8.5, with different Hys values, the condition (8.1) is satisfied earlier for faster UEs, than slower UEs, so that the handover procedure can be triggered in time for faster UEs, and the radio link failure due to a too-late handover is avoided.

Next, we conduct simulations to test the performance of our solution. We consider a 19-cell scenario, where each cell consists of three sectors. The simulation setup is listed in Table 8.2.

We first show the system performance in terms of radio link failure rate when a uniform Hys is used for different mobilities. Figure 8.6 shows that as Hys value increases, the radio link failure rates decrease, and reach the optimal value and then increase for both the normal and medium mobility. For the high mobility, the radio link failure rate increases as Hys value increases. This is because when Hys is small, the handover will be triggered early where too-early handovers may happen; in the meanwhile, when it is large, the handover will be triggered late where too-late handovers occur. Since the handover is triggered earlier when Hys is smaller, a smaller Hys is required for UEs with higher speed. Figure 8.6 shows that the optimal Hys values for high, medium and normal mobility UEs are 2, 3 and 4, respectively. Note that the optimal Hys values depend on the settings of the network such as the cell size, the topology of cells, TTT and $Offset$, and vary in different settings.

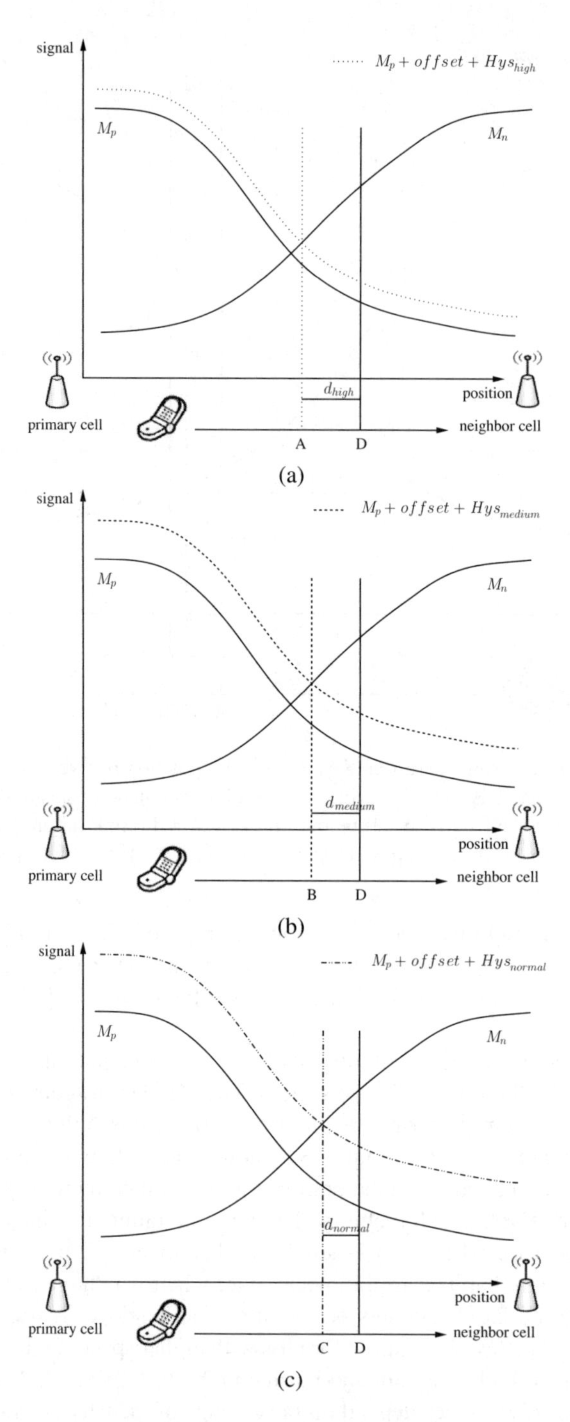

Figure 8.5 Different values of Hys are used for UEs at different speeds. Radio link failures are avoided. Adapted from Z. Wei, "Mobility robustness optimization based on ue mobility for LTE system", Proceedings of 2010 International Conference on Wireless Communications and Signal Processing (WCSP), pages 1–5, October 2010.

Table 8.2 Simulation setup for mobility robustness algorithm. Adapted from Z. Wei, "Mobility robustness optimization based on ue mobility for LTE system", Proceedings of 2010 International Conference on Wireless Communications and Signal Processing (WCSP), pages 1–5, October 2010

Parameter	Value
Distance between origins	500 m
Path loss model	$pl(d) = 136.3 + 39.1\log_{10}(d)$, d in km
Shadowing standard deviation	6 dB
Shadowing correlation	0.5
Qin	−6 dB
Qout	−8 dB
N310	1
N311	1
T310	1 s
HO delay	0.234 s
HO Check Time	2 s
#UEs at a normal speed	3333
#UEs at a medium speed	3333
#UEs at a high speed	3334
Simulation time	500 s
TTT	0.3 s
CIO	0 dB

By obtaining the optimal Hys for different mobility states, we are able to conduct experiments to show the performance of our solution. As a benchmark, in the test of uniform Hys, the value of Hys is set to be 4. In Figure 8.7, it shows the number of radio link failures drops from 398 to 96 after different Hys values are applied to the handover trigger event, and the number of handovers increases from 9568 to 9872. It means that our solution significantly reduces the number of too-late handovers. It also shows that the numbers of radio link failures immediately after a handover are 34 and 32 for the uniform Hys and the different Hys,

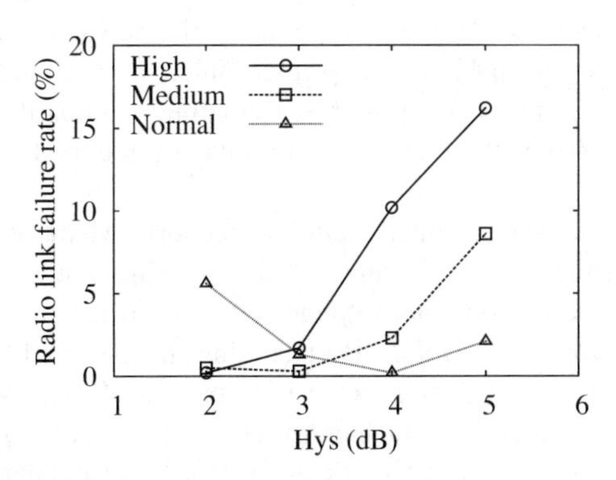

Figure 8.6 Radio link failure rate when a uniform Hys is used for different mobility.

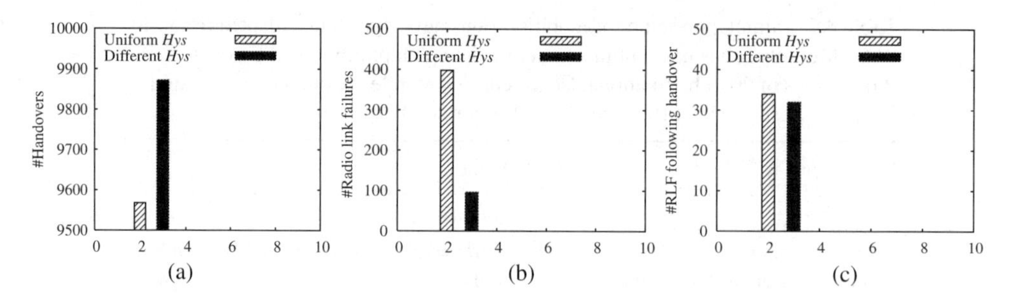

Figure 8.7 For the uniform Hys test, $Hys = 4$. For different Hys, $Hys = 2$ for UEs at a high speed, $Hys = 3$ for UEs at a medium speed and $Hys = 4$ for UEs at a normal speed. (a) #Handovers (b) #Radio link failures (c) #Radio link failures immediately following handover.

respectively, which are almost comparable. This is because a uniform Hys of 4 is a relatively large value, and hence very few too-early handovers happen when it is used.

8.4 Mobility Load Balancing Optimization

Efficient resource utilization is an important issue for the next generation networks to accommodate various applications with different QoS and rate requirements, such as voice, data and streamed multimedia, and to deliver services at much higher data rates compared to previous generations. In cellular networks, the arrivals of mobile users and the resulting traffic load are random, time-varying and probably unbalanced, which makes cell loads in the system unequal. There might be a huge number of users in some cells which are overloaded, while there are a fewer users in other cells where resources are not fully utilized.

The inefficient resource utilization may be mitigated by optimal network management and planning. However, the current network planning strategy is far from solving the load balancing problem completely. There are several facts which make the load balancing problem a critical challenge faced by 4G networks. First, the network applications and services are developing rapidly and the demand for resources is increasing very fast, which makes resource shortage in cellular networks very common. Second, the traffic is time-varying and unpredictable, so the static and pre-fixed network planning cannot make the network adapt to the varying load dynamically, in a timely fashion. Third, one primary interest of mobile operators is to reduce capital and operational expenditures if possible. Remaining competitive at a reduced cost by utilizing resources efficiently is the motivation for studying the load balancing problem from the market side [24].

We consider a downlink-constrained cellular network where channel conditions are time-varying. One possible way to balance load is to shift some users at the border of overlapping or adjacent cells from more congested cells to less congested cells, which is often referred to as *handover* or *handoff*. By changing the assigned base stations for users, the load is balanced and system performance is improved at the cost of system overhead caused by handovers. The procedure for handovers consumes substantial system resources, and the involved users may experience significant system delay and performance degradation.

Hence, handovers cannot happen arbitrarily often; otherwise, the performance improvement gained by the more balanced traffic load cannot compensate the performance loss caused by handovers.

In this section, we aim to balance the unequal traffic load, improve the system performance and reduce the number of handovers needed to achieve load balancing. The importance of minimizing the number of handovers to reach the balanced load has also been recognized by the standardization body 3rd Generation Partnership Project (3GPP). 3GPP has proposed the mobility load balancing optimization problem as a self-organizing network (SON) use case in its technical report TR 36.902 [2], where its proposed objective overlaps with ours.

8.4.1 Related Works

The mobility load balancing problem has attracted much attention from both industry and academia. Both 3GPP and the operators' lobby Next Generation Mobile Networks (NGMN) have foreseen the importance of load balancing in 4G networks. Several projects, such as SOCRATES [13], Monotas [22] and ANA [12], have been conducted to investigate the load balancing problem. In [24], NEC Corporation presents its self-organizing network architecture for 4G networks, which contains the load balancing component.

The load balancing problem has been addressed in various works [6–8, 10, 11, 14, 18, 23, 28–30, 33]. We now give a brief summary of prior works. The works in [10, 14, 33] investigated the base station assignment problem in circuit switched cellular networks. In [11], the load balancing is achieved through a centralized scheduling, which is too complicated to be practical. The work in [6] examined the load balancing problem in CDMA-based systems and proposed a complicated centralized algorithm as well. Both [11] and [6] have not considered the cost of handovers. In [7, 28, 30], the problem of load balancing was studied under the framework of maximizing system utilities, which has been proved to be a powerful model in wired networks in the milestone work [17] and [21]. In [28], a three-stage cross-layer framework was proposed to coordinate opportunistic packet-level scheduling, load-aware handoff and cell-site selection, and system-level cell coverage based on cell breathing. In [7], Bu et al. considered the load balancing problem with the proportional fairness objective, and proposed an optimal offline algorithm and a heuristic online algorithm. The work in [30] is an extension of [7] for 4G networks. However, in [7, 28, 30], all practical online algorithms are heuristic with no intention of decreasing the number of handovers and performance analysis.

Kim et al. [18] investigated the vertical handover strategy in heterogeneous networks, where the objective is different from ours. Casey [8] proposed a base-station-initiated load balancing procedure in WiMAX networks. The works in [23] and [29] studied how to auto-tune the handover parameters in 3GPP LTE systems. The work in [23] proposed dynamically adapting the handover margins based on the loads of adjacent cells. The work in [29] developed a trial-and-error method based on the gradient direction to tune the handover margins.

8.4.2 Problem Description

Consider a down-link-constrained packet-switched multi-cell system shown in Figure 8.8, where there is one channel in each cell shared by mobile users (i.e., mobile stations, MS).

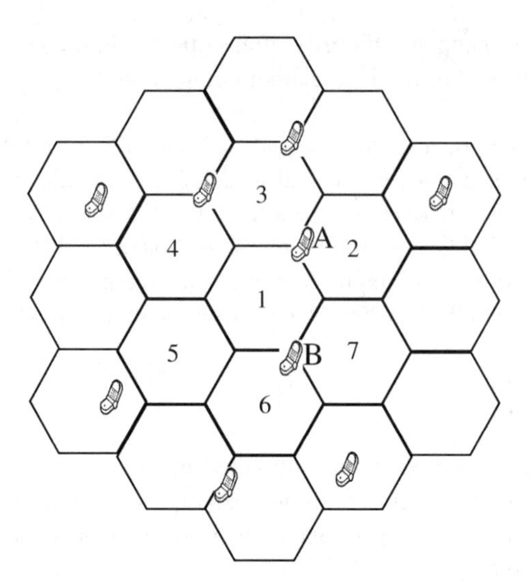

Figure 8.8 A cellular network.

Each cell is managed by its base station (BS) and is referred to as a BS interchangeably.[2] Let us denote the set of BSs by B, and the set of MSs by U. All BSs are connected to a wired core network with sufficient bandwidth. For each MS $u \in U$, if it is able to communicate with a BS $b \in B$, we say that there is a directed link e from BS b to MS u. Thus, the multi-cell system can be represented by a directed expanded graph $G(N, E, L)$, where $N = B \cup U$ is the set of nodes, E is the set of directed links originating from a BS b to an MS u, and L is the set of directed pseudo-links fully connecting all the BSs. Let $o(e)$ and $t(e)$ denote the origin and end of e, respectively. Similarly, the origin and end of any link $l \in L$ are denoted by $o(l)$ and $t(l)$, respectively. For any MS $u \in U$, $E_u = \{e \in E : t(e) = u\}$ is the set of directed wireless links ending at MS u. For any BS $b \in B$, $E_b = \{e \in E : o(e) = b\}$ is the set of directed wireless links originating at BS b.

In cellular networks, a BS is assigned to an MS, which is called the MS's *assigned* BS. An MS only communicates with its assigned BS, though the MS may be able to communicate with other BSs. Let the variable $I_{b,u}$ indicate whether BS b is assigned to MS u, that is, $I_{b,u} = 1$ if BS b is assigned to MS u; $I_{b,u} = 0$ otherwise. Let Θ represent the indicating matrix of the base station assignment as:

$$\Theta = [I_{b,u}]. \tag{8.3}$$

Let the set of all possible base station assignments be denoted by $A_\Theta = \{\Theta\}$.

The channel conditions between BSs and MSs are time-varying due to random environmental conditions, local mobility and wireless fading. Let w_e denote a random variable describing the current channel condition from BS $o(e)$ to MS $t(e)$. Let $\mathbf{w} = (w_e)_{e \in E}$ be the vector of random variables. Due to the shared nature of the wireless medium, the transmission rate c_e of a

[2]The model is ready to be extended to cover the case that a BS has multiple channels and can transmit data to multiple MSs concurrently. The extension is straightforward and is omitted for ease of presentation.

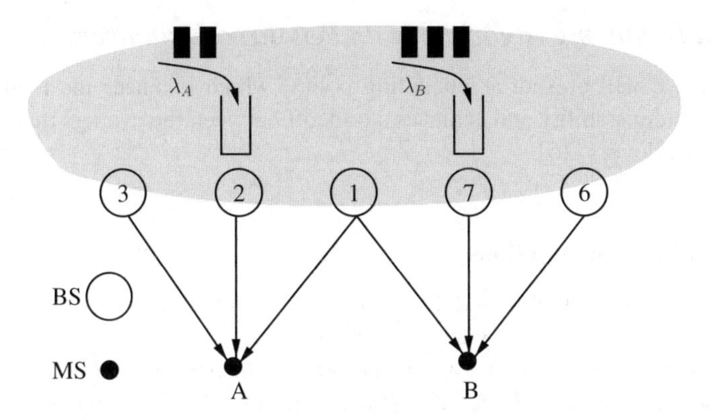

Figure 8.9 The topology of the cellular network in Figure 8.8 with MS A and MS B. The mean data arrival rates of MS A and MS B are λ_A and λ_B, respectively.

wireless link e depends not only on its own modulation/coding scheme, power assignment, the channel condition and the ambient noise but also on the interference from other transmitting links, which in turn depends on their power assignments. Let $\mathbf{s} = (s_e)_{e \in E}$ denote a vector of power assignment, which is also referred to as a *schedule*. Note that, since an MS can only communicate with its assigned BS, $s_e > 0$ only if the BS $o(e)$ is the assigned BS of the MS $t(e)$. Let $S(\Theta)$ denote the set of all possible schedules when the base station assignment vector is Θ, and $S = \bigcup_{\Theta \in A_\Theta} S(\Theta)$ is the set of all possible schedules. Let $\mathbf{c} = (c_e)_{e \in E}$ denote the vector of the wireless link transmission rates. We assume that the link rate vector $\mathbf{c}$ is completely determined by $\mathbf{w}$ and $\mathbf{s}$ together, which means there exists a rate-power function f such that $\mathbf{c} = f(\mathbf{s}, \mathbf{w})$ [20]. The rate-power function f is determined by the interference model. Let us denote the transmission rate of link e by the shorthand $c_e(\mathbf{s}, \mathbf{w})$, where $c_e(\mathbf{s}, \mathbf{w}) = [f(\mathbf{s}, \mathbf{w})]_e$.

For instance, let us inspect MS A and MS B in the cellular network shown in Figure 8.8, and study the corresponding network topology. Suppose MS A is able to communicate with BS 1, 2 and 3, and MS B is able to communicate with BS 1, 6 and 7. The topology of the cellular network and the resulting expanded graph are shown in Figures 8.9 and 8.10, respectively.

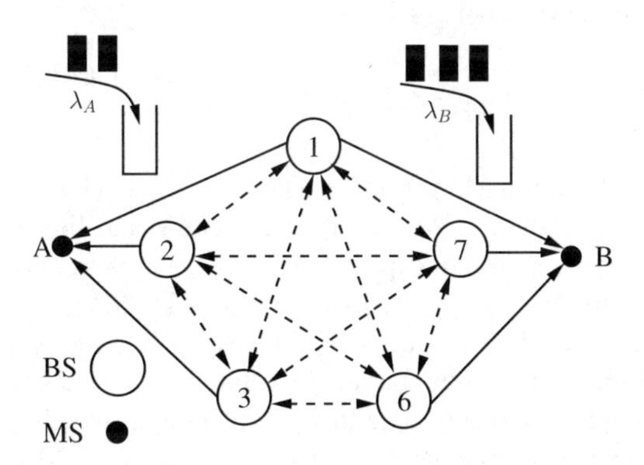

Figure 8.10 The expanded graph of the example in Figure 8.9.

8.4.3 Load Balancing Algorithm with Penalized Handovers

In this section, we will present a scheduling policy, which balances the load among cells, maintains the system stability and achieves a tradeoff between the average queue backlog and the number of handovers [15].

8.4.3.1 Definitions and notations

We first define some variables and functions, and describe the concept of approximate algorithms required by our policy later.

Given two base station assignments Θ and Θ', let $h_{l,u}(\Theta, \Theta')$ be a function indicating whether a handover from BS $o(l)$ to BS $t(l)$ is required for MS u to switch from assignment Θ to Θ', that is, $h_{l,u}(\Theta, \Theta') = 1$ if a handover is required; $h_{l,u}(\Theta, \Theta') = 0$ otherwise. Let us define a function $h(\Theta, \Theta')$ as:

$$h(\Theta, \Theta') = \sum_{u \in U} \sum_{l \in L} h_{l,u}(\Theta, \Theta'), \tag{8.4}$$

that is, the function $h(\Theta, \Theta')$ denotes the number of handovers required to switch from the base station assignment Θ to Θ'.

Let $\{v_{l,u}(k)\}$ be a sequence of control parameters for each pseudo-link l and each MS u. The sequence $\{v_{l,u}(k)\}$ satisfies the following assumption.

- $AS1$: There exists an arbitrary control parameter $0 < V < \infty$ such that:

$$0 \leq v_{l,u}(k) \leq V, \forall l, \forall u, \forall k.$$

At any time slot k, given the initial base station assignment $\Theta(k-1)$, the current channel states $\mathbf{w}(k)$, the control parameter vector $\mathbf{v}(k)$, the current queue backlog vector $\mathbf{q}(k)$, the selected schedule $\mathbf{s}(k)$ and the resulted base station assignment $\Theta(k)$, we define a link weight $\omega_e(k)$ for link $e \in E$ as:

$$\begin{aligned} \omega_e(k) = {} & q_u(k)c_e(\mathbf{s}(k), \mathbf{w}(k)) \\ & - \sum_{l \in L: t(l) = o(e)} v_{l,u}(k) h_{l,u}(\Theta(k-1), \Theta(k)), \end{aligned} \tag{8.5}$$

where $u = t(e)$ is the destination MS for link e. The first term $q_u(k)c_e(\mathbf{s}(k), \mathbf{w}(k))$ in the right-hand side of (8.5) is the product of MS u's queue backlog and link e's data rate, which is used in [31] to achieve the maximum capacity region for wireless networks. We modify the link weight by subtracting $\sum_{l \in L: t(l) = o(e)} v_{l,u}(k) h_{l,u}(\Theta(k-1), \Theta(k))$, which takes the value of 0 if a handover is not required to activate e, and takes $v_{l,u}(k)$ otherwise, where link l is the pseudo-link from the source BS to the target BS if the handover takes place. We will show later that our algorithm prefers to activate a link e with a larger weight $\omega_e(k)$, so the term $v_{l,u}$ can be interpreted as the penalty if a handover is required to switch MS u's assigned BS from BS $o(l)$ to BS $t(l)$.

The definition of a ρ-*approximate* solution is: Let $\rho \geq 1$ be an approximate ratio. Let $g^* \geq 0$ denote the optimal objective value of a maximization problem. A solution with objective value $g_\rho \geq 0$ is said to be a ρ-approximate solution, if:

$$\frac{1}{\rho} g^* \leq g_\rho \leq g^*. \tag{8.6}$$

8.4.3.2 An inter-cell scheduling policy with penalized handovers

The goal of our study is to balance the traffic load among cells and as well as to reduce the number of required handovers to reach the balance. We will present an inter-cell scheduling policy where handovers are penalized.

Inter-cell scheduling with penalized handovers (ISPH):

At each time slot k, based on the current queue backlog $q_u(k)$ for each MS u, the previous base station assignment $\Theta(k-1)$ and the current channel conditions $\mathbf{w}(k)$, a base station assignment vector $\Theta(k)$ and a schedule $\mathbf{s}(k)$ are selected to activate, where $\Theta(k)$ and $\mathbf{s}(k)$ are a ρ-approximate solution of the following optimization problem:

$$\begin{aligned} &\max \textstyle\sum_{e \in E} \omega_e(k) \\ &\text{s.t.} \quad \Theta \in A_\Theta, \\ &\qquad \mathbf{s} \in S(\Theta). \end{aligned} \tag{8.7}$$

Remark 1: For cellular networks in consideration, a schedule $\mathbf{s}$ is feasible if the activated wireless links are node-exclusive. Therefore, the optimization problem (8.7) is in fact a maximum weighted matching (MWM) problem [19,20]. There is a centralized algorithm to solve MWM precisely with the time complexity of $O(|N|^3)$ [27], and a greedy algorithm to solve it approximately with an approximation ratio $\rho = 2$ [19,20]. The greedy algorithm is more useful to our problem because it is decentralized.

Remark 2: If (8.7) is solved precisely, we call the resulting ISPH algorithm a *perfect ISPH algorithm*; otherwise, we call it an *imperfect ISPH algorithm*. Note that, in the perfect ISPH algorithm, a coordinator is needed to observe and collect the information of the network, and run the centralized solver of (8.7). By contrast, the imperfect ISPH algorithm, (8.7) can be solved approximately in a distributed fashion, thus the imperfect algorithm is also distributed.

Remark 3: Regarding the sequence of control parameters $\{v_{l,u}(k)\}$, we choose the following in our simulation.

$$v_{l,u}(k) = \min\{V, \beta q_u(k) c_{t(l),u}(\mathbf{s}(k), \mathbf{w}(k))\}, \tag{8.8}$$

where $0 \leq \beta \leq 1$. Note that the control parameters $\{v_{l,u}(k)\}$ used in the simulation depend on the current queue backlogs and channel conditions.

The key component of the ISPH algorithm, the maximum weighted matching problem (8.7) should be solved at each time slot. Even with the distributed approximate solution, the system overhead is still higher than localized policies, where BSs manage the intra-cell scheduling independently.

In practice, we propose to solve the MWM problem (8.7) only periodically. Every $T > 0$ time slots, the imperfect ISPH algorithm is activated to perform handovers. At all other time slots, each BS b selects an MS u which maximizes $q_u(k)c_{b,u}(\mathbf{w}(k))$ from the set of associated MSs, and transmits data to the selected MS, where $c_{b,u}(\mathbf{w}(k))$ is the transmission rate from b to MS u. Obviously, at most time slots, BSs manage their intra-cell scheduling locally and independently.

The simulation results will show that the periodical scheme works almost as well as the scheme that the ISPH algorithm runs at every time slot. Note that since the ISPH algorithm runs only every T time slots, the system overhead is low and the overall scheme is practicable.

8.4.4 Numerical Examples

8.4.4.1 Performance evaluation metrics

We briefly introduce two scheduling policies which are well known and adopted in practice. The two policies will be used as benchmarks in the simulation results analysis.

- Signal to interference plus noise ratio (SINR) based scheduling policy: Let $sinr_{u,b}(k)$ denote the SINR strength of the channel from BS b to MS u measured at time k. The SINR-based policy greedily schedules MS-BS pairs with the largest $sinr_{u,b}(k)$ until no more MS-BS pairs are able to be scheduled. Higher SINR strength usually means higher data rate, thus the SINR-based scheduling aims to maximize system throughput. However, the SINR-based policy can be very unfair, and users with relatively bad channel conditions can be starved [7].
- Proportional fair (PF) scheduling policy: At each time slot, a weight for each MS-BS pair is defined as $\frac{r_{u,b}(k)}{R_u(k)}$, where $r_{u,b}(k)$ is the instantaneous data rate that MS u receives from BS b if it succeeds in competition for transmission, and $R_u(k)$ is the average data rate that MS u receives. $R_u(k)$ is computed by a sliding window as follows:

$$R_u(k) = (1-\alpha)R_u(k-1) + \alpha r_u(k),$$

  where $0 \leq \alpha \leq 1$ is a control parameter, and $r_u(k)$ is the data rate received at slot k. If MS u is not scheduled by any BSs at k, then $r_u(k) = 0$; otherwise, $r_u(k) = r_{u,b}(k)$, where b is MS u's assigned BS. In our simulation, $\alpha = 0.001$. The PF policy greedily schedules MS-BS pairs with the largest $\frac{r_{u,b}(k)}{R_u(k)}$ until no more MS-BS pairs are able to be scheduled. The PF policy maintains proportional fairness among all users [7].

8.4.4.2 Simulation setup

A two-tier multi-cell network composed of 19 hexagonal cells as in Figure 8.8 is considered, where the distance between origins of adjacent cells is 1732 m. Each cell has three sectors (i.e., base stations), and each sector has only one channel. The frequency reuse factor is 3. All BSs have the same transmission power of 43 dBm. For generating channel gains, the path loss model, $pl(d) = 35.63 + 35\log_{10}(d)$ is adopted, where d is the distance between MS and BS. For each downlink $e \in E$, its channel condition varies among *Poor, Medium and Good* randomly and independently. The probability to be in each condition is 1/3. In each cell, we assume that the MSs are uniformly distributed. The number of MSs per sector (i.e., per BS) is 9, 18 or 36. About 1/3 of BSs have 36 MSs, 1/3 of BSs have 18 MSs and the remaining 1/3 have 9 MSs. There are 1242 MSs in total.

The time is slotted into 1 ms lengths. For each MS, the data arrives randomly according to a Poisson process, where the mean data arrival rate is generated according to a uniform distribution. The simulation lasts 40,000 time slots (40 s) and MSs are initially allocated to BSs based on their SINR strength.

We first run the SINR-based policy and the PF policy at each BS locally and independently, where handovers are absolutely not allowed. The results can be served as the lower bounds of system performance. Then we run the SINR-based and the PF policies globally as if there is a coordinator which can schedule the channels of all the base stations globally, and handovers can happen arbitrarily often. In the two test cases, the procedure of handovers is assumed to be instantaneous, and an MS can participate in the competition for transmission at the target cell immediately, which is not possible in practice. However, the results of the two test cases are served as the upper bounds of system performance in terms of throughput and the maximum queue backlog. Note that the upper bounds are loose since they are under an impractical assumption.

Next we test the perfect ISPH algorithm which runs globally at each time slot. We first set the control parameter β to be 0, which means no handover penalty at all. In this test, we still assume the instantaneous handover assumption in order to see how good the ISPH algorithm could be under the ideal condition that handovers can happen arbitrarily often. Then we set β to be 0.2 to penalize handovers, and set the processing time of a handover to be 100 ms (i.e., 100 time slots), which means that an MS will not compete for transmission during the following 100 ms after it decides to perform a handover. Finally, we test the imperfect ISPH algorithm and the periodical scheme. The penalty parameters are 0.2 for both the tests, and 100 ms idle time is required for a handover.

We have eight tests summarized as follows:

- Test A: the SINR-based policy runs locally at each BS at each time slot.
- Test B: the SINR-based policy runs globally in a distributed fashion at each time slot. The instantaneous handover assumption is assumed.
- Test C: the PF policy runs locally at each BS at each time slot.
- Test D: the PF policy runs globally in a distributed fashion at each time slot. Instantaneous handover is assumed.
- Test E: the perfect ISPH algorithm runs globally at each time slot. We set $\beta = 0$, which means handovers can happen arbitrarily often. Instantaneous handover is assumed. Remember that β is defined in (8.8).
- Test F: the perfect ISPH algorithm runs globally at each time slot. We set $\beta = 0.2$ to penalize handovers. A handover takes 100 ms to complete.

Table 8.3 Comparison of Tests A to H

Test	Throughput	#Handover	Maximum backlog
A	100.00%	0	100.00%
B	118.98%	439,173	99.80%
C	91.93%	0	85.02%
D	114.19%	119,513	76.74%
E	112.92%	62,654	77.09%
F	113.09%	455	80.50%
G	113.11%	456	81.22%
H	112.17%	409	83.90%

- Test G: the ρ-approximate ISPH algorithm runs globally in a distributed fashion at each time slot, where $\rho = 2$. We set $\beta = 0.2$ to penalize handovers. A handover takes 100 ms to complete.
- Test H: we test the periodical scheme. The ρ-approximate ISPH algorithm runs globally every 1000 time slots with $\beta = 0.2$. At other slots, each BS schedules its associated MSs locally and independently. A handover takes 100 ms to complete.

8.4.4.3 Results analysis

In Table 8.3, we show the results of Tests A to H in terms of the system throughput, the number of handovers and the maximum queue backlog of MSs. The throughput and maximum queue backlog are normalized with respect to Test A.

Regarding the maximum system throughput, Test B shows that the throughput can be increased by as much as 18.98% under the instantaneous handover assumption, which is the upper bound of the throughput improvement. Tests E to H (all ISPH algorithm based) improve the system throughput by 12% to 13%, which is significant considering the upper bound of 18.98%. In addition, the improvement achieved by Tests E to H is very close to that achieved by Test D (i.e., 14.19%). Note that Test D assumes instantaneous handovers and allows infinitely frequent handovers. Thus the ISPH algorithm based tests improve system throughput significantly. The improvement is even larger, if compared to Test C, which performs the local PF policy and is often adopted in practice.

Regarding the number of handovers that happened, in Tests F, G and H where handovers are penalized, the values drop dramatically compared to Tests B, D and E (in Tests A and C, handovers are not allowed). Note that the performance of Tests F, G and H is very close to that of Tests B, D and E in terms of the throughput and the maximum queue backlog.

Consider the maximum queue backlog, which is proportionally related to the worst-case system delay. Test D gives the smallest backlog as we expected, since the global PF policy aims to maintain proportional fairness, which can be served as the lower bound of the worst-case backlog. The worst-case backlogs achieved by Tests E to H are only slightly larger than that of Test D, which are even smaller than what a local PF test obtains. Therefore all ISPH algorithm based tests maintain good fairness.

Table 8.4 Comparison of different parameters β

β	Throughput	#Handover	Maximum backlog
0.1	100.00%	3,509	100.00%
0.2	100.32%	456	102.20%
0.3	100.50%	348	106.00%
0.4	100.58%	304	109.37%
0.5	100.63%	282	111.12%
0.6	100.73%	264	112.16%
0.7	100.56%	244	112.37%
0.8	100.51%	250	113.06%
0.9	100.06%	270	113.30%

Now we discuss the complexity of all tests. Tests A and C use localized scheduling and have the lowest complexity and system overhead. Tests E and F run the perfect ISPH algorithm globally at each slot, hence require the highest complexity and system overhead. Tests B, D and G can run in a distributed fashion and have the medium complexity and overhead. Finally, in Test H, the distributed approximate ISPH algorithm takes place only every 1000 time slots. Its complexity and overhead are relatively low.

We summarize the improvement that our ISPH algorithms obtained. First, the approximate ISPH algorithm works almost as well as the perfect one, while it requires less complexity and overhead. Second, the approximate ISPH algorithm achieves a significant improvement compared to the local SINR or PF policy, at the cost of an affordable complexity and system overhead. Third, we can run the approximate ISPH algorithm periodically, which has relatively low complexity and a good performance in practice.

8.4.4.4 Tests of different control parameters β

The control parameter β defined in (8.8) is used to penalize handovers. The larger the parameter β, the more penalty a handover will receive. We test different β with the approximate ISPH algorithm running at each slot. Table 8.4 shows that the number of handovers drops while the maximum queue backlog increases as β increases, which can be interpreted as the tradeoff between the number of handovers and the maximum queue backlog.

8.5 Cooperation of MRO and MLB

In LTE systems, the mobility robustness management and mobility load balance are coupled with each other due to the fact that they both adjust the mobility parameters. If the two mobility entities work in an isolated fashion, it is possible that their respective decisions may conflict with each other. For instance, when a cell is overloaded and plans to offload to one of its neighbouring cell, the load transferring is implemented by modifying the mobility parameters between the two adjacent cells. An improper adjustment may result in a sudden increase of radio link failures or unnecessary handovers (ping-pong handovers). Consequently, the mobility robustness management entity is triggered to correct the handover setting, and a loop of actions results. Therefore, the system resources have been wasted and the user experiences have deteriorated.

The parameter conflict problem has been discussed in the 3GPP document [26], where the mobility load balance entity is assigned a higher priority. The mobility robustness entity is suspended temporarily when the load balancing is in operation, or it should avoid adjusting parameters that would oppose load balancing actions. However, the rationale to assign a priority to the load balancing entity has not been justified in [26]. It may lead to a failure of the robustness management and a large number of call drops, which are severe indications of bad performance. In the 3GPP document [9], it proposes to solve the conflict by informing the load balancing entity of the range of parameters that the robustness management allows. The allowed range of parameters or how to get the range has not been specified in the document.

In this section, we first introduce a mobility load balancing approach proposed in [9], which adjusts the cell individual offset between adjacent cells. We then develop the coordination rules between MRO and MLB, and hence jointly consider the mobility robustness and the mobility load balance [34]. We aim to investigate a load balancing approach which avoids the conflict with the robustness management. Finally, we conduct experiments and compare our solution to the approach in [9].

8.5.1 Achieve Load Balance by Adjusting CIO

The work in 3GPP document [9] proposed a load balancing approach which adaptively adjusts the handover parameters between adjacent cells. Consider two adjacent cell, where c_p denotes the cell with high load and c_n denotes the cell with low load. By manipulating the entering condition of Event A3 defined in (8.1), we have:

$$M_n > M_p - Oc_n + Hys + Of_p - Of_n + Oc_p + Off. \tag{8.9}$$

In [9], it proposed to adjust the parameters Oc_n and keep all other mobility parameters constant. For clarity, let us denote the Oc_n of the neighbouring cell by $CIO_{s,n}$ and the Oc_n of the source cell by $CIO_{n,s}$. There are three options to divert UEs from the high-load cell c_p to the low-load cell c_n and hence balance the load by modifying the parameters CIO:

- Option I: the neighbouring cell c_n increases its $CIO_{s,n}$, and therefore UEs in the overload cell c_p are urged to transfer to the cell c_n.
- Option II: the source cell c_p decreases its $CIO_{n,s}$, and therefore the handovers from the underload cell c_n to the overload cell c_p are postponed.
- Option III: it is a combination of Option I and Option II. The parameter $CIO_{s,n}$ increases and the parameter $CIO_{n,s}$ decreases at the same time.

8.5.2 Coordination Rules between MRO and MLB

Since the mobility robustness and load balance entities share the same set of handover parameters and they are closely coupled, we propose a set of coordination rules between

Table 8.5 Coordination rules between MRO and MLB

MRO performance indicator	Cause	Rules
$r_{early}(s, n) \geq \lambda_{early}$	$CIO_{s,n}$ is too large	Stop increasing $CIO_{s,n}$
$r_{late}(s, n) \geq \lambda_{late}$	$CIO_{s,n}$ is too small	Stop decreasing $CIO_{s,n}$
$r_{pp}(s, n) \geq \lambda_{pp}$	$CIO_{s,n}$, $CIO_{n,s}$ are too large	Stop increasing $CIO_{s,n}$ and $CIO_{n,s}$

the two entities. For a pair of source cell c_s and neighbouring cell c_n, we define the number of handovers, the number of too-early handovers, the number of too-late handovers and the number of ping-pong handovers observed in a time interval t as $N(s, n)$, $N_{early}(s, n)$, $N_{late}(s, n)$, and $N_{pp}(s, n)$, respectively. The rate of too-early handovers, the rate of too-late handovers, and the rate of ping-pong handovers are defined as the following:

$$r_{early}(s, n) = \frac{N_{early}(s, n)}{N(s, n)}, \tag{8.10}$$

$$r_{late}(s, n) = \frac{N_{late}(s, n)}{N(s, n)}, \tag{8.11}$$

$$r_{pp}(s, n) = \frac{N_{pp}(s, n)}{N(s, n)}. \tag{8.12}$$

As discussed in Section 8.3, $r_{early}(s, n)$, $r_{late}(s, n)$ and r_{pp} are important performance indicators for the mobility robustness management entity. When the rate of too-early handovers $r_{early}(s, n)$ is above a predefined threshold λ_{early}, it means that the cell individual offset $CIO_{s,n}$ is too large, and the MLB procedure should not increase it further; when the rate of too-late handovers $r_{late}(s, n)$ exceeds a predefined threshold λ_{late}, it means that the cell individual offset $CIO_{s,n}$ is too small and should not be decreased; when $r_{pp}(s, n)$ is above a predefined threshold λ_{pp}, both $CIO_{s,n}$ and $CIO_{n,s}$ are too large and cannot be increased further. We summarize the rules for MRO and MLB coordination in Table 8.5.

8.5.3 Jointly Consider MRO and MLB

With the set of coordination rules, we can jointly consider the mobility robustness management and load balance together. Figure 8.11 shows the joint procedure in a mobility management round. For each cell, both the mobility load balancing entity and the mobility robustness entity collect data and compute the performance indicators, that is, cell load λ_{thr}, r_{early}, r_{late}, and r_{pp}, respectively. For any cell, if its load is above the predefined threshold λ_{thr}, it will check with its neighbouring cells to see whether there exists any neighbouring cell that can accept load (i.e., its load is no more than $\lambda_{thr} - \delta_{thr}$). If such a cell exists, a coordinator of MRO and MLB checks the mobility robustness performance indicators and selects the coordination rules for the load balance actions from Table 8.5. Finally, following the rules, the load balancing entity adjusts the mobility settings and makes the load balanced.

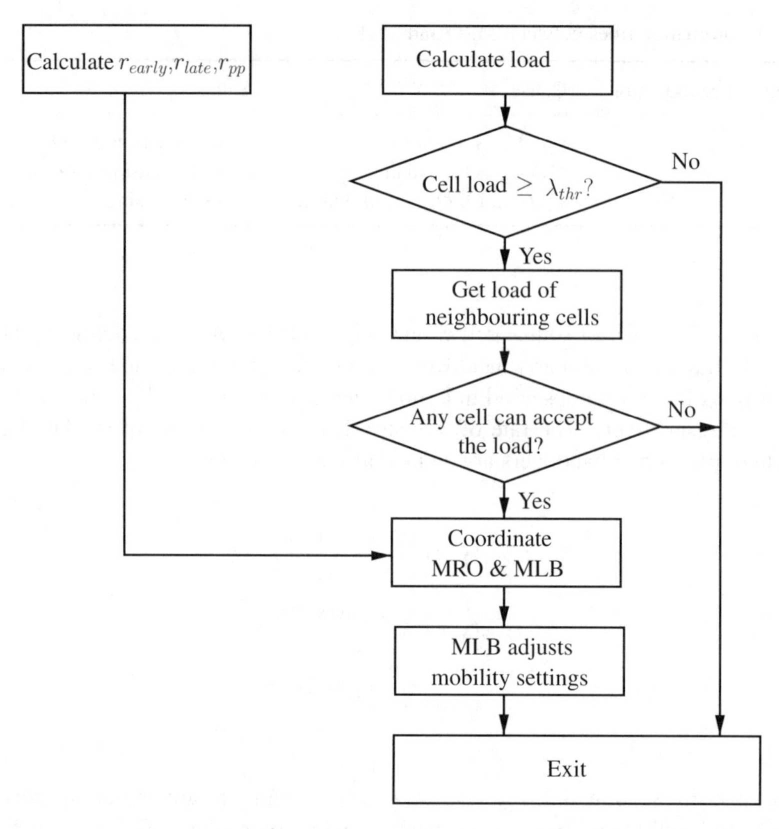

Figure 8.11 A coordinator of MRO and MLB checks the performance indicators of mobility robustness and sets rules for the load balance entity.

8.5.4 Simulation Results

8.5.4.1 Simulation setup

Consider the three cells shown in Figure 8.12, where a two-tier interference model is adopted. There are three types of UE mobility: high speed (30 km/h), low speed (3 km/h) and stationary (0 km/h). In our simulation, the UEs with high and low speed will distribute uniformly in the round area, while the stationary UEs can only stay at the three rectangle areas (hotspot areas) as shown in Figure 8.12. The detailed simulation setup is summarized in Table 8.6.

8.5.4.2 Calculate cell load

The system performance will be sampled every T_m seconds and the samples are indexed by k. Given the SINR received by UE u at time k, $SINR_{u,k}$, the throughput of the downlink can be approximated by (8.13) [3].

$$Thr_{u,k}(bps/HZ) = \begin{cases} 0, & SINR_{u,k} < SINR_{min} \\ Thr_{max}, & SINR_{u,k} > SINR_{max} \\ \alpha \log_2(1 + SINR_u), & SINR_{min} \leq SINR_{u,k} \leq SINR_{max}. \end{cases} \tag{8.13}$$

The parameters presented in (8.13) are summarized in Table 8.7.

Table 8.6 Simulation setup

Parameter	Value
Distance between origins	500 m
BS power	46 dBm
Path loss model	$136.3 + 39.1(log_{10}d)$
TTT	400 ms
Hyst	2 dB
CIO	0 dB
#UEs at high speed	300
#UEs at low speed	90
#Stationary UEs	$Cell_1 : Cell_2 : Cell_3 = 100 : 0 : 0$
Data rate of UEs (high speed)	32 kbps
Data rate of UEs (low speed)	64 kbps
Data rate of UEs (stationary)	96 kbps
Duration of MLB cycle	10 s
Stepsize	0.5 dB
λ_{thr}	5 MHz
δ_{thr}	1 MHz
λ_{early}	2%
λ_{late}	2%
λ_{pp}	5%

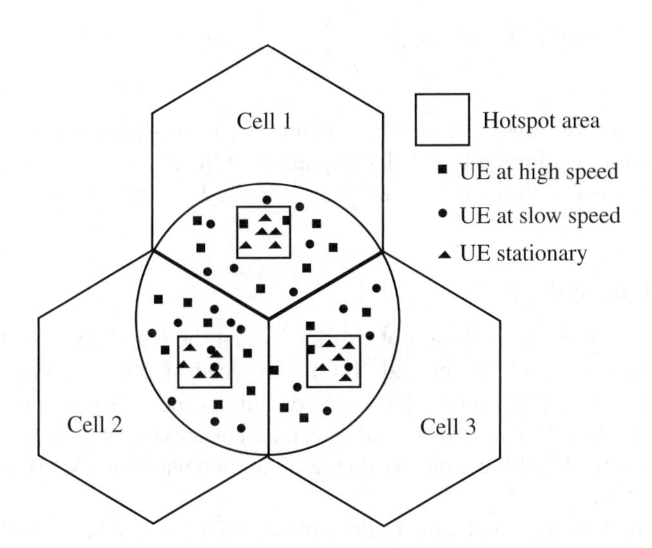

Figure 8.12 Simulation scenario.

Table 8.7 Parameters used in (8.13)

Parameter	Value	Description
α	0.6	attenuation factor, representing implementation losses
$SINR_{min}$	-10 dB	minimum SINR of the codeset
$SINR_{max}$	22 dB	SINR at which max throughput is reached
Thr_{max}	4.4 bps/Hz	maximum throughput of the codeset

For each UE u, a task arrives at a random time t_u, the duration of the task is a fixed value T_u, and the data rate of UE u is D_u. Therefore, the bandwidth consumed by UE u at time k is

$$B_u(k) = \begin{cases} D_u/Thr_u, & t_u \leq kT_m \leq t_u + T_u \\ 0, & \text{otherwise.} \end{cases} \tag{8.14}$$

The cell load of cell c at time k is the sum of bandwidth consumed by all the UEs in cell c (i.e., $u \in U_c$, where U_c is the set of UEs in cell c) as follows:

$$Load_{c,k} = \sum_{u \in U_c} B_u(k) \tag{8.15}$$

The performance of load balancing functionality is evaluated by calculating the average offload for a cell c during the simulation interval as follows.

$$AvgOffLoad_c = \frac{1}{K_c} \sum_{k:Load_{c,k} \geq \lambda_{thr}} Load_{c,k} - Load^*_{c,k}, \tag{8.16}$$

where $Load_{c,k}$ is the cell load of c at time k before the load is balanced and $Load^*_{c,k}$ is the cell load of c at time k after the load balancing entity shifts the load. The variable K_c is the number of samples where the cell c is overloaded, that is, $K_c = |\{k : Load_{c,k} \geq \lambda_{thr}\}|$.

8.5.4.3 Results analysis

In Figure 8.13, we show the evolution of cell load when the load balancing function is off. During the interval between 100 ms and 300 ms, the load of cell c_1 is above the overload threshold (i.e., $\lambda_{thr} = 5$ MHz), and both the load of cells c_2 and c_3 is less than the load of cell c_1 by more than 1 MHz. Hence, cells c_2 and c_3 can accept some load from cell c_1.

Next, we turn on the MLB function. We do the simulation with three MLB options and with the coordinator off/on.

In Table 8.8, we show the simulation results with MLB Option I. When cell c_1 is overloaded, the MLB entity increases $CIO_{1,2}$ and $CIO_{1,3}$. Therefore, the entering condition of Event A3 (8.1) is triggered earlier and UEs are prompted to shift from cell c_1 to cell c_2 or c_3 to balance the load. When the coordinator is off, the MLB entity repeatedly adjusts $CIO_{1,2}$ and $CIO_{1,3}$ to make them too large, which results in a large amount of too-early handovers and ping-pong handovers. When the coordinator is on, it will observe both the actions and performance of

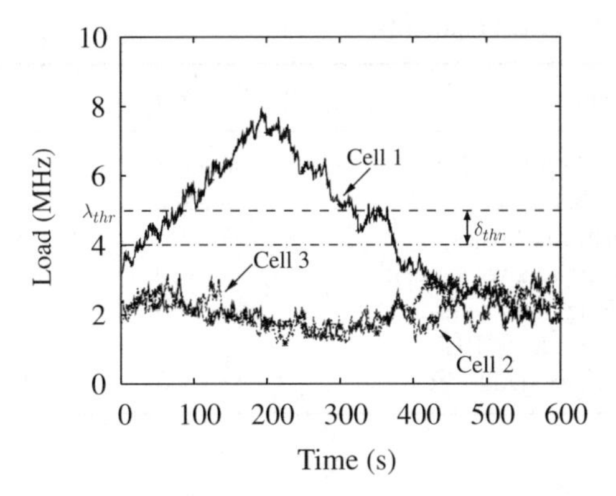

Figure 8.13 The evolution of cell load when the coordinator is off.

the MLB and MRO entities. When the too-early handover rate r_{early} is over the threshold, the coordinator will not allow the MLB entity to further increase $CIO_{1,2}$ and $CIO_{1,3}$. Hence too-early handovers and ping-pong handovers are avoided. When the coordinator is on, the range of the mobility parameters in which the MLB entity can set is more restricted. Hence the amount of load shifted from cell c_1 to cells c_2 and c_3 with the coordinator on is less than that with the coordinator off.

In Table 8.9, we show the simulation results with MLB Option II. When cell c_1 is overloaded, the MLB entity increases $CIO_{2,1}$ and $CIO_{3,1}$. Therefore, UEs in cell c_2 or c_3 are postponed to shift to cell c_1. When the coordinator is off, the MLB entity repeatedly adjusts $CIO_{2,1}$ and $CIO_{3,1}$, which results in a large amount of too-late handovers. With the coordinator on, when the too-late handover rate r_{late} is above the threshold, the coordinator will not allow the MLB entity to further decrease $CIO_{2,1}$ and $CIO_{3,1}$ to avoid too-late handovers. Similar to the MLB Option I, when the coordinator is on, the allowed adjusting range of the mobility parameters is more restricted. Hence the amount of shifted load is less.

In Table 8.10, we show the results with MLB Option III. Option III is a combination of Options I and II. Regarding mobility robustness, Option III has a small number of handover related radio link failures. Regarding the amount of shifted load, Option III makes the load of the three cells more balanced even with the coordinator on.

Table 8.8 MLB Option I

Coordinator	off	on
$AvgOffLoad_1$ (MHz)	0.6369	0.3001
#too-late handovers	0	0
#too-early handovers	120	6
#handovers to a wrong cell	0	0
#ping-pong handovers	643	21
#handover related RLFs	120	6
#unnecessary handovers	643	21

Table 8.9 MLB Option II

Coordinator	off	on
$AvgOffLoad_1$ (MHz)	0.5532	0.2433
#too-late handovers	122	4
#too-early handovers	0	0
#handovers to a wrong cell	0	0
#ping-pong handovers	0	0
#handover related RLFs	122	4
#unnecessary handovers	0	0

Table 8.10 MLB Option III

Coordinator	off	on
$AvgOffLoad_1$ (MHz)	0.9494	0.6671
#too-late handovers	138	2
#too-early handovers	4	6
#handovers to a wrong cell	0	0
#ping-pong handovers	0	9
#handover related RLFs	142	8
#unnecessary handovers	0	9

8.6 Mobility Enhancement for Femtocells

The large-scale deployment of femtocells to enhance indoor coverage has been promised by LTE/LTE-Advanced systems. A femtocell is a low-cost and low-power access point with limited functionality, which is often installed at home or in an office environment to provide voice and data services to mobile users. The unique character of femtocells compared with the conventional cellular base stations makes the mobility management for a heterogeneous network consisting of both macro and femtocells challenging.

A conventional base station is an expensive professional product, which provides much more capacity and coverage; a femtocell is a low-cost and low-power access point designed for end users, and is only able to support a small number of mobile users concurrently with limited capacity and pico-size coverage. The base stations are inter connected by the mobile-provider-owned dedicated core networks; a femtocell connects to the mobile provider's network via the internet. The deployment of cellular base stations is performed and optimized by the mobile providers; the deployment of femtocells is usually carried out randomly by the consumers. Since the coverage of femtocells is small and installation is easy and can be implemented by the consumers, the density of femtocells is much larger than that of the cellular base stations, which makes the management of femtocells difficult.

3GPP started to work on a femtocell standard in 2008. Some agreements about mobility management in a heterogeneous network consisting of femtocells and macrocells have been reached. In LTE systems, a closed subscriber group (CSG) femtocell broadcasts its CSG ID

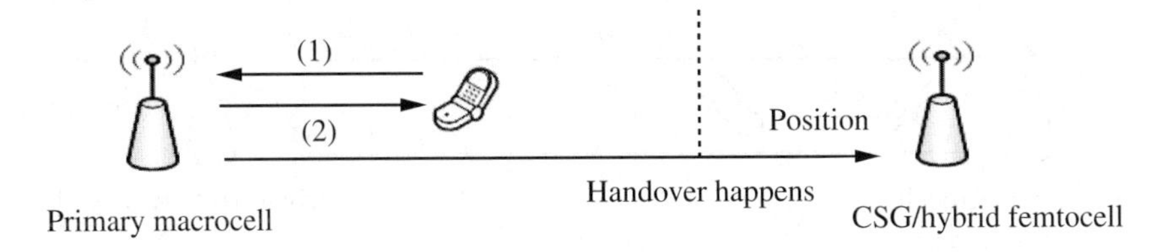

(1) Proximity indication

(2) Measurement configuration

Figure 8.14 In-bound handover: the UE sends the proximity indication message to the source base station.

and a CSG indicator to identify itself as a CSG cell, where the CSG indicator is set to be true. If only the CSG ID is broadcast without the CSG indicator, then the femtocell is a hybrid cell. All other fetmocells are normal or open cells. A CSG/hybrid cell is accessed as a CSG cell if its CSG ID is in the UE's CSG whitelist. If a cell's CSG ID is not in the UE's CSG whitelist, a hybrid cell can be accessed as a normal cell while the access to the CSG cell is not allowed. Based on the access mode, femtocell-involved handovers are classified into two categorises as in-bound and out-bound handovers. An in-bound handover refers to a handover towards a CSG/hybrid cell. An out-bound handover is defined as a handover from a CSG/hybrid cell to an open cell, which follows the normal handover procedure. In the meanwhile, the in-bound handover is more complicated, and hence new issues arise for mobility robustness management.

In a normal handover procedure, the UE keeps measuring the signal from nearby cells according to the measurement configuration set by the source base station. In the case of in-bound handover in LTE systems, when the UE is approaching a CSG/hybrid cell which it is allowed to access, if the measurement configuration is not present for the frequency/RAT of the concerned cell, the UE will not measure the signal from the concerned cell and no handover will be triggered. Hence, as shown in Figure 8.14, in the in-bound handover procedure, the UE needs to inform the source base station that it is near a desired CSG/hybrid cell by sending a proximity indication message to the source base station. The proximity indication message indicates whether the UE enters or leaves the proximity of the CSG/hybrid cell and also the frequency/RAT of the CSG/hybrid cell. Upon receiving the proximity indication message, the source base station updates the measurement configuration of the UE, and therefore the UE is able to measure the signal from the target CSG/hybrid femtocell and initiate a handover if needed.

Now the problem arises when the proximity indication message does not arrive at the source base station in time or does not arrive at all, or the frequency of the concerned CSG/hybrid cell changes after the proximity indication is sent to the base station. In any of the above cases, the measurement configuration of the UE cannot be set correctly to trigger the measurement for the target CSG/hybrid cell. Therefore, a handover towards the concerned CSG/hybrid cell will not be initiated, even if the signal from the CSG/hybrid cell is good enough and the mobility parameters between the source cell and the target CSG/hybrid cell are set correctly. Consider a scenario that the concerned CSG/hybrid cell is deployed to enhance the indoor coverage

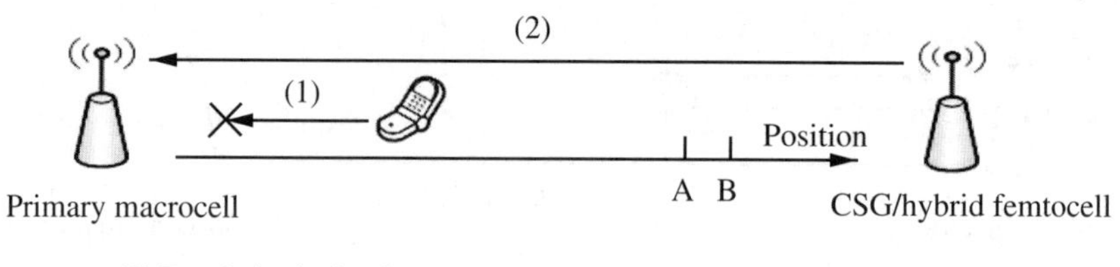

Figure 8.15 The proximity indication message is lost and the UE experiences a radio link failure. Position A: RLF happens. Position B: UE connects to the CSG/hybrid cell by RRC reestablishment.

where the signal from a macrocell is poor. As shown in Figure 8.15, when the UE enters the indoor environment, as already mentioned, the handover towards the CSG/hybrid femtocell will not happen since the measurement configuration is not properly set. Thus eventually the UE loses its connection with the source macro cell and a radio link failure happens. After the UE successfully connects to the CSG/hybrid cell by RRC reestablishment, it will follow the mobility robustness management procedure to report the radio link failure as one resulting from a too-late handover. The mobility robustness management will count it as an indication of the improper mobility setting and may further seek to adjust the setting. However, in this case, the radio link failure is not due to the improper mobility setting but due to the delayed, missing or mis-configured proximity indication message. Hence the source base station should be able to distinguish this too-late handover report from other too-late handover reports and will not take it as an indication of improper mobility setting and adapt the handover parameters consequently.

One possible solution is to add more information to the radio link failure indication message as shown in Figure 8.16. As defined in 3GPP TS 36.423 [4], the radio link failure indication message is sent from the base station where the RRC reestablishment attempt is made or the RLF report is received by the base station that the UE has connected to prior to the failure. In order to help the source base station to judge whether a radio link failure indication message is a true indication of improper setting, the frequency of the CSG/hybrid cell and the access mode of the UE can be included in the radio link failure indication message. Once the source base station receives the indication, it can check whether it has properly configured the measurement for the UE, given the information carried in the message. Hence, false radio failure indication can be figured out, and no further actions will be taken to modify the handover setting.

8.7 Conclusion

In this chapter, we study mobility management for the next generation mobile networks, in particular, the LTE/LTE-Advanced systems, where we are facing serious challenges brought by higher data rates, higher mobility and more heterogeneous networks. We investigate the mobility robustness optimization and the mobility load balance optimization. Since the two functionalities are linked, we propose a coordination scheme to avoid conflict between the two entities. Finally, as femtocells will be deployed in a large scale to enhance the indoor

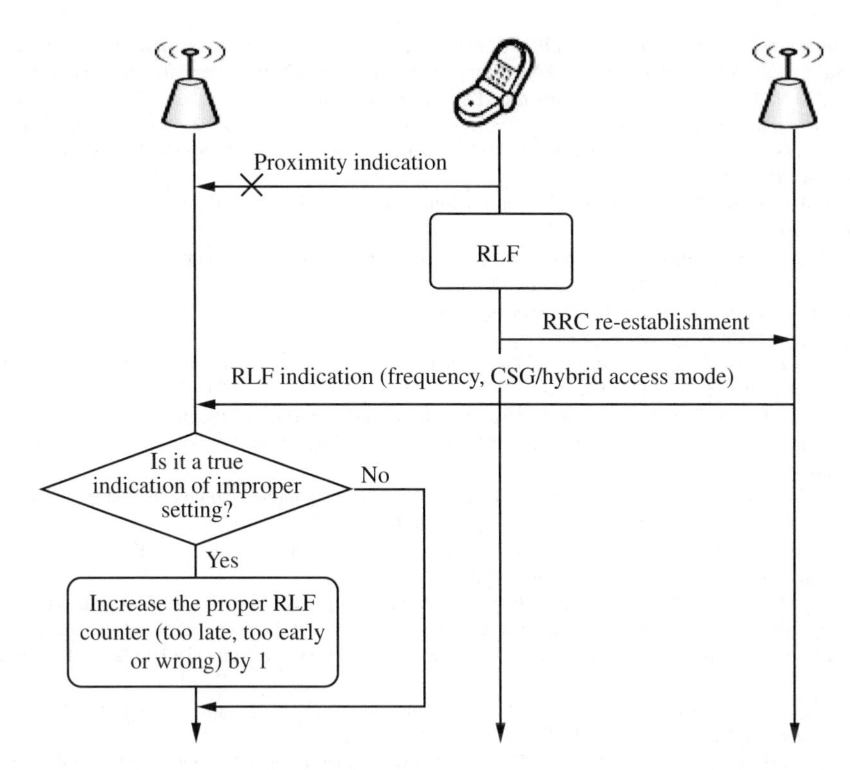

Figure 8.16 The proximity indication message is lost and the UE experiences a radio link failure. After the source base station receives the RLF indication message, it can judge whether the message is a true indication of improper setting and take action accordingly.

coverage, the mobility management of femtocell involved heterogeneous networks becomes even more difficult. We propose a solution to solve the mobility robustness issue caused by the introduction of femtocells.

Acknowledgements

This work was partially supported by the NSFC under grant number 61100238.

References

1. 3rd Generation Partnership Project (3GPP). Evolved universal terrestrial radio access (E-UTRA); user equipment (UE) procedures in idle mode. Technical Report 3GPP TR 36.304 v8.5.0, 3rd Generation Partnership Project (3GPP), March 2009.
2. 3rd Generation Partnership Project (3GPP). Self-configuring and self-optimizing network use cases and solutions. Technical Report 3GPP TR 36.902 v1.2.0, 3rd Generation Partnership Project (3GPP), May 2009.
3. 3rd Generation Partnership Project (3GPP). Radio frequency (RF) system scenarios (release 9). Technical Report 3GPP TR 36.942 v9.0.1, 3rd Generation Partnership Project (3GPP), 2010.

4. 3rd Generation Partnership Project (3GPP). Technical specification group radio access network; evolved universal terrestrial radio access network (E-UTRAN); X2 application protocol (X2AP)(release 10). Technical Report 3GPP TS 36.423 V10.2.0, 3rd Generation Partnership Project (3GPP), June 2011.
5. 3rd Generation Partnership Project (3GPP). Evolved universal terrestrial radio access (E-UTRA); radio resource control (RRC) protocol specification (release 10). Technical Report 3GPP TR 36.331 v10.5.0, 3rd Generation Partnership Project (3GPP), March 2012.
6. T. Bonald, S. Borst and A. Proutiere. Inter-cell scheduling in wireless data networks. In *Proceedings of European Wireless*, 2005.
7. T. Bu, L. Li and R. Ramjee. Generalized proportional fair scheduling in third generation wireless data networks. In *IEEE INFOCOM 2006*, pp. 1–12, April 2006.
8. T. Casey. Base station controlled load balancing with handovers in mobile WiMAX. Master's thesis, Helsinki University of Technology, 2008.
9. ZTE Corporation and China Academy of Telecommunications Technology. The correlation between MLB and MRO. Technical Report R3-092294, 3GPP TSG RAN WG3 Meeting 65bis, 2009.
10. S. K. Das, S. K. Sen, R. Jayaram and P. Agrawal. A distributed load balancing algorithm for the hot cell problem in cellular mobile networks. In *The Sixth IEEE International Symposium on High Performance Distributed Computing*, pp. 254–263, 1997.
11. S. Das, H. Viswanathan and G. Rittenhouse. Dynamic load balancing through coordinated scheduling in packet data systems. In *IEEE INFOCOM 2003*, pp. 786–796, 2003.
12. European Union Information Society Technologies Framework Programme 6 (EU IST FP6). ANA: Autonomic network architecture. http://www.ana-project.org/.
13. European Union under the 7th Framework Program. The socrates project. http://www.fp7-socrates.eu/.
14. S. V. Hanly. An algorithm for combined cell-site selection and power control to maximize cellular spread spectrum capacity. *IEEE Journal on Selected Areas in Communications*, 13:1332–1340, September 1995.
15. H. Hu, J. Zhang, X. Zheng, Y. Yang and P. Wu. Self-configuration and self-optimization for LTE networks. *IEEE Communications Magazine*, 48(2):94–100, February 2010.
16. T. Jansen, I. Balan, J. Turk, I. Moerman and T. Kurner. Handover parameter optimization in LTE self-organizing networks. In *the 72nd IEEE Vehicular Technology Conference*, pp. 1–5, 2010.
17. F. Kelly, A. Maulloo and D. Tan. Rate control for communication networks: shadow price, proportional fairness and stability. *Journal of the Operational Research Society*, 49:237–252, 1998.
18. J.-S. Kim, E. Serpedin, D.-R. Shin and K. Qaraqe. Handoff triggering and network selection algorithms for load-balancing handoff in CDMA-WLAN integrated networks. *EURASIP Journal on Wireless Communications and Networking*, 2008.
19. X. Lin, N. B. Shroff and R. Srikant. A Tutorial on Cross-Layer Optimization in Wireless Networks. *IEEE Journal on Selected Areas in Communications*, 24(8):1452–1463, August 2006.
20. X. Lin, N. B. Shroff and R. Srikant. The Impact of Imperfect Scheduling on Cross-Layer Rate Control in Wireless Networks. *IEEE/ACM Transaction on Networking*, 14(2):302–315, April 2006.
21. S. H. Low and D. E. Lapsley. Optimization flow control – I: Basic algorithm and convergence. *IEEE/ACM Transactions on Networking*, 7(6):861–874, 1999.
22. Multiple Access Communications Limited. Project MONOTAS (mobile network optimisation through advanced simulation). http://www.macltd.com/monotas/index.php.
23. R. Nasri and Z. Altman. Handover adaptation for dynamic load balancing in 3GPP Long Term Evolution systems. In *MoMM 2007*, pp. 145–154, 2007.
24. NEC Corporation. Self organizing network. NEC's proposals for next-generation radio network management, February 2009. http://www.nec.com/global/solutions/nsp/mwc2009/images/SON_whitePaper_V19_clean.pdf.
25. INC NTT DOCOMO. Evaluation model for rel-8 mobility performance. Technical Report R1-091578, 3GPP TSG RAN WG1 Meeting 56bis, March 2009.
26. China Academy of Telecommunications Technology, China Academy of Telecommunication Research, and ZTE Corporation. Conflict avoidance between MLB and MRO. Technical Report R3-091565, 3GPP TSG RAN WG3 Meeting 65, 2009.
27. C. H. Papadimitriou and K. Steiglitz. *Combinatorial Optimization: Algorithms and Complexity*. Dover Publications, 1998.
28. A. Sang, X. Wang, M. Madihian and R. D. Gitlin. Coordinated load balancing, handoff/cell-site selection, and scheduling in multi-cell packet data systems. In *MobiCom 2004*, pp. 302–314, 2004.

29. A. Schroder, H. Lundqvist and G. Nunzi. Distributed self-optimization of handover for the Long Term Evolution. *Lecture Notes In Computer Science*, 5343:281–286, 2008.
30. K. Son, S. Chong and G. de Veciana. Dynamic association for load balancing and interference avoidance in multi-cell networks. In *The 5th International Symposium on Modeling and Optimization in Mobile, Ad Hoc and Wireless Networks and Workshops*, pp. 1–10, April 2007.
31. L. Tassiulas and A. Ephremides. Stability properties of constrained queuing systems and scheduling policies for maximum throughput in multihop radio networks. *IEEE Transactions on Automatic Control*, 37(12), December 1992.
32. Z. Wei. Mobility robustness optimization based on UE mobility for LTE system. In *Wireless Communications and Signal Processing (WCSP), 2010 International Conference on*, pp. 1–5, October 2010.
33. R. D. Yates and C.-Y. Huang. Integrated power control and base station assignment. *IEEE Transactions on Vehicular Technology*, 44:638–644, August 1995.
34. J. Yu, H. Hu, S. Jin and X. Zheng. Conflict coordination between mobility load balancing and mobility robustness optimization. *Computer Engineering (in Chinese)*, 38(1):37–41, January 2012.
35. H. Zhang, X. Wen, B. Wang, W. Zheng and Z. Lu. A novel self-optimizing handover mechanism for multi-service provisioning in LTE-Advanced. In *International Conference on Research Challenges in Computer Science, 2009*, pp. 221–224, 2009.

9

Connected-mode Mobility in LTE Heterogeneous Networks

Carl Weaver and Pantelis Monogioudis
Alcatel-Lucent, USA

9.1 Introduction

In this chapter we consider networks with open subscriber group (OSG) heterogeneous nodes of two types: macro and metro eNBs. It is worth reviewing some definitions that the reader may not be familiar with.

OSG cells can be selected by everyone with a subscription to the mobile operator. This contrasts to closed subscriber group (CSG) cells, such as femtocells, that can be selected by only specific subscribers.

Metros are OSG eNBs with lower maximum transmission power than macros, usually employing antenna configurations that are simpler than those of the macro eNBs but exhibit certain compactness that is usually achieved with increased integration with other space-consuming radio components such as in active antenna arrays.

We consider a mixture of these nodes where the operator macro network is already in place, already achieving some capacity and coverage targets, and metros are deployed as an underlay, usually aiming to fill coverage holes or local throughput improvements. Only simple antenna configurations are considered because the aim in this chapter is to highlight the effect of inter-eNB mobility procedures in connected mode. The connected mode is the mode where the UE has established a connection with its peer radio resource control (RRC) layer at the serving eNB and is actively engaged in user-plane traffic transmission/reception with this eNB. In HTNs, two important mobility procedures determine the extent to which metros will contribute to the overall system performance: the initial selection and reselection of the macros/metros and the handover (HO) while in connected mode. Let us have a closer look to the cell selection problem and the tools that the radio network planners have to introduce metros in existing macro networks.

Heterogeneous Cellular Networks, First Edition. Edited by Rose Qingyang Hu and Yi Qian.

9.2 Cell Selection and Problem Statement

During the cell selection procedure, the UE will determine the best cell that can be used to establish an RRC connection. Conventionally, the UE will select a cell based on the received signal reference power (RSRP) that is determined based on measurements it performs on the cell reference signal (CRS) broadcasted by the cell with a nominal transmission power.

$$CellID_{serving} = \arg\ \max_{\{i\}}\{RSRP_i\}$$

In Figure 9.1, the RSRP slopes of the macro and metro are shown simplistically. Because of the substantially higher power emitted by the macro, the two received power levels will be equal at a very small distance from the metro. This shrinks the DL coverage of the metro and in addition, it can create asymmetry in the UL coverage that is unaffected by the transmission power differences of the cells. R8 specifications allow to bias the cell selection, and the UE can then determine the serving cell, based on the relationship:

$$CellID_{serving} = \arg\ \max_{\{i\}}\{RSRP_i + bias_i\}.$$

The cell selection bias can range from 0 to 20 dB, according to 3GPP specifications, with the optimal setting determined by proprietary SON algorithms outside the scope of this chapter. It is communicated to the UE via higher layer signalling. The effect of the bias is to extend the DL metro coverage as the UE artificially increases the measured RSRP from the metro by the bias value. At the same time, the DL signal to interference and noise ratio (SINR) when an RRC connection establishment is attempted is degraded. As shown in Figure 9.2, in cases where the metro cell is sufficiently far away from the macrocell, in coverage-challenged areas, the range of the metro is adequate without bias, especially if the transmission power of the metro is increased to 5 W. Both biased and unbiased cases have been studied, but here we focus primarily on the handover procedure without bias.

The handover procedure in LTE has been covered adequately in the existing literature and the interested reader can consult the 3GPP Radio Resource Control (RRC) standards specification TS 36.331 [1]. In summary, the UE initiates a handover procedure, based on signal and time thresholds defined and transmitted to it by the serving cell. These are the HO hysteresis, time to trigger (TTT) and the parameter cellIndividualOffset (O_c). The last is a *vector* quantity sent by the serving cell that contains the offsets of the serving cells ($O_c(s)$) and neighbour cells ($O_c(n)$) that all UEs in this cell must apply in determining whether the measured RSRPs satisfy

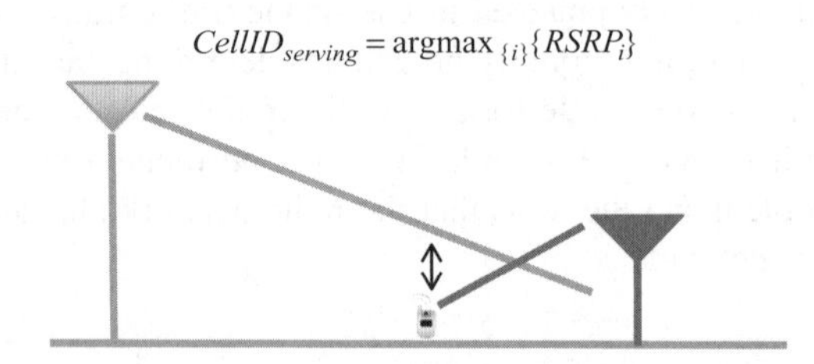

Figure 9.1 RSRP from the macro (blue) and from the metro (red) in the UE association procedure.

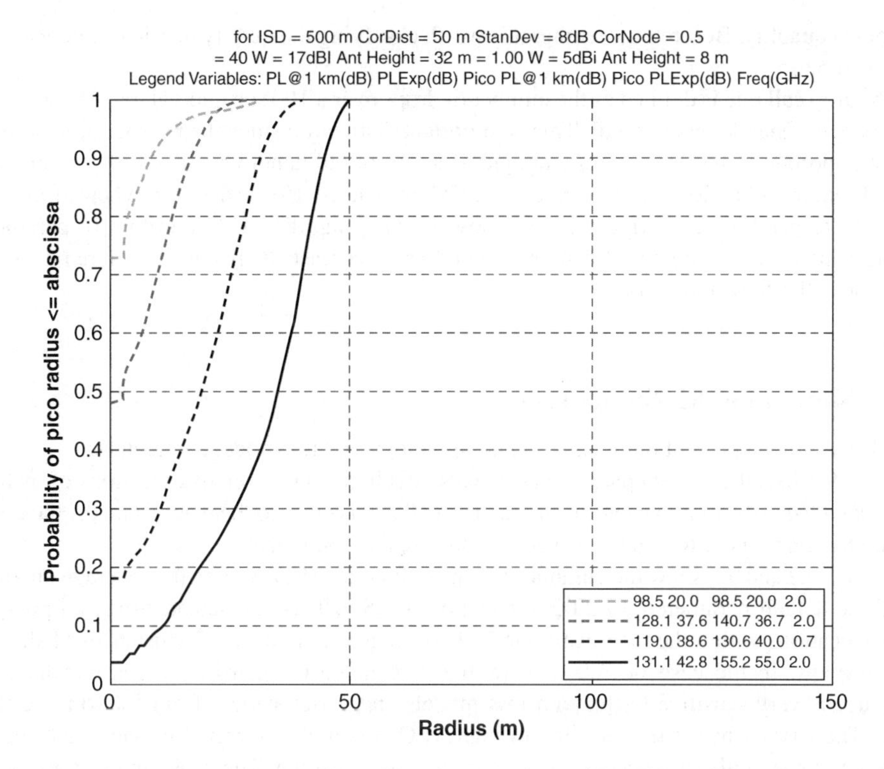

Figure 9.2 Metro cell coverage for all possible metro locations within 500 m ISD macro coverage and over 25 instances of shadow fading maps.

the conditions of the so-called 'event A3', that results in the UE sending a measurement report (MR) that triggers the handover. O_c values can be positive or negative. The measured RSRPs are filtered and O_c values are added to determine the cell id {k} according to the relationship:

$$CellID_k = \arg \max_{\{i\}} \{RSRP_i + O_c(i)\}$$

The primary objective of this study is to compare the HO rate and the dropped call rate performance of the macro network with an HTN network, in the high UE mobility regime. In the high UE mobility case the HO rate for a given deployment environment is quite high with very short time allowed to complete such handovers. Additionally, we consider the handover (HO) procedure defined in TS 36.331 [1], and the tuning of its parameters necessary to obtain less than 1% dropped call rate without reestablishment. This objective is more stringent than necessary for best effort traffic on LTE, but it may be required for VoIP where reestablishments to correct handover failures degrades VoIP QoS. It is important to look at the macro system dropped call rate and the HO rate to calibrate with familiar results from macro-only studies. It would also be useful to compare with field results, but the driving times required for statistical significance are long and service measurements must be conditioned on UE speed, a usually unknown quantity. Before proceeding, it is worth clarifying the quality metric of interest – the drop call rate.

The drop call rate is defined as the number of drops over a 100 sec interval, rather than as the percentage of handovers that fail. This is important distinction since the number of handovers is not important to the UE; so for example, if a UE did 20 handovers in a 100 sec period, it would require a handover failure rate of 0.05% instead of 1% in the case where it did just 1 handover per 100 sec period. This is also especially significant if algorithm or parameter changes increase the number of handovers but do not decrease the handover failure rate, since the drop call rate would worsen.

9.3 Simulation Methodology

In this section we present our evaluation methodology and the model parameters.

Table 9.1 lists the system parameters. We also distinguish two relative locations of metros: the *embedded* in the macro location where the small cell is close to the macro, and the handover region location where the metro is placed at the edge of the macro.

Figures 9.2 and 9.3 show the cumulative density function (CDF) of metro coverage obtained for these default parameters for 500 m and 1700 m ISD. This is calculated over all possible metro locations within the macrocell nominal coverage area and over 25 instances of shadow fading maps. In the case of 500 m ISD, the percentage of metro locations that have no coverage is very sensitive to the path loss models and varies widely from 5% to more than 50%. The cases where there are line of sight (LOS) conditions between macro and metro, are even worse, with the majority of cases having no coverage. The problem of very limited metro coverage can be addressed by increasing the maximum metro transmission power level, increasing the directionality of its antennas or with inter-cell interference coordination (ICIC) where blanking of the macrocell transmissions with almost blank subframes (ABS) would be required to provide some usable coverage. ABS with appreciable cell selection bias is not feasible in the R8 and R9 LTE specifications, and we expect systems to operate optimally with ABS beyond R10. Therefore we limit the discussion to unbiased cell selection without ABS-based macro interference blanking. This limits the placement of metros in the handover regions and especially so for dense macro deployments such as in the 500 m ISD case. Therefore we place 12 metros uniformly spaced in a circle around the centre node of a 21-macrocell (7-site) cluster. The placement radius is half the ISD. A map showing the coverage area, defined by strongest RSRP, of macros and metro is shown in Figure 9.4 while Figure 9.5 shows the best SINR on the same map.

The high level analysis process that is used to study HO performance for HetNet is as follows: first we perform drive tests on a reference system with only macrocells to determine the HO and dropped call rate at high speed. In a second step, we place metros in good coverage areas and determine HO rates and dropped call rates at high speed.

It is important that the driving time be long enough to determine statistical significance and to experience all the possible handover scenarios by covering a high density of locations within a drive test zone. For example, movement with respect to the metro should include both driving through areas where HO procedures will be executed and a grazing route where HO may not be necessary. Statistical significance can be estimated by assuming a binomial distribution for dropped call events and the density requirement for drive routes can be checked

Table 9.1 Default system parameters

network topology	LTE R8 macro: seven sites with three cells per site LTE R8 metro: twelve sites with one omni cell per site.
carrier freq./sys. BW	700 MHz and 2 GHz / 10 MHz
path loss models	We have used the following models unless otherwise specified: suburban model for 700 MHz macro: path loss (dB) = 119.65 + 38.63log10(d_km) metro: path loss (dB) = 130.57 + 40.0log10(d_km) suburban model for 2 GHz macro: path loss (dB) = 128.1 + 37.6log10(d_km) metro: path loss (dB) = 140.7 + 36.7log10(d_km) LOS (free space) model for 2 GHz path loss (dB) = 128.1 + 20log10(d_km) note: over all models path loss (dB) = max(path loss (dB), free space loss (dB))
site-to-site distance (ISD)	macro-to-macro 500 m or 1750 m, macro-to-metro 0 to maximum macro coverage distance
shadow fading	macro; $\sigma = 8$ dB, metro: $\sigma = 8$ dB, site-to-site correlation $\rho_s = 0.5$ shadow correlation distance $\rho_d = (25,50,100,200)$ shadow map area $4 \times \text{ISD}^2$ with $2^9 \times 2^9$ pixels
BS antenna	macro: 17 dBi antenna gain; pattern: horizontal 70° beamwidth and vertical 10° beamwidth 20 dB max. attenuation; 32 m height metro: 5 dBi antenna gain; omni; 8 m height
BS TX power	macro: 46 dBm, metro: 30 dBm
antenna configuration	2 Tx, 2 Rx (assume coherent gain only)
cable loss	3 dB macro 0 dB metro
penetration loss	0 dB
UE noise figure (NF)	10 dB
fading	No multi-path is modelled as the aim here is the transitions of the UE between lognormal slow fading states.
metro cell placement	12 metros at 0.5 ISD equally spaced in azimuth around the centre macro.
UE dropping	continuous driving one UE with random variation in azimuth uniformly distributed over a circular region of 0.8 × ISD radius cantered on centre macro eNB.
traffic model	There is no traffic modelled in this simulation; only the reliability of signalling is considered
scheduling algorithm	since there is no traffic modelled, no scheduling is applied as we look at fully loaded conditions.

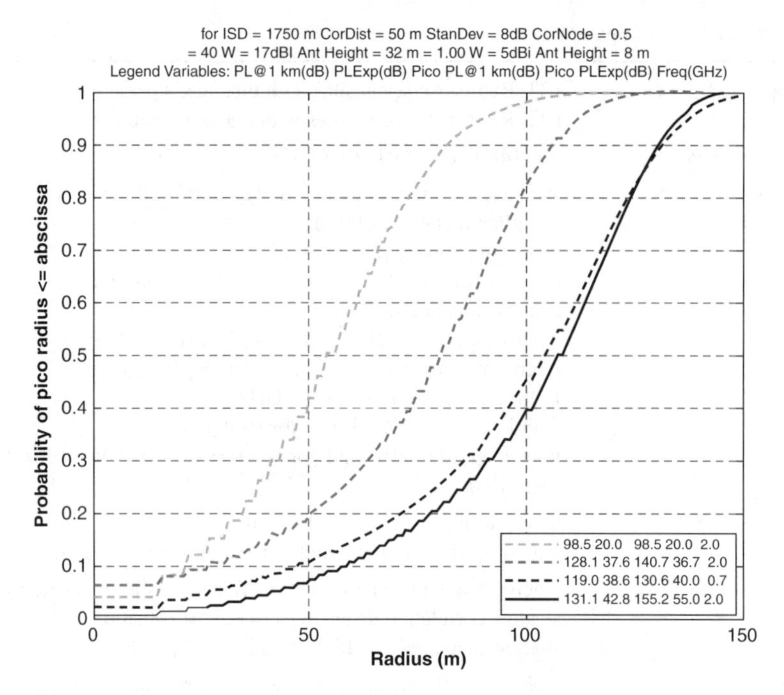

Figure 9.3 Metro cell coverage for all possible metro locations within 1750 m ISD macro coverage area and over 25 instances of shadow fading maps.

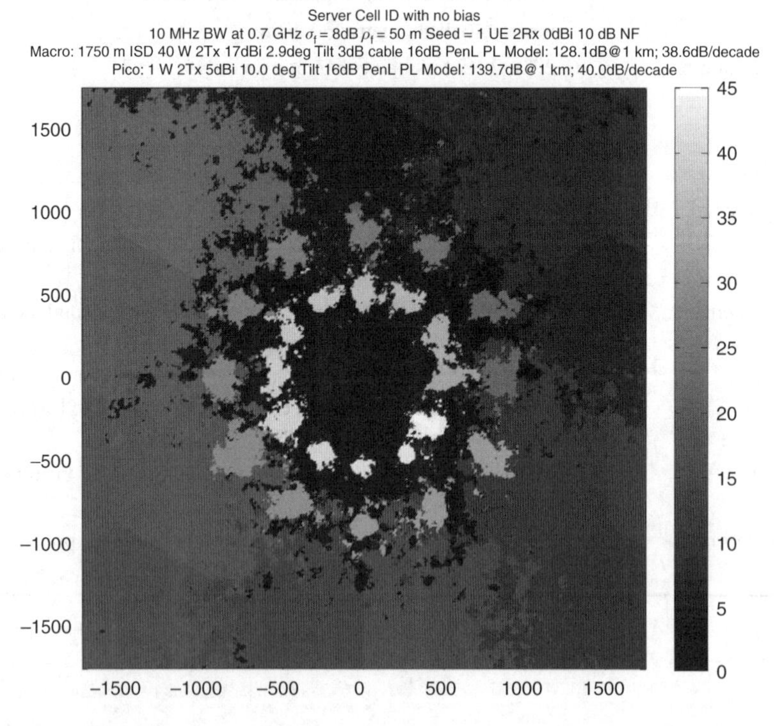

Figure 9.4 Strongest RSRP by cell number. Cells 1–21 are macrocells and 22–45 are metros. Cells 1–3 are in centre node. Metros are at 250 m from centre macrocell.

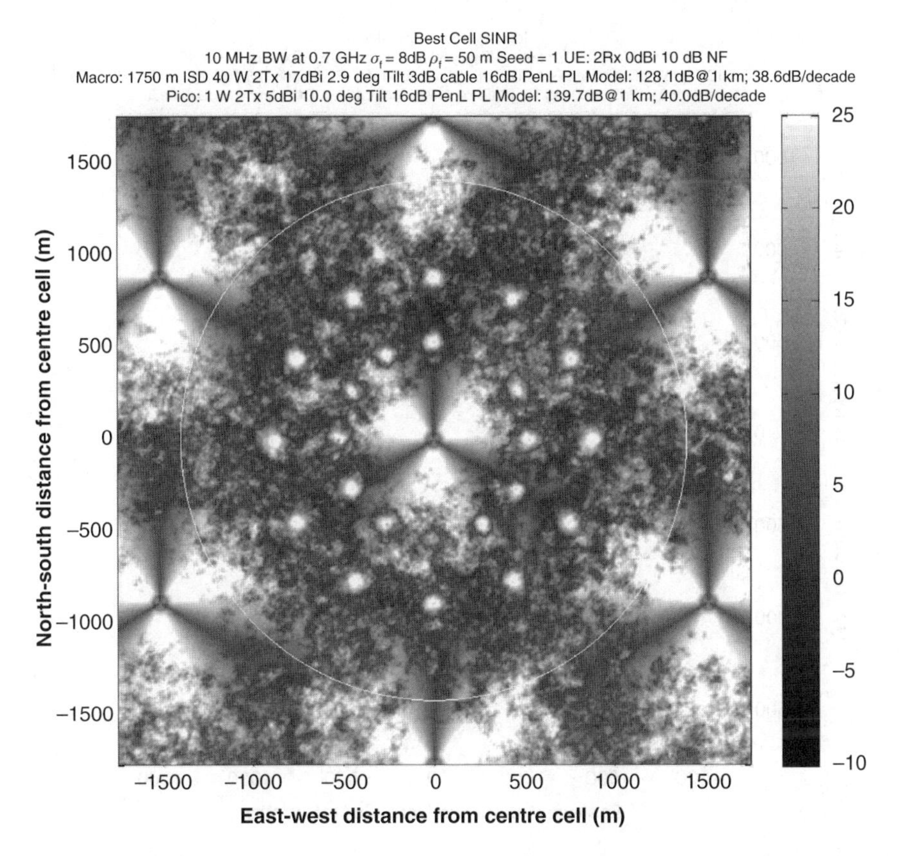

Figure 9.5 Best SINR calculated for two transmit antenna and two receive antenna.

by inspection of the drive route. If driving is restricted to the coverage area of one eNB, then the statistical significance requirement is probably sufficient. It is also necessary to test the metro placement for several different shadowing instances. Either many metros can be placed within one drive zone or one metro can be placed in one drive zone with the experiment repeated over several different instances of shadow fading. In this work, 24 metros are placed within an eNB coverage zone, with 12 at 0.3 ISD and 12 at 0.5 ISD and four instances of shadowing are done, resulting in 36 uncorrelated instances of shadowing for 0.3 ISD and 0.5 ISD. The drive zone used in this simulation is outlined by the white circle in Figure 9.5, and the obtained ideal handover locations are shown in Figure 9.6.

If it is desired to measure dropped call rates of between 0.8% and 1.2% with 95% confidence, then we would require 220 hours of driving time. If 11 hours are driven and no dropped calls are observed, then the dropped call rate is less than 1% with approximately 95% confidence. In the interest of shorter run times, typically the simulated drive time is 80,000 seconds or 22.2 hours. In addition, the UE drive vectors are not straight lines but include a small amount of random variation in azimuth. This is done to ensure that spatial uniformity is obtained.

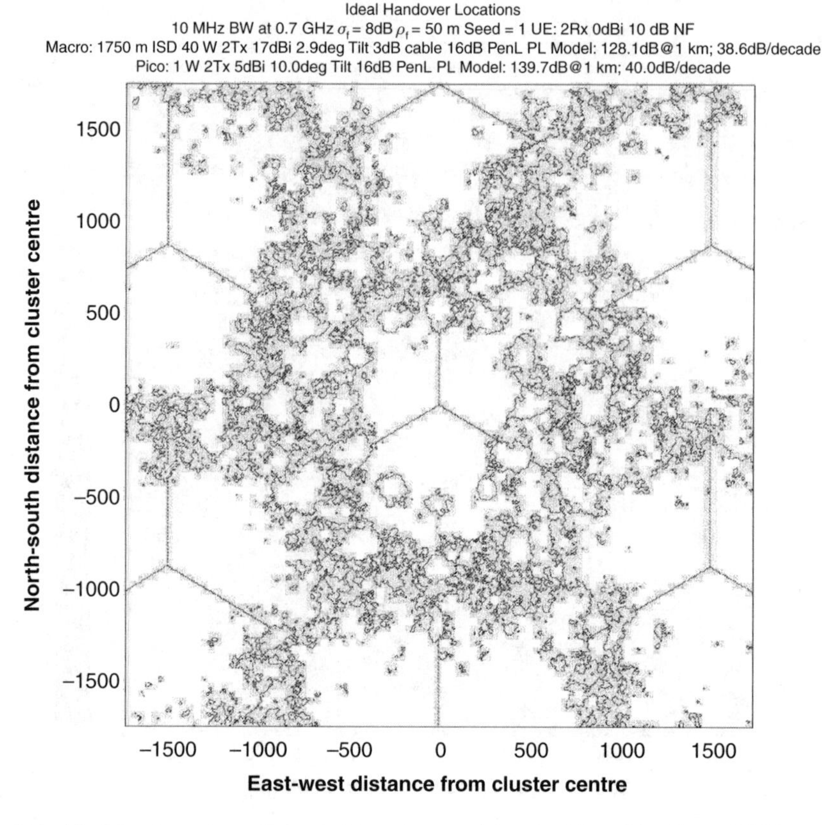

Figure 9.6 Ideal handover boundaries (without hysteresis) and without bias for an HTN system with 24 metro cells in the vicinity of one macro Node-B. Ideal boundaries reflect the largest RSRP.

To summarize, the important parameters for selecting the parameters of the drive routes are:

- the drive duration to obtain the necessary statistical significance,
- the number of independent instances of shadowing with respect to macro and the metros,
- the degree of drive path diversity through and near the metros, and
- the degree of uniform spatial distribution of UEs within the drive route.

Generally, satisfying the required drive times is the most challenging requirement. Also it is worth noting that in this study we implicitly consider uniform distribution of UEs, but non-uniform distributions are also important, and could be studied by setting up circular or rectangular subzones within the drive test zone that could contain additional UEs to model hotspots.

In addition, mixed mobility can similarly be modelled with a good percentage of the UEs being indoors and therefore pseudo-stationary. Notably, even for stationary UEs, environment motion can still cause non-zero HO rates, depending on the Doppler of the environment motion. Without considering the environment motion, the scenario described here would be a worst case scenario, since all the UE are moving in and out of the metro coverage very quickly.

In reality, some of the metros need to be embedded in the macros, and in such high interference zones ICIC and larger cell selection bias values must be used and optimized together with the handover algorithm parameters. In addition, multi-path fading and low-mobility or stationary UEs must also be included to find the speed-dependent tuning of time-to-trigger (TTT) and the other HO parameters.

9.4 Handover Modelling

In this section we detail the modelling of the event-triggered HO procedure that is initiated whenever the event-A3 entrance condition is satisfied, that is:

$$RSRP_{serving} - RSRP_{target} > HYS$$

for more than *TTT* seconds. The UE will then send a measurement report (MR) to the eNB and the eNB will initiate the handover.

The UE applies certain filtering to the measured RSRP quantities to remove multi-path fading, noise and any other high-frequency error terms. The filtering requirement adds the majority of latency in the UE and eNB handover process. The layer-3 filter processing done at UE specified in TS 36.331 [1] is:

$$F_n = (1 - a) \cdot F_{n-1} + a \cdot M_n,$$

where $a = (1/2)^{K/4}$, M_n = current measured value, $F_n = n^{th}$ filtered value, $F_{n-1} = (n-1)^{th}$ filtered value, $t_s = t_n - t_{n-1} = 200$ ms.

Although the filtering process is specified to be performed at a 200 ms sampling interval, the standard allows for other sampling intervals, as long as the filtering is of the same infinite impulse response (IIR) nature. Presumably, this means that either the bandwidth or the latency of the filter is the same. The filter mean latency, the average group delay more specifically is related to the 3dB bandwidth as follows:

$$\tau \approx \frac{1}{2\pi f_3}$$

In addition, the layer-3 filter alone is not able to remove the high frequency components of multi-path at high mobility due to aliasing, and this necessitates a layer-1 filter. Without the layer-1 filter the aliased responses of multi-path may look more like a superposed error process. So why would we want a lower layer-3 sampling rate if the mean latency is not affected? The reason is that with larger sampling rates there is a random component in the step response of the filter. The tail of that distribution would be worse for larger sample intervals and could increase the dropped call rates. This has been verified in other studies in the industry, with 100 ms sampling interval providing most of the improvement possible.

The next modelling decision is the layer-1 filter. The standard does not specify the layer-1 filter, and instead requires a given measurement accuracy on the RSRP. A good layer-1 filter would be a moving-average filter with window equal to the layer-3 sampling interval.

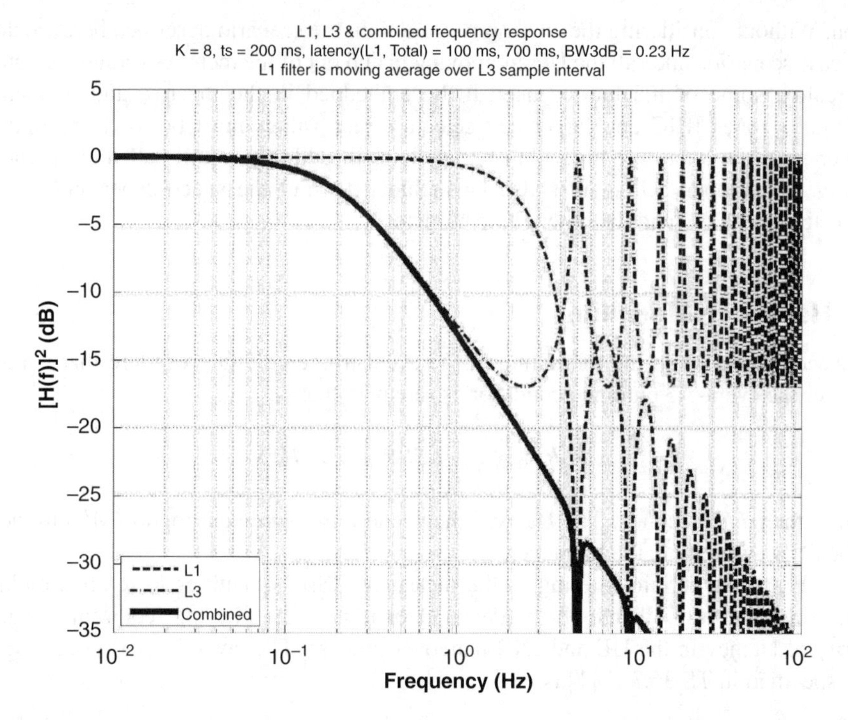

Figure 9.7 Filter responses of modelled layer-1 and layer-3 filters and combined response for 200 ms sampling interval (default in 3GPP specs) and K = 8. Other sample intervals are allowed with requirement that they have equivalent filtering to the 200 ms sampling interval case.

This means that the layer-1 filter consists of averaging all the per-TTI RSRP samples once per layer-3 sampling interval. This is the simplest processing possible that on the one hand provides attenuation of multi-path responses at all stop-band frequencies, and on the other hand, the layer-1 FIR filter places a transmission zero at the transmission pole of the layer-3 filter.

With 200 ms sampling interval and K = 4 the bandwidth of the layer-3 filter is approximately 0.55 Hz and the mean latency is 200 ms. The combined layer-1 and layer-3 latency is 300 ms. Figure 9.7 shows the frequency response of the layer-1, layer-3, and combined filters. Figure 9.8 shows K values of layer-3 filter for other sampling intervals the same bandwidth and latency. Figures 9.9 and 9.10 show the effect on K and the sampling interval on the latency and bandwidth of the combined filter.

Note that the RAN processing time in LTE – the time to send an HO message in response to the received MR – is assumed to be less than 50 ms and is such a short interval that it doesn't have much impact on the HO reliability. Consequently, it is not included in the simulation of this contribution. The biggest factor is the reliability of the HO control messaging on DL and UL and in particular the PDCCH grants that assign UL resources for the transmission of the MR. This simulation does not go into the detail of the physical layer and simply models a failure as having a DL SINR< -8 dB at the time the MR was sent.

Since this work is focusing on how the metros affect the HO rate and dropped call rate, rather than on optimizing parameters for HO, it is enough to provide results primarily for

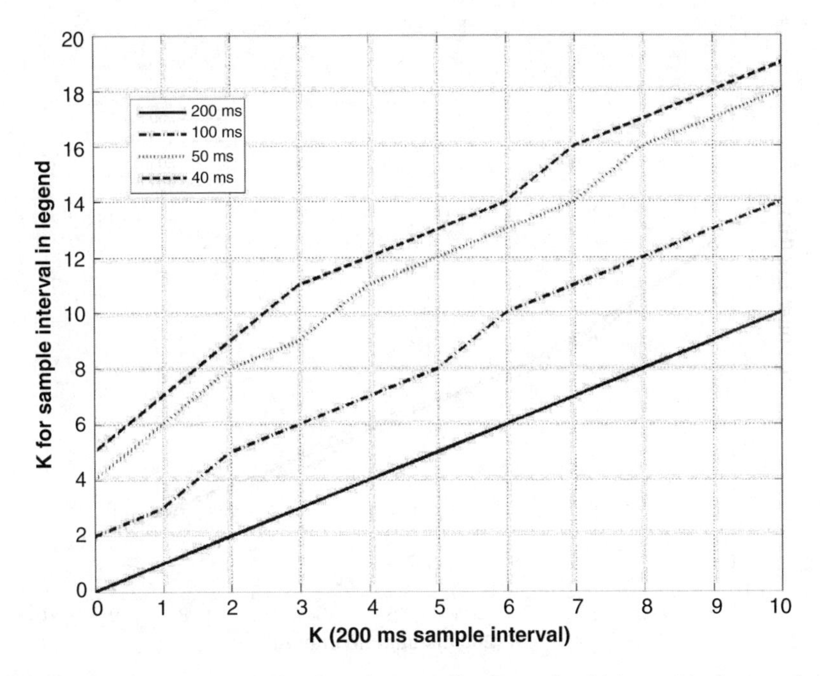

Figure 9.8 K values of the layer-3 filter for other sampling intervals which provide the same bandwidth and latency.

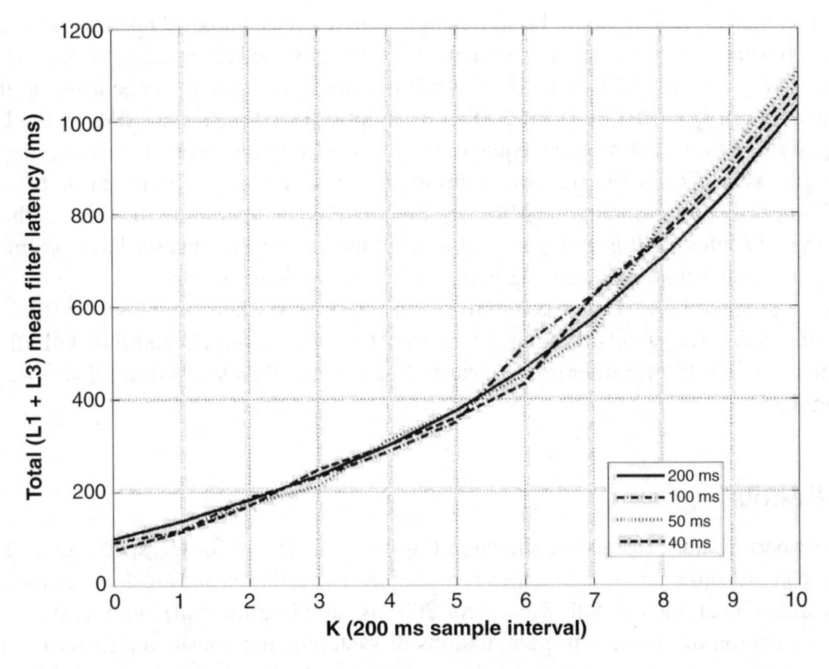

Figure 9.9 Combined filter latency for various K values and sampling intervals.

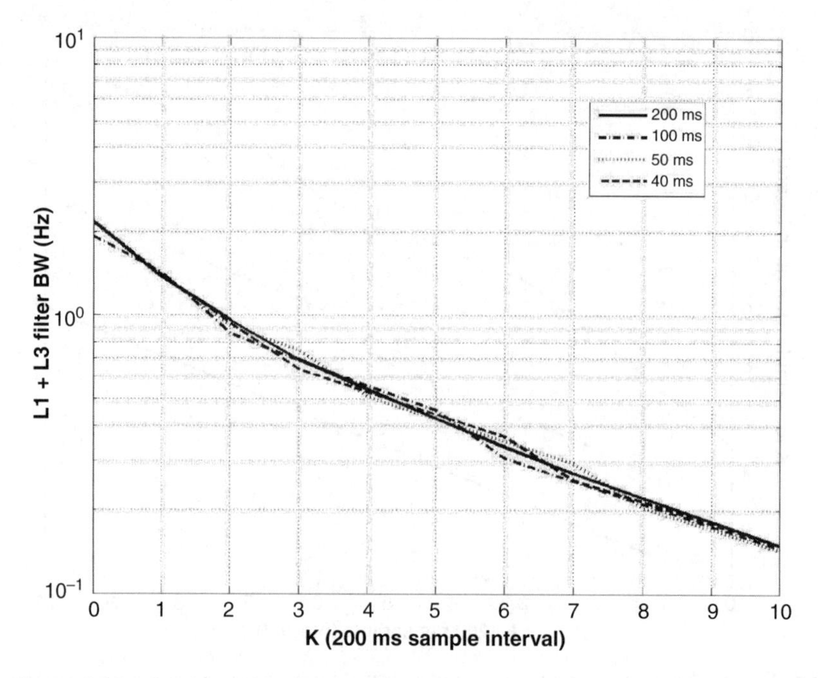

Figure 9.10 Combined filter bandwidth for various K values and sampling intervals.

default parameters, which should be sufficient to provide reasonable HO performance at high mobility. To summarize, the default parameters assumed for the HO model in the high mobility case are $TTT = 40$ ms, $HYS = 1$ dB, $t_s = 100$ ms (the exact value is dependent on the UE implementation), $K = \{2,4,6,8\}$ (for 200 ms assumption), and the layer-1 filter at the UE is a moving average filter with window equal to the layer-3 sampling interval. It is also important to note that when *TTT* is less than the sampling interval there is really no reason to wait for the *TTT* interval before sending the MR since there will be no updates to the layer-3 filter state during the *TTT* interval. It is ambiguous as to what the UE may do, but we have assumed that any *TTT* less than sampling interval is equal to zero in implementation.

Additionally, the layer-3 filter is initialized and runs only when neighbour RSRP $E_s/I_{ot} >$ *DetectThreshold*. The standard requires a $DetectThreshold < -6$ dB and this will affect the performance of the layer-3 filter in some cases. Some cases of $DetectThreshold = -12$ dB are also simulated.

9.5 Results

Handover performance results are shown in Figures 9.11–16, all for UE speeds of 120 km/h. Figures 9.11–13 show important metrics recorded from simulation and are done for the default system and HO parameters with $K = 5$ (for 200 ms sampling interval), with a 50 m shadow fading correlation distance, with permutations of system being considered (macro to metro, metro to metro, macro to macro and metro to macro), shadow fading standard deviation of 0 or 8 dB and metro power of 1 W (30 dBm) or 5 W (37 dBm). Figure 9.11 shows the HO

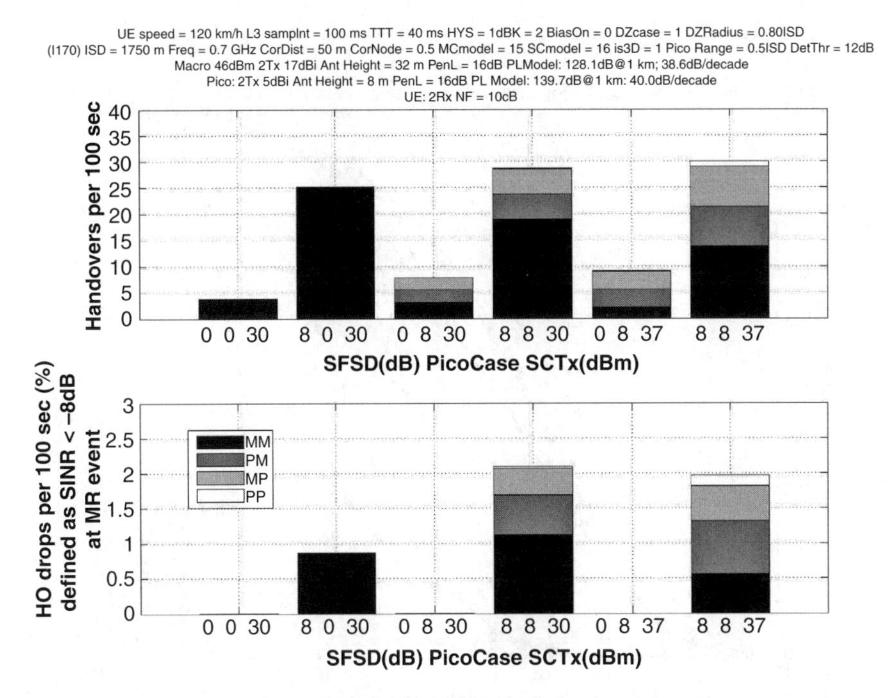

Figure 9.11 The presence of shadow fading has a much greater impact on handover rate and dropped call rate than does the presence of 24 metro cells in a Node-B coverage area. Variables in the legend (left to right) are shadow fading standard deviation, system type (0 to macro; 8 to HTN) and metro cell transmission power in dBm.

rate and dropped call rate and Figure 9.12 shows the observed server dwell time CDF. It can be seen that the lowest tails of server dwell time distribution are mostly for macro to macro handovers.

Figure 9.13 shows the geometry distributions for the four system permutations considered. It is clear in all of these metrics (HO rate, dropped calls, dwell times and geometry) that shadow fading is the single most important parameter which degrades performance and more than the presence of metro cells. For this reason, results in Figures 9.14 and 9.15 show performance versus shadow fading correlation distance with shadow fading correlation distance constant at 8 dB. It is clear that even a small variation of correlation distance has a large impact on dropped call and handover rates. This is a very important observation, since although 3GPP typically uses 50 m correlation distance, probably as a typical worst case, in real field scenarios the correlation distance is believed to vary over a large range, with dense urban having the lowest correlation distances of 10 m to 25 m and rural with US inter-state highways exhibiting 100 m to 1000 m correlation distance. Additionally, recognizing that mobility should be lowest in dense urban and highest in rural environments, designing for the highest mobility in the dense urban areas with low correlation distances may result in extreme over-design. Figure 9.14 shows also the impact of neighbour detection threshold. It appears that the minimum 3GPP requirement may not be ideal in the case of $K = 2$. Figure 9.14 shows that a small adjustment in K is sufficient to provide HTN performance as good as macro system.

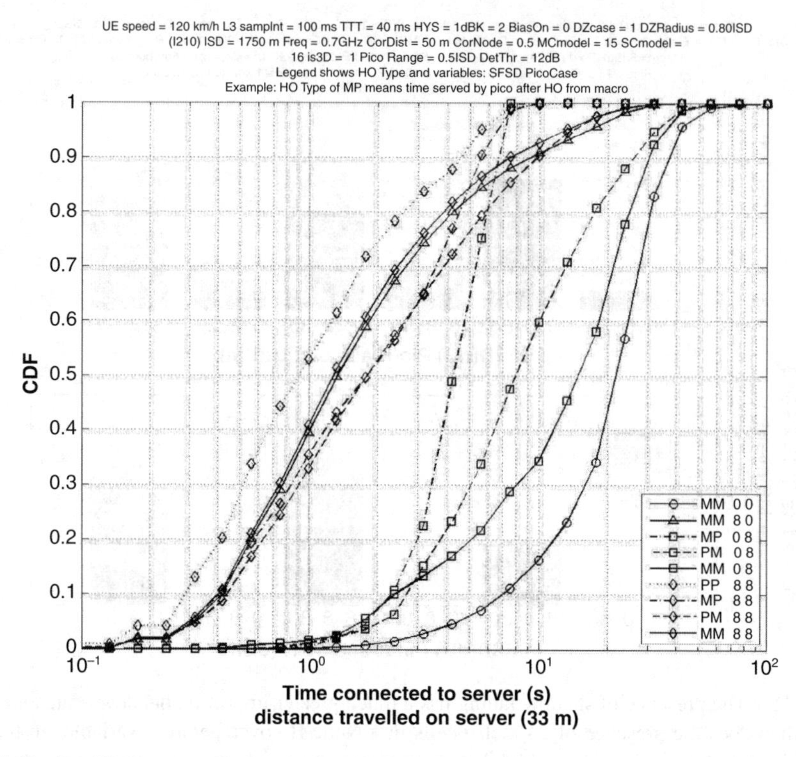

Figure 9.12 Dwell times on servers are dominated by shadow fading and not by metro cell presence.

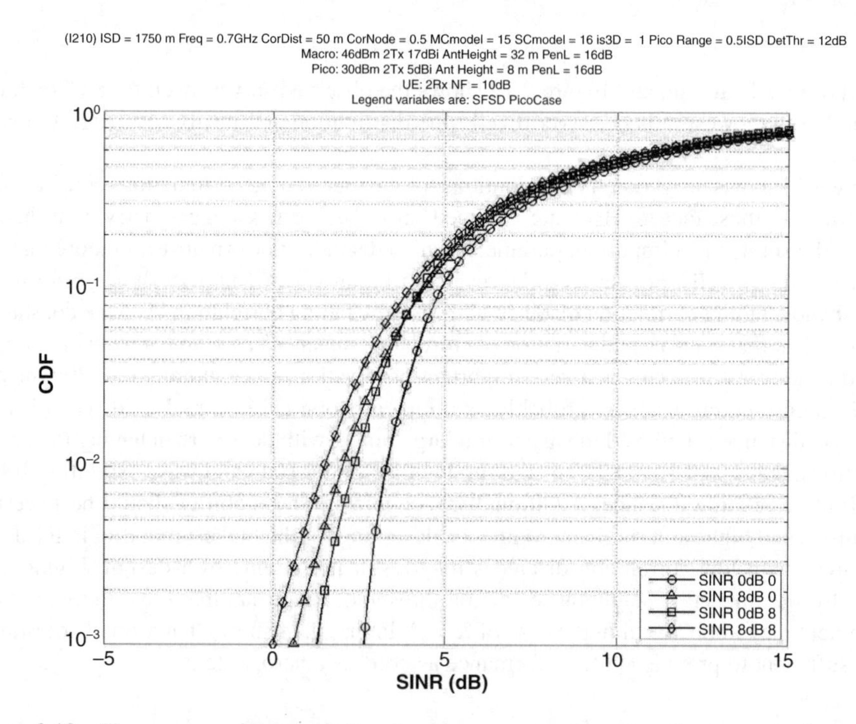

Figure 9.13 The presence of shadow fading degrades geometry more than the presence of metro cells.

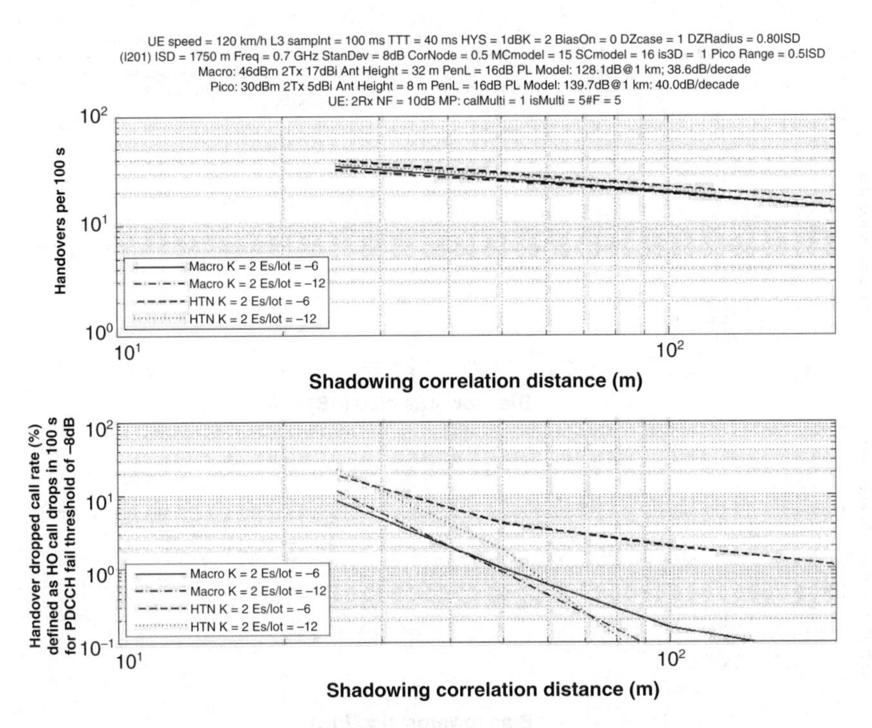

Figure 9.14 Impact of shadow fading correlation distance and neighbour detection threshold on macro and HTN performance.

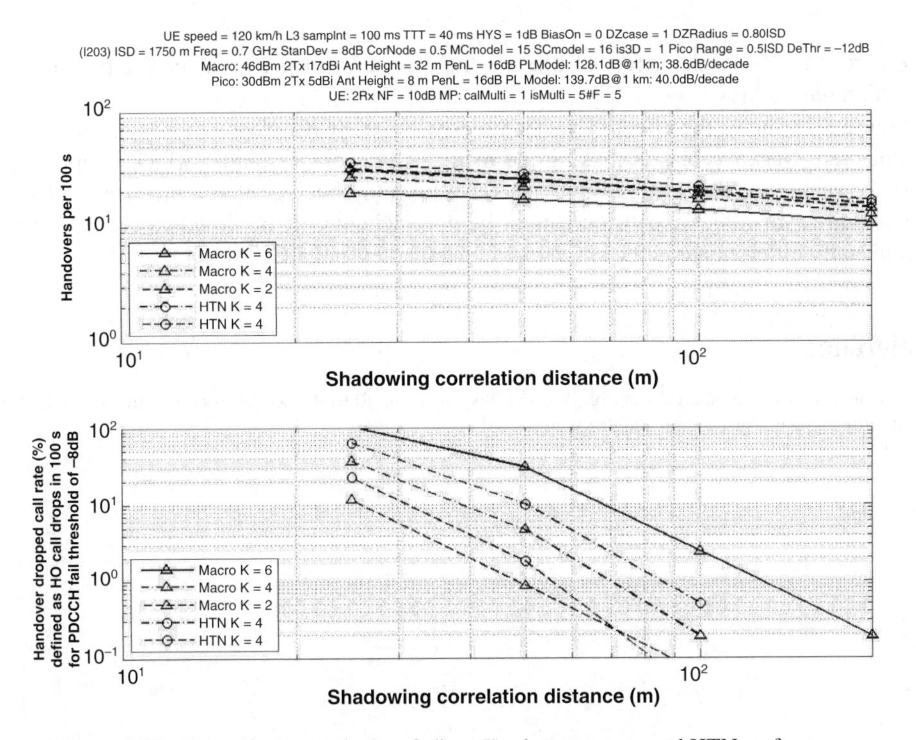

Figure 9.15 Dependence on shadow fading, *K* value on macro and HTN performance.

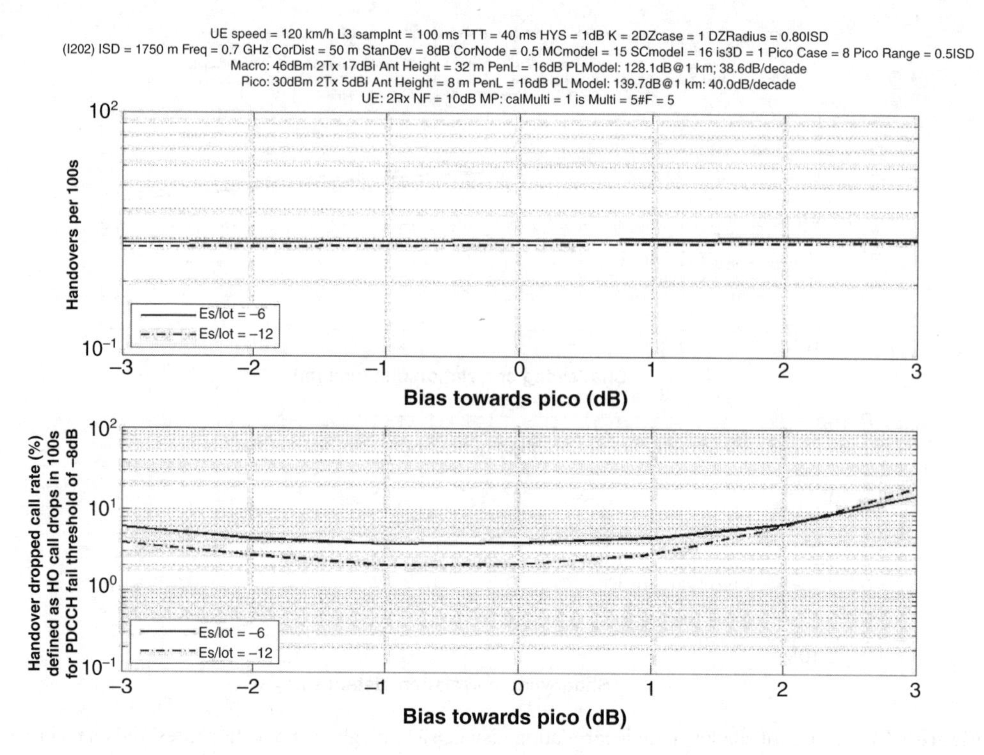

Figure 9.16 Effect of bias on dropped call performance for two detection thresholds.

Figure 9.16 shows the effect of bias on the dropped call performance for two different detection thresholds.

Figure 9.12 shows the effect of varying bias without eICIC. Negative bias of 1 dB is best in this case. Negative bias is making the metro cell coverage smaller which is opposite to what is wanted for range expansion. Thus, we have to conclude that a practical amount of bias required for range expansion cannot be done without eICIC. Vendors are expected to significantly differentiate in the space of HO parameters and eICIC design.

Reference

1. TS 36.331 Radio Resource Control Protocol Specification, Release-specific versions are available from: ftp://ftp. 3gpp.org/specs/html-info/36331.htm

10

Cell Selection Modes in LTE Macro–Femtocell Deployment

Meryem Simsek and Andreas Czylwik
University of Duisburg-Essen, Germany

10.1 Introduction

Long Term Evolution (LTE) is the next major step in mobile radio communications and was introduced in 3rd Generation Partnership Project (3GPP) Release-8. In addition to LTE, the 3GPP has also defined IP-based flat network architecture. This architecture is defined as part of the System Architecture Evolution (SAE) effort. Since the LTE/SAE is based on a flat all-IP architecture, the architecture and interfaces are the same for femtocells (HeNBs) as for macrocells (eNBs). LTE femtocells require no new interfaces to be defined and no changes are required to the Evolved Packet Core (EPC), which is the main element of the LTE/SAE network. The EPC is defined in 3GPP Release-8 and is integral to the industry drive towards flatter, all-IP networks. The EPC supports local breakout of traffic whether a roaming subscriber is accessing the EPC via a 3GPP-based or a non-3GPP-based access network according to the design principles described in [1]. Specific EPC components include the Mobility Management Entity (MME) in the control plane and the Serving Gateway (SGW) and Packet Data Network Gateway (PGW) in the bearer plane. The MME is the control node that processes the signalling between the User Equipment (UE) and the Core Network (CN). The protocols running between the UE and the CN are known as the Non Access Stratum (NAS) protocols. Among other functions, the NAS control protocols handle Public Land Mobile Network (PLMN) selection, tracking area update, paging, authentication and the Evolved Packet System (EPS) bearer establishment, modification and release. The NAS interacts with the Access Stratum (AS). The control plane of the AS handles radio specific functionality. The applicable AS-related procedures largely depend on the Radio Resource Control (RRC) state of the UE, which is either RRC_IDLE or RRC_ CONNECTED.

Heterogeneous Cellular Networks, First Edition. Edited by Rose Qingyang Hu and Yi Qian.

Heterogeneous network deployment applies to deployment scenarios where macro eNBs are deployed with other low power nodes in an overlay manner to provide additional capacity needs, especially in indoor scenarios. These non-macro eNBs could be picocells, relays or femtocells. The discussions on topics related to heterogeneous networks are ongoing, with numerous contributions studying the feasible gains of heterogeneous deployments. Gains in system throughput through coverage extension as well as capacity enhancements have been shown using preliminary simulation studies. Factors such as co-channel deployment scenarios or different transmission powers make it essential to mitigate interference on heterogeneous network deployments in order to reduce their impact on system performance. A network operator would find it desirable for the overlay macrocellular network not to experience any degradation in throughput or performance due to the deployment of HeNBs. This can be accomplished by different interference reduction schemes such as frequency selective scheduling, power control and partial frequency reuse or by introducing different cell selection mechanisms.

A UE in RRC_IDLE state performs cell selection and cell reselection, that is, it decides on which cell to camp depending on its preferences, authorities and access mode. The cell (re)selection process takes into account the priority of each applicable frequency of each applicable Radio Access Technology (RAT), the radio link quality and the cell status. Before performing cell selection, it is essential that a UE is able to distinguish macrocells from femtocells and whether the targeted femtocell is accessible or not. Additionally, access control mechanisms, including their specific parameters – which were introduced in 3GPP LTE Release-8/9 – have to be taken into account. It is important to note that access control is invoked whenever a UE is trying to camp on a femtocell in order to prevent the unauthorized use of that femtocell. Depending on this information, a UE will perform cell selection when it is first powered on or after having lost its previous coverage or while camping on a cell but will need to reselect a better cell.

The distinction of cells, the access modes and cell (re)selection procedures will be detailed in the following sections.

10.2 Distinction of Cells

Due to the network architecture and the deployment characteristics of femtocells, it is impossible to treat femtocells like macrocells. Signalling overhead between macrocell UEs and femtocells will impact the overall network performance. Hence, identifiers and mechanisms are necessary to reduce this impact on network performance. Especially, in the case of cell (re)selection it is essential to distinguish femtocells from macrocells, so that mobility procedures can be enhanced. These identifiers and mechanisms are introduced in 3GPP in order to prioritize the cells for UEs to camp on once UEs enter the coverage area of the allowed femtocell. Additionally, these parameters make it easier to prevent the UEs that are not allowed to connect to a femtocell from scanning or reading the information from the femtocell layer, and this leads to a longer battery life and shorter cell selection periods. Besides, the ability to distinguish femtocells from macrocells it is also important to enable UEs to find and distinguish among neighbouring femtocells during cell (re)selection and mobility procedures. Here, a list of parameters is given that a UE can use to distinguish femtocells from macrocells or neighbouring femtocells from each other.

Hierarchical cell structure

A Hierarchical Cell Structure (HCS) is a system where different cell types and sizes, such as femto-, pico- and macrocells, can coexist. These cell types are called layers/tiers. Within hierarchical cell structures, cells in different layers can constitute different localized service areas. In 3GPP LTE, HCS can additionally be used as a parameter to better distinguish femtocells from macrocells. Assigning each layer a different priority will allow different cell (re)selection criteria and thus more flexibility for femtocells. Using HCS it is possible to set cell (re)selection parameters so that UEs that camp on macrocells can automatically find femtocells and UEs that camp on femtocells will prioritize these cells.

Separate HeNB PLMN ID

In 3GPP it is proposed that HeNBs would have separate PLMN IDs that are distinct from those of the eNBs. These different PLMN IDs are another parameter to distinguish femtocells from macrocells [2]. The advantage of this approach is that UEs that are not allowed to get access to an HeNB could be configured not to access the HeNB PLMN ID, which will result in a better battery performance and less signalling load towards the CN. This approach will also tell the UE that it camps on an HeNB and will allow this UE to display the right network identifier. However, this will cause some compatibility problems with some older Subscriber Identity Module (SIM)/Universal Integrated Circuit Cards (UICC). Thus, an update of SIM/UICC is required. A further drawback of using separate PLMN IDs for HeNBs is that operators might not have additional PLMN IDs. Thus, the introduction of these additional PLMN IDs will be costly.

Reserve frequency and PCI

Radio networks need to handle non-unique local physical identifiers of cells to support efficient measurement and reporting procedures. In LTE each cell must be assigned a Physical Layer Cell Identity (PCI). It is important that the allocation of 504 PCIs is free of conflict and confusion, so that a UE can properly detect reference signals and discover neighbouring cells. These 504 PCIs may be split into subgroups for eNBs and HeNBs separately in order to simplify the introduction of new cells (e.g., activation of new HeNBs) in one layer without impacting other layers. However, due to the small cell size of HeNBs, multiple HeNBs with the same PCIs can be within the coverage of a common source eNB. This will lead to a PCI confusion, wherein the source eNB is unable to detect the correct target cell. PCI confusion is solved by the UE by reporting the global cell identity of the target HeNB.

In LTE, a range of PCI values is reserved by the network for Closed Subscriber Group (CSG) femtocell use [3]. These are cells that provide access to a limited set of authorized UEs. The PCIs reserved for CSG cells are signalled in the Information Element (IE) *SystemInformation-BlockType4* (SIB4). SIB4 contains neighbouring cell-related information for intra-frequency cell (re)selection and indicates the set of PCIs that are reserved for CSG cells. This field is optional for non-CSG cells and mandatory for CSG cells in LTE.

In the case of hybrid cells, which were introduced in 3GPP LTE Release-9, the network can reserve similarly to CSG cells a list of PCIs in order to distinguish hybrid cells.

UE autonomous search

The introduced parameters enable the UEs to distinguish HeNBs from eNBs. However, in order to effectively find HeNBs, further functions have been presented in 3GPP Release-8.

The autonomous search function is introduced for the UE to find CSG cells [4–6], which supports a CSG cell manual search and selection procedure during both UE idle mode and connected mode. To detect the allowed CSG cell(s), a UE applies an autonomous search function, regardless of which RAT the UE is camping on. The autonomous search function may be used only on the frequencies that are listed in the system information and in case of dedicated CSG frequencies the autonomous search function may be used on the dedicated frequencies.

If the CSG whitelist – the list of CSG IDs for CSG cells maintained by a UE/Universal Subscriber Identity Module (USIM) – does not exist or is empty, the autonomous search procedure is disabled.

In the case of hybrid cells, 3GPP Release-9 proposed that UEs should use the autonomous search function to detect at least the hybrid cells previously visited, whose CSG IDs and associated PLMN IDs have been stored in the UE's CSG whitelist. In this case, the UE should treat the detected hybrid cells as CSG cells.

Finding and selection of allowed cells

The introduced parameters and mechanisms like HCS, separate HeNB PLMN ID, reserve frequency and PCI and UE autonomous search, enable the UE to distinguish HeNBs from eNBs. Using these parameters and mechanisms, a macrocell UE can avoid trying to get access from a femtocell and femtocell UEs can favour access from femtocells over macrocells. However, in the case of closed access femtocells, all femtocell UEs will try to camp on the target femtocell regardless of whether they have the authority to access it or not. Hence, methods are necessary to infer whether a femtocell is accessible or not. Using these methods (such as usage of CSG IDs), unnecessary signalling overhead can be avoided.

A CSG ID is a unique identifier within the range of a PLMN. It identifies a CSG in the PLMN associated with a CSG cell or a group of CSG cells [7]. The CSG ID has a fixed length of 27 bits and is a field in *SystemInformationBlockType1* (SIB1). SIB1 contains information that is relevant when evaluating whether a UE is allowed to get access to a certain cell or not. According to 3GPP Release-9, SIB1 has the following two fields: CSG ID and CSG-Indication. The CSG ID is an identity of the CSG within the PLMN which the cell belongs to. If CSG-Indication is set to TRUE the UE is only allowed to access the cell if in addition the CSG ID of the cell exists in the whitelist of the UE.

For closed access cells the CSG-Indication and CSG ID are both set to TRUE. Hybrid access cells have both CSG-Indication and CSG ID set to FALSE. Open access cells do not have any CSG IDs. The CSG-Indication is set to FALSE for them.

Using the CSG-Indication and CSG ID a femtocell UE will be able to find out whether a femtocell is accessible or not. The UE will not try to camp on a femtocell if it does not have the authority to access this cell by comparing the CSG ID against the entries in its whitelist.

Allowed list in CN

After distinguishing femtocells from macrocells and finding out the accessible femtocells, it needs to be decided whether a UE is allowed to access a target femtocell or not. In 3GPP Release-8, an Allowed CSG List (ACL) has been provided by NAS to provide all CSG identities and their associated PLMN IDs of the CSGs to which the subscriber belongs. The NAS forms the highest stratum of the control plane between UE and MME at the radio interface. The main functions of the NAS are to support the mobility of the UEs and to support the session management procedures to establish and maintain IP connectivity between the UEs and PGW. In 3GPP Release-9 the name of ACL provided by NAS has been changed to CSG whitelist.

In [8] the conflict of the name 'allowed CSG list' that is provided by AS to NAS has been resolved. In AS, the allowed CGS list includes all CSG IDs of the cells to which the UE is authorized. In NAS, on the other hand, the permitted CSG IDs are shared in an operator CSG list and an allowed CGS list. However, a combination of the operator CSG list and the allowed CSG list will be provided by the NAS to the AS. The combined list received by the AS should have a different name from the two constituent lists maintained by NAS. Therefore, the name of this combined list has been changed to CSG whitelist, and its contents are provided to the AS by NAS. The CSG whitelist may be stored either in the USIM or in the Mobile Equipment (ME), depending on the USIM capability [9].

The UE will maintain a list of allowed CSG IDs. According to [10] it will be possible for both the operator and the UE to modify the ACL. In case of manual CSG selection, the UE will be allowed to introduce new CSG IDs to the ACL. The UE will also maintain an operator controlled list of allowed CSG IDs (operator CSG list). The modification of this list will be possible for the operator. These two lists are independent from each other, so that a change in the operator CSG list should not trigger the UE to modify the ACL.

In the case of manual CSG selection the UE will perform a scan for available CSGs, their CSG IDs and their HeNB names. The UE should be informed of all the CSGs that have an entry in the ACL or operator CSG list. The available CSGs will be displayed in the UE in the following order [10]:

1. CSGs with identities contained in the ACL,
2. CSGs with identities contained in the operator CSG list, and
3. CSGs whose identities are not included in the ACL or in the operator CSG list.

If the UE has been accepted via an attachment procedure by a CSG cell, the UE will check whether the CSG ID and the associated PLMN ID of the cell are in the ACL or not. If it is not in the ACL, the UE will add that CSG ID and associated PLMN ID to the ACL. The UE may also add the HeNB name to the ACL if it is provided by the lower layers.

In the case of hybrid cells, the UE will be informed whether it is a member of the hybrid cell or not. If it is a member of the hybrid cell, the UE will add the CSG ID of this cell to its ACL.

10.3 Access Control

In a macrocell network, access control takes place either during Location Area Update (LAU) and Tracing Area Update (TAU), or when a UE requests data transmission services. However, in a mixed femtocell–macrocell deployment scenario, access control has to be done whenever a UE tries to get access from a femtocell in order to prevent unauthorized access; that is, access control is defined as the process that checks whether a UE is allowed to access and to be granted services in a closed cell [10].

Since interference remains one of the major problems in heterogeneous networks, the impact of access control for femtocells is particularly significant in the mixed femtocell–macrocell deployment scenario. The access control method is also an important concern for femtocell customers, because they pay for the femtocell themselves and also for the broadband internet connection being used for backhaul. So the question arises as to whether all the UEs that are within a femtocell's coverage should have the right to access the femtocell or not, if not all of them pay for the service. To address that, there are basically three access control modes in which a femtocell could be operated: open, closed and hybrid. In the following, further details on these access modes are provided.

10.3.1 Access Control Scenarios

Access control by mobility management signalling

One of the common assumptions is that access control will be performed within LAU/TAU if it is initiated by the UE [11]. In this access control scenario, each HeNB is assigned an HeNB specific Location Area Identity (LAI), which differs from the eNB's LAI. When UEs carry out inbound mobility, access control can be initiated so that non-subscriber UEs will receive a negative response upon location registration, which means that these UEs are not allowed to camp on this HeNB. Reject causes for these UEs could be either that the location area is not allowed or that roaming is not allowed in this location area.

An advantage of this access control scenario is that it is an immediate and a quite simple solution, but this method has a high signalling overhead when the UE is at high mobility. Another side effect of using LAU rejects is that a rejected UE will have to wait for 300 seconds before it can try to reselect another target HeNB. The only alternative to reduce the time of the out-of-service or limited camped state is that there should be always a non-HeNB frequency layer.

Access control by redirection and handover

Allowing non-subscriber UEs to use an HeNB for roaming and camping is an alternative approach [2]. Here, access control will be done by redirecting or handing over these UEs to an eNB when receiving data transmission service requests. In this approach, a number of HeNBs are configured in the same Location Area (LA), so that the allocation of LAIs for HeNBs can follow the same procedure as that for eNBs. While the coverage of an HeNB is approximately constant within such an LA, no LAU is necessary when the UE is moving from one HeNB to another HeNB.

An advantage of this access control scenario is the reduced mobility management signalling and consequent battery use, especially in dense HeNB deployments. The drawback of this method is that when there is no eNB coverage, redirection and handover might not be possible, which will lead to radio connection failures. A possible solution to handle the access control for the HeNBs that are out of eNB's coverage is to use mobility management procedure.

10.3.2 Access Control Executor

In 3GPP LTE, access control executors are defined depending on available information and the current scenario. Access control can be done by HeNB Gateway (HeNB-GW), CN/MME and UE.

Since femtocells are normally of very small size, the handover frequency has to be much higher than that in macrocells. This can potentially cause a significant ping-pong effect and delay to the overall UE handover procedure which, in the worst case scenario, could lead to the UE dropping the ongoing call. Therefore, the following ideas have been considered when performing access control [14]:

- early rejection in the case of non-allowed UEs;
- minimum involvement of the CN;
- minimum signalling;
- avoidance of changes in existing system architecture, if not absolutely necessary; and
- security aspects of validating information.

In 3GPP LTE, access control is performed either in case of UE registration or in case of inbound mobility from macrocell to femtocell [12]. The same procedures for inbound mobility are also applicable to the handover between HeNBs under the same HeNB-GW. Depending on the type of target cell (CSG, open or hybrid cell – see Section 10.3.3 for further details), the node that performs access control could be different.

For a CSG UE that wants to handover to a CSG cell or a hybrid cell, initial access control is performed by the UE itself in order to accelerate the mobility procedure. Figure 10.1 shows access control steps performed by the UE [12]. Since the CSG UE can distinguish whether access to the target femtocell is allowed or not, the UE can either exclude home cells which are not in its ACL in the RRC measurement report or can report all home cells but indicate which are available, so that it can avoid attaching to a femtocell that is not in its ACL. Access control requires the UE to frequently read and decode the System Information Block (SIB) of the target femtocell candidate containing CSG ID. A drawback of UE access control is that the UE might not be synchronized with the ACL and the UE may try to access a cell for which the CSG subscription has expired, so that the network cannot completely trust the information reported by the UE. An additional access control has to be performed in the network by the CN. In addition, as the network cannot trust the CSG ID reported by the UE completely, a

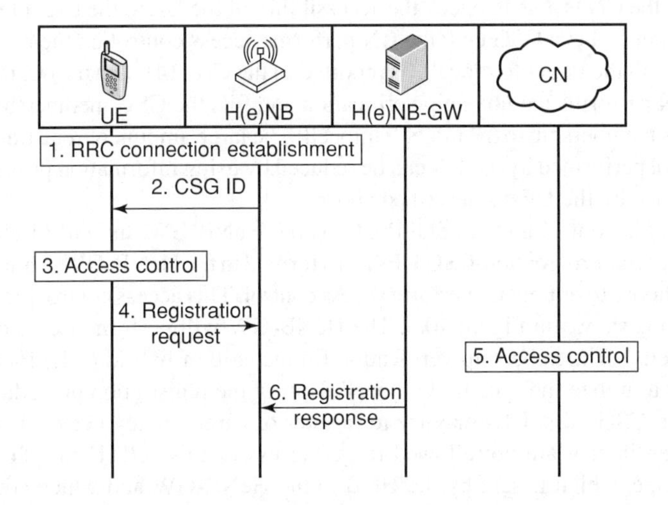

Figure 10.1 Access control performed by UE during UE registration.

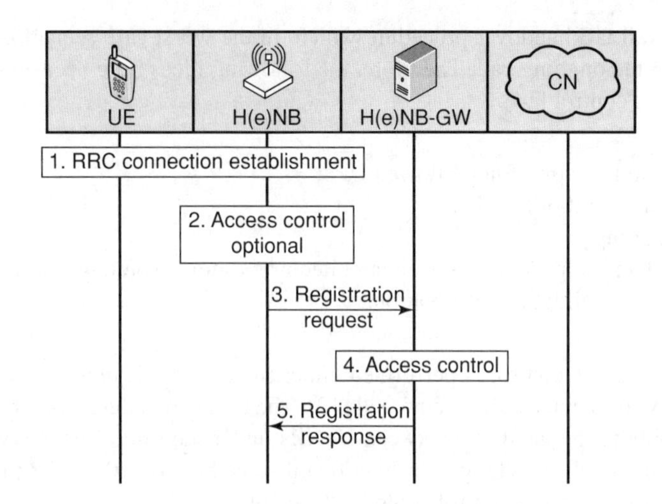

Figure 10.2 Access control performed by HeNB-GW during UE registration.

CSG ID check at the target femtocell is needed. The network should pass the CSG ID reported by the UE to the target femtocell. If there is no match, the target femtocell should reject the access of that UE.

In 3GPP LTE Release-8, the CN performs access control for CSG UEs trying to access the network through an LTE CSG cell during attach, combined attach, detach, service request and TAU procedures. The principle is the same as that applied to the access control of UTRAN CSG cells. However, in this case, Mobile Switching Centre (MSC)/visitors location register (VLR) and serving GPRS support (SGSN) perform access control [13]. In Figure 10.2 the access control steps performed by the CN during UE registration is depicted [12]. It is essential that the HeNB-GW forwards the initial UE message, including the CSG-ID of the femtocell to the CN, so that the CN is able to check the accessibility of the UE to the target femtocell using ACL. If the target cell is a CSG cell, the CN performs access control on the basis of the CSG ID associated with the target femtocell, as reported to the CN [14]. Otherwise, if the target is a hybrid cell, CN performs membership verification and fills the CSG membership status IE to reflect the UE's membership to the target femtocell. The large amount of registration signalling in access control performed by the CN can be reduced by using information provided by access control performed by the UE, as described above.

The third access control node in 3GPP LTE is the HeNB-GW. In 3GPP Release-8, it was agreed that access control for non-CSG UEs is performed in the HeNB-GW, while the femtocell (HeNB) may choose to optionally perform access control. This access control procedure during UE registration is shown in Figure 10.3. The HeNB-GW will perform access control (in the case of CSG cells) or membership verification (in the case of hybrid cells) for the particular UE attempting to utilize the specific femtocell. During the registration procedure or inbound mobility of non-CSG UEs, UEs may try to register or camp on any detected cell (including femtocells) even if they are not allowed to get access of this cell. Using the UE REGISTER REQUEST message, which is sent by the HNB to the HeNB-GW and which contains the UE identity, UE capabilities and also the registration cause [12], the HeNB-GW will be able to

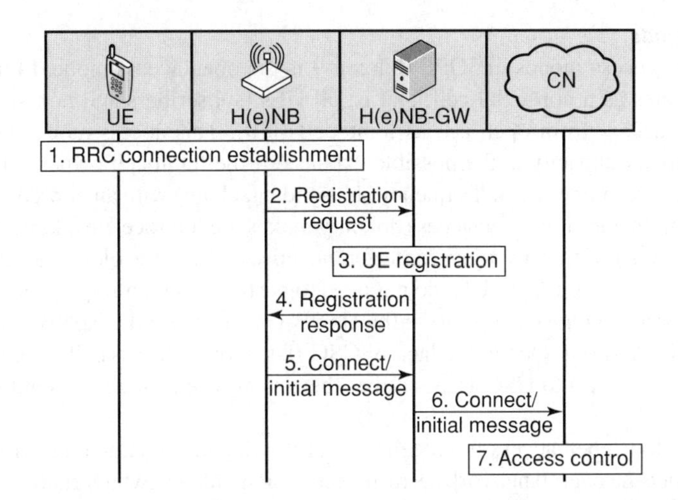

Figure 10.3 Access control performed by CN during UE registration.

perform access control for non-CSG UEs and for all UEs trying to access non-CSG cells. The advantage of locating the access control in HeNB-GW is that it can significantly reduce signalling overhead to the CN. A drawback, however, is that the HeNB-GW needs to call the ACL from the CN.

10.3.3 Access Control Mechanisms

In 3GPP Release-8, femtocells are fully closed access cells. In 3GPP Release-9, two additional access mechanisms have been introduced: open access cells and hybrid access cells. In the following, a brief summary of each access control mechanism is given.

Closed access mode

In closed access mode femtocells, also known as CSG cells in 3GPP, only femtocell subscribers are allowed to connect to their corresponding femtocells. CSG cells are expected to be widely used in individual home deployment. In this case, the list of registered UEs is decided by the femtocell owners, because it is not in the interest of femtocell owners to share their cells, because of the limited resources of the backhaul.

Access control should always be performed whenever a UE tries to camp on a femtocell. In urban scenarios including CSG cells, non-CSG UEs will be rejected by the CSG cells. This may cause a strong interference between femtocell and macrocell, or so-called cross-layer interference [27]. Cross-layer interference is still an important concern to be addressed.

In order to register a non-CSG UE in the HeNB-GW and find out its International Mobile Subscriber Identity (IMSI), the femtocell performs an identity request. After performing access control for that non-CSG UE, the HeNB-GW may either accept or reject the UE to camp on the femtocell.

From the business point of view using CSG cells is the favourite method for femtocell owners, because most of the customers will only be interested in paying for a femtocell if they are sure that they will have full control over the list of authorized users.

Open access mode

One of the new access modes in 3GPP Release-9 is the open access mode. In this mode, the femtocell operates as a non-CSG cell, that is, all UEs (subscribers and non-subscribers) are allowed to get access from open access femtocells [10]. This access mode should increase the overall network capacity at the possible expense of the femtocell owner, who must share her/his femtocell resources (time/frequency slots and backhaul) with an unpredictable number of cellular users. In addition, open access not only makes the femtocell available to more users, but also avoids most of the cross-layer interference discussed in the closed access mode.

In this access mode non-CSG UEs do not need to perform an identity request to find out the UE's IMSI in order to register this UE in the HeNB-GW, because the HeNB-GW will always accept any access request. Due to the lack of CSG IDs in open access cells, the femtocell will not include any CSG ID to CSG UEs, so that there is no need for access control in the CN, either.

One drawback of open access femtocells is that it will cause a large number of handovers, which will reduce the overall network performance. Soft handover, which enables simultaneous connection of a UE to several cells, is not supported in femtocells. Instead, hard handover is supported. Therefore, a UE will always handover between femtocells or between femtocell and macrocell, when it is passing an open access femtocell. This will substantially increase the signalling in the network.

From the business point of view it is not clear who should cover the costs of femtocell maintenance, when open access femtocells are deployed in homes. A commercialization of open access femtocells might be aimed at the enterprise market, where a company, for example, will be able to provide low-cost calls to employees, when they are in the office. In hotspot scenarios, open access femtocells might be deployed by the operator, in order to increase the coverage and capacity of the network.

Hybrid access mode

The second access mode introduced in 3GPP Release-9 is the hybrid access mode. The main drawback of CSG cells is that femtocells suffer as well as cause high interference. In case of open cells, however, femtocell owners will not be willing to pay for resources which they have to share with other UEs. 3GPP introduced, in Release-9, a tradeoff between CSG cells and open cells, namely hybrid cells. Hybrid cells allow the connectivity of non-CSG UEs, while limiting the resources that can be shared. They are similar to open access cells, except that here femtocell owners can select CSG UEs and give them preferential treatment over non-CSG UEs. With hybrid cells, the level of access to femtocells can be controlled and the connection can be configured to guarantee a minimum performance, so that non-CSG UEs are only authorized a limited Quality of Service (QoS).

Although a hybrid cell allows all UEs to camp on it and offers services to any type of UEs, there is still a need to distinguish CSG UEs from non-CSG UEs in order to provide them with different levels of QoS. Consequently the HeNB has to perform an identity request. In this case the HeNB-GW has to inform the HeNB about the membership status (based on the CSG ID received from the target cell and the UE's CSG whitelist). In case of a CSG UE, the femtocell has to include hybrid mode information, while the CN may inform the femtocell about the UE membership status for the accessed cell.

A possible deployment scenario for a hybrid cell might be a shopping mall. The network operator might provide a special contract to the shopping mall/femtocell owner, where the

Table 10.1 Overview of HeNB access control mechanisms

	HeNB access control mode		
	Closed access	Open access	Hybrid access
CSG UE	Access to services	Access to services	Preferential access to services
Non-CSG UE	Access to services	No access	Access to services

employees of the shopping mall get priority access when requesting services. The shopping mall/femtocell owner will, at the same time, allow the public to use the femtocell to access normal network operator services. Hence, hybrid cells are an interesting solution for shopping mall deployment scenarios.

However, from the business point of view a commercial model has still to be proposed by the network operators for hybrid cells.

In Table 10.1 the HeNB access control mechanisms and allowed accesses for CSG and non-CSG UEs are illustrated.

10.3.4 Performance of Access Control Mechanisms

In order to analyse the performance of wireless systems, system-level simulation environments are used by both industry and research institutions. To understand the impact of femtocells deployed within an existing macrocellular network in terms of coverage, capacity and interference, system-level simulations can be carried out. This enables the development of new algorithms to face and solve the technical challenges of new wireless technologies.

In this section, the main components of a system-level simulation environment, including macrocell and femtocell scenarios, are introduced. In addition, simulation results for all possible cell selection scenarios, showing the advantages and drawbacks of these scenarios, are given.

System-level simulation environment

Network simulations can be done based on system-level simulation environments modelled by software. Within system-level simulations the behaviour of a network is analysed, considering Media Access Control (MAC) related issues such as mobility, scheduling, radio resource management and hybrid automatic repeat request. The target of system-level simulations is to study an overall system's performance, that is network capacity, QoS and delay time within the network, instead of considering the performance of a single link.

Simulating transmission of all individual bits of all radio links between UEs and eNBs of a whole network is an impractical way of performing system-level simulations due to high required computational power and time. Therefore, in system-level simulations the physical layer (link level) is abstracted by simplified models that capture its essential characteristics with high precision. This abstraction, which acts as an interface between both simulation levels, is a Look-Up Table (LUT). The simulated link level performance as a function of metrics reflecting the performance of the studied network for any instantaneous channel realization, are generated off-line and saved in LUTs. The LUTs are a set of tables that represent the results of link level simulations and provide bit error rate or block error rate values in terms of

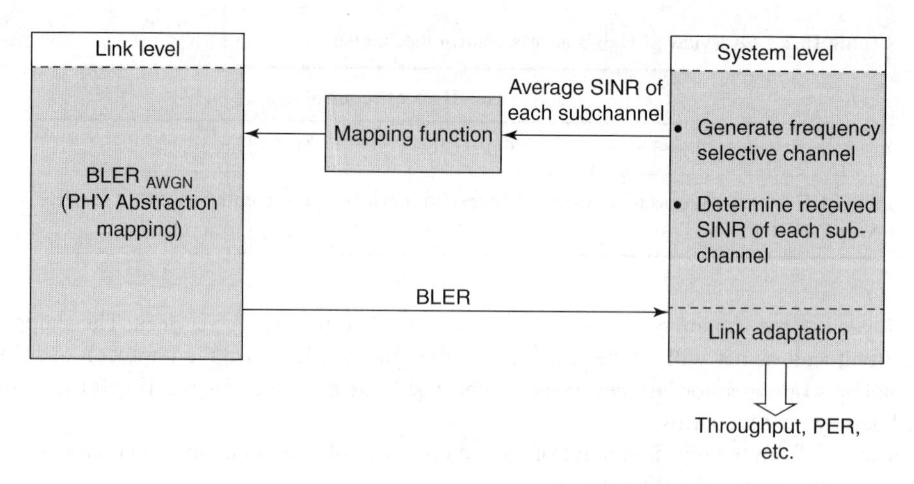

Figure 10.4 Link level to system-level mapping procedure.

Signal-to-Noise-Plus-Interference-Ratio (SINR). The LUTs are used in system-level simulations to precisely predict the link layer performance in a computationally simple and effective way. Figure 10.4 shows the link-level to system-level mapping procedure as defined in [15].

In addition to the simulations of the physical layer and the MAC layer, it is necessary to model accurately the radio channel to predict the performance of a wireless system. Since the radio channel depends strongly on the behaviour of the environment, system-level simulation tools consist of additional components, so-called simulation blocks. These simulation blocks are the deployment scenario/geometry, the mobility and traffic model of mobile stations and the channel model including indoor propagation and fading models. An additional simulation block is the radio resource management block, which enables the implementation of scheduling and power control algorithms that are essential to analyse and control the interference, especially in two-tier networks.

Performance of cell selection scenarios

It is expected that femtocells will extend the indoor coverage and enhance the network capacity. Femtocells will also help to manage the exponential growth of traffic within the macrocells, due to the handover of the indoor traffic to the backhaul connection. To achieve these benefits, interference management/coordination between macrocells and femtocells is essential. Besides radio resource control and power control, cell selection is one possibility to manage this interference. Enabling a non-CSG UE that strongly suffers from interference caused by a nearby HeNB or permitting a CSG UE to connect to an eNB due to better channel conditions, will reduce cross-tier interference.

In the following, LTE based system-level simulations have been performed by the authors to obtain network-wide statistics about the impact of femtocells deployed within existing macrocell networks for different cell selection scenarios. In this way the benefits and drawbacks of the simulated cell selection scenarios will be demonstrated in order to address the challenges that network operators must face.

The scenario that is used to perform these evaluations looks as follows: Within an existing macrocellular environment consisting of a total of 19 sites (corresponding to 57 macrocells), femtocells that operate in the same frequency band (co-channel deployment) are randomly dropped. A suburban scenario, as introduced in [16], is simulated with ten femtocells per macrocell. In total, 10 macro UEs and one to four femto UEs are randomly dropped within each macrocell and femtocell, respectively. Note that these UEs will change their status depending on the simulated cell selection scenario. In the performed simulations, this is the status before any base station assignment or cell selection has been performed, that is, a macro UE can get access to a femtocell and become a femto UE and vice versa. This is a terminology that is used to explain the different cell selection scenarios and to emphasize that no different service types or priorities will be considered for these UEs as is the case in hybrid cells. Table 10.2 summarizes all system-level simulation parameters.

The examined cell selection scenarios are illustrated in Figure 10.5. The lower right box shows an overview of all possible connections represented by encircled numbers. The

Table 10.2 System level simulation parameters

Parameter	Value
Simulation direction	Downlink
Cellular layout	Hexagonal grid, three cells per site
Number of snapshots	1 (50) (each snapshot: 1 ms)
Number of random scenarios	500 (20)
Number of sites	19 (7)
Number of femtocells per macrocell	10
Inter-site distance	500 m
Carrier frequency	2000 MHz
System bandwidth	10 (3) MHz
Duplexing scheme	FDD
Min/max eNB transmission power	0 dBm/46 dBm
eNB antenna gain	14 dBi
eNB noise figure	5 dB
eNB thermal noise density	−174 dBm
eNB antenna pattern	$A(\theta) = -\min[12(\frac{\theta}{70^\circ})^2 \text{ dB}, 20 \text{ dB}]$
Min./max. HeNB transmission power	0 dBm/20 dBm
HeNB antenna gain	0 dBi
HeNB antenna pattern	Omnidirectional
UE antenna gain	0 dBi
UE noise figure	9 dB
UE thermal noise density	−174 dBm
UE antenna pattern	Omnidirectional
Shadowing standard deviation	8 dB
Shadowing correlation	Between macrocells: 0.5 fixed Between macro sites: 1.0 fixed
Path loss model	see [17]
Traffic model	Full buffer
Scheduling algorithm	Proportional fair
Link-to-System mapping	Exponential effective SINR mapping

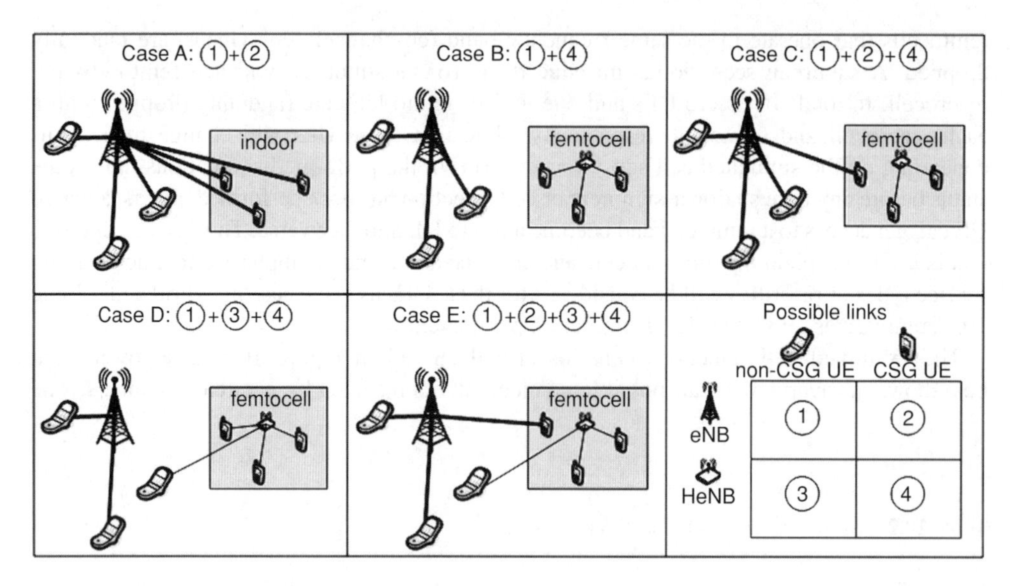

Figure 10.5 Cell selection scenarios.

connection of a macro UE to an eNB is described by an encircled 1, whereas an encircled 2 reflects the connectivity of a femto UE to an eNB. The numbers 3 and 4 describe the connection of macro UEs and femto UEs to HeNBs, respectively. Note that this is the terminology used for the UEs before activating any cell selection scenario.

The following cell selection scenarios are considered similar to [19]:

- **Case A**: HeNBs are assumed to be inactive, so that each UE gets access from the eNB. This is a baseline scenario, which is used to express the benefits and drawbacks of femtocells when introducing them into a macrocellular network.
- **Case B**: This is the closed access cell scenario. HeNBs only grant access to a particular set of authorized UEs (femto UEs). In this case no status change of the UEs takes place, that is, femto UEs remain femto UEs and get access to HeNBs that they are subscribed to, and macro UEs remain macro UEs and select the eNB providing the strongest link as their serving BS. With closed access, macro UEs can receive severe interference from nearby femtocells. Even if the received power from the HeNB is larger than that of the nearest eNB, non-subscribers are not allowed to connect to femtocells. In [17, 18] it was shown that in closed access femtocell networks macro UEs lying in a femtocell's coverage area greatly suffer from high interference in the downlink and that such UEs cause destructive interference to HeNBs in the uplink.
- **Case C**: In this cell selection scenario, UEs that were initially defined as femto UEs are not forced to get connected to their corresponding HeNB, that is, subscribers are free in becoming non-subscribers and getting access to eNBs if their received power is larger. This may occur especially if a femtocell is very close to an eNB. Since the transmission power of an eNB is larger (46 dBm) than that of an HeNB (20 dBm) the received power from an eNB at a femto UE might be larger, although this user is indoors and the received power suffers

from penetration losses. In this case a UE, which was preliminarily defined as femto UE, is allowed to become a macro UE. This scenario has not been introduced in 3GPP Release-9, but reflects a realistic scenario, in which an indoor UE need not connect to the femtocell that is in the same building.

- **Case D**: This scenario is introduced as open access in 3GPP LTE Release-9. It is an approach to reducing interference by simply handover macro UEs that cause (in the uplink) or experience (in the downlink) strong interference to the femtocell. In this access mode macro UEs that lie within a femtocell's coverage area are allowed to get access from the corresponding HeNB. Intuitively, this should increase the overall network capacity at the possible expense of a given femtocell owner, who must now share his femtocell resources with an unpredictable number of cellular users.
- **Case E**: After introducing these access modes an upper bound is also considered. The last cell selection method we analysed is case E in Fig. 10.2. In this case no matter how the status of a UE was initially defined, each UE is free to select the base station (BS) it wants to get access to. It is expected that this case shows the best result concerning the wideband SINR and link gain distribution. It can also be expected that the average throughput of UEs with bad channel conditions/low SINR values will be improved in case E.

For the described scenarios (cases A–E) the same cell selection criterion is used. The UE's selection criterion is the so-called long-term received power, which is the transmission power level of a BS minus all the losses, including path loss, antenna gain, shadowing and penetration loss. As long as the selected access mode gives the UE permission to access a certain BS type, the UE will always choose the BS with the largest long-term received power. This is also known as the strongest cell selection criterion in 3GPP LTE.

One of the most important parameters when designing the characteristics of a new mobile communication system is the downlink wideband SINR. This is the frequency- and time-averaged power received from the serving BS in relation to the average interference power from all other BSs plus noise. For a UE m connected to BS i, the wideband SINR Γ_{WB} is defined as:

$$\Gamma_{\text{WB},m} = 10 \cdot \log \left(\frac{p_{\text{rx},i(m)}}{\sum_{\forall j \neq i} p_{\text{rx},j} + \sigma} \right) \quad [\text{dB}], \qquad (10.1)$$

where $p_{\text{rx},j}$ is the received power from BS j and σ is the noise power. In order to show the impacts of each cell selection scenario, calibrations have been carried out based on system-level simulations. The cumulative distribution function of the wideband SINR for each cell selection scenario is depicted in Figure 10.6.

The curve, which shows the case that no femtocell is activated (case A), provides a basis for comparison of the results obtained from the introduced cell selection scenarios. This reference curve shows that with a probability of 50% (median) a UE receives a wideband SINR of more than +3.6 dB as shown in Table 10.3. By using cell selection methods the median of the wideband SINR can be increased to 12.0 dB. This can be interpreted as the benefit of femtocells, because activating femtocells leads to higher received powers particularly for indoor UEs. As a drawback of the investigated cell selection scenarios, it can be observed that low wideband SINRs occur with a larger probability than for the reference system. This is

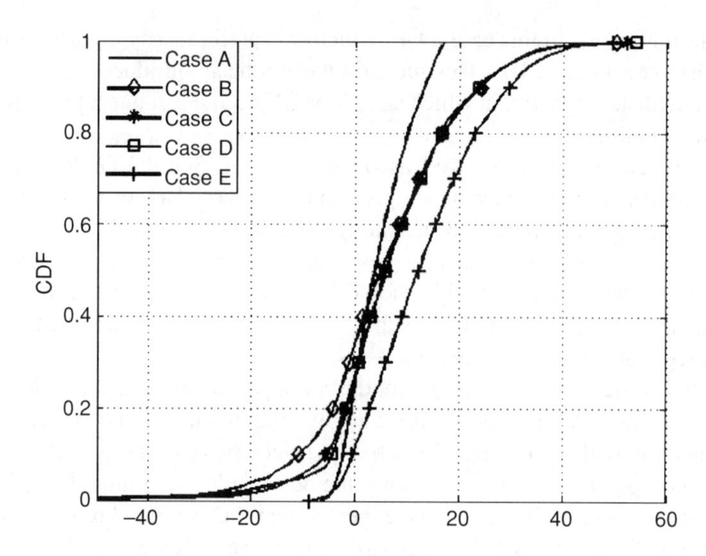

Figure 10.6 Simulation results for the downlink wideband SINR [dB].

due to the cross-layer interference that arises when activating femtocells. Here interference mitigation techniques, for example power control or resource allocation, become more important. However, from Table 10.3 it can be seen that the mean values are larger than the one of the reference model. As expected, case E provides the highest mean value, because this cell selection scenario selects the best access for each type of UE.

Additionally, the average normalized UE throughput is considered, which is defined as the number of information bits that the user successfully receives, divided by the simulation time and the bandwidth. Simulation parameters that differ from those for downlink wideband SINR simulations are given in brackets in Table 10.2. In Figure 10.7 the CDF of the average normalized throughput of all UEs in the system is depicted for each cell selection scenario. Obviously the curves of all cell selection scenarios are quite close to each other besides the curve of case A. The difference of case A to the curves of all other cell selection scenarios shows the impact of femtocells in a cellular system. Introducing femtocells significantly improves the average normalized throughput of UEs in a system.

Table 10.3 Mean values of wideband SINR for each cell selection scenario

Cell selection scenario	Wideband SINR [dB]
Case A	3.6
Case B	4.8
Case C	5.7
Case D	5.8
Case E	12.0

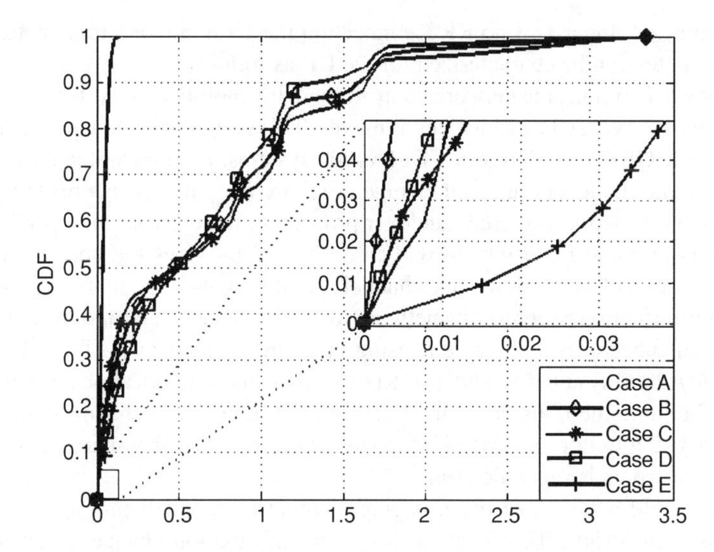

Figure 10.7 Average normalized throughput per UE [bps/Hz].

It can be outlined that for higher rates (high average throughput) the poorest performance is obtained for case A. If no femtocells are deployed in the system, high rates cannot be achieved, because indoor UEs have bad channel conditions. Although the curves for the cell selection scenarios are close to each other, some differences can be noted. At higher rates, case D shows the weakest performance. This is because the number of femto UEs increases rapidly for this case. Case E shows a better performance than case D for higher rates, because in case E femto UEs may get access from eNBs. Comparing cases B and C, it can be seen that C outperforms B for higher rates. This is because in case C femto UEs with bad channel conditions get access to eNBs, that is a smaller number of remaining femto UEs get higher rates due to the small distance to their HeNBs and the reduced number of UEs in their femtocell. Case C also outperforms case E for higher rates, because in case E macro UEs can get access to HeNBs. This increases the number of UEs within a femtocell and consequently decreases the number of high rate femto UEs.

For cell-edge UEs (zoomed in Figure 10.7), the cell selection scenarios show different performances compared to high rate UEs. Case B shows for cell-edge UEs the worst performance, because macro UEs are interfered by femtocells, and femto UEs will suffer from interference caused by eNBs. Allowing macro UEs to get access to femtocells (case D) enhances the performance, whereas for case C even more enhancement can be achieved. For cell-edge UEs, case E outperforms all cell selection scenarios including the reference case A.

10.4 Cell Selection and Cell Reselection

The applicable AS-related procedures largely depend on the UE's RRC state, which is either RRC_CONNECTED or RRC_IDLE. A UE is in RRC_CONNECTED state when an RRC connection has been established. In case of no RRC connection the UE is in RRC_IDLE state. According to [3], the two states can be characterized for the UE as follows.

In RRC_CONNECTED state, the network controls the UE's mobility, that is the network decides when the UE will move, and to which cell it moves. This cell may be on another frequency or RAT, for example GERAN, UTRA or CDMA2000 systems. For network controlled mobility in RRC_CONNECTED, handover and cell change order are the only procedures that are defined [3]. To facilitate a network triggered handover procedure, the network may configure the UE to perform measurement reporting. In this case, the UE provides the network with reports of its buffer status and of its downlink channel quality, as well as of its neighbouring cell measurement information to enable the network to select the most appropriate cell for this UE. These measurement reports include cells using other frequencies or RATs such as GERAN, UTRA or CDMA2000 systems. In RRC_CONNECTED state, the network allocates radio resources to the UE to facilitate the transfer of (unicast) data via shared data channels. To support this operation, the UE monitors a control channel associated with the shared data channel to determine whether data is scheduled to it.

In RRC_IDLE state a UE monitors a paging channel to detect incoming calls. A paging message is used to inform the UE about a system information change. Even if the UE is informed about changes in system information, no further details are provided to the UE regarding which system information will change. In RRC_IDLE state a UE also acquires system information, which mainly consists of parameters by which E-UTRAN can control the cell (re)selection process. Additionally the UE performs neighbouring cell measurements and cell (re)selection in RRC_IDLE state. Hence, the UE decides on which cell to camp. During cell (re)selection process the priority of applicable RATs, the radio link quality and the status of each cell – that is whether it is accessible, barred or reserved – is taken into account.

In this section a description of idle mode states and state transitions will be given to describe the processes of idle mode tasks such as PLMN selection, cell selection and cell reselection for 3GPP Release-9.

10.4.1 UE in Idle Mode

The UE procedures in idle mode have been specified in general in [4] and specifically for E-UTRAN in [6]. Here, idle mode tasks are subdivided into the following four processes:

- PLMN selection;
- cell selection and cell reselection;
- location registration; and
- support for manual CSG selection.

The overall idle mode process, and the relationship between the four idle mode processes, are illustrated in Figure 10.8.

When a UE is switched on, a PLMN is selected, and within this selected PLMN the UE searches for a suitable cell to camp on. A suitable cell in this context is a cell on which a UE with a valid USIM may camp to obtain normal service in E-UTRA. This cell is either part of the selected PLMN or part of the registered PLMN. In the case of CSG cells, the CSG ID must be present in the CSG whitelist associated with the selected or registered PLMN.

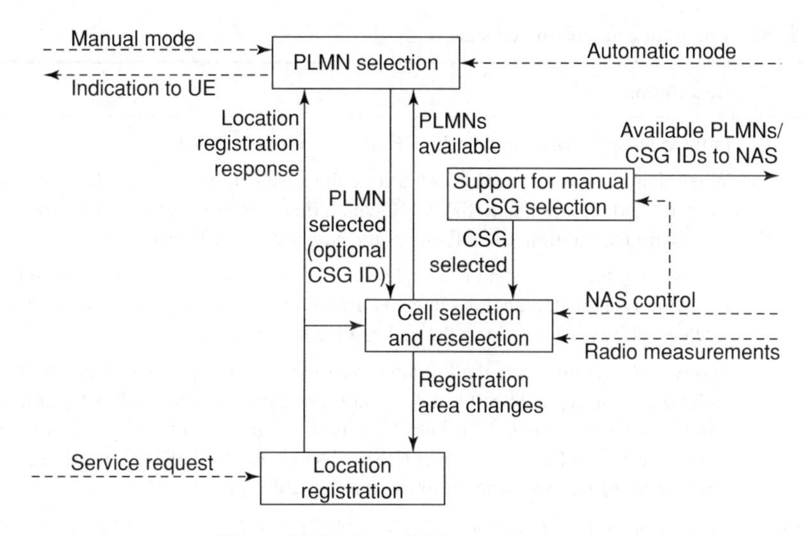

Figure 10.8 Overall idle mode process.

After searching for a suitable cell of the selected PLMN, the UE will choose that cell to camp on. A location registration procedure will be carried out by the UE. The UE will register itself in the registration area of the selected cell. Finally, the selected PLMN will be a registered PLMN as outcome of the location registration.

A description of idle mode states together with possible state transitions including conditions for these transitions is given in Figure 10.9.

Table 10.4 provides an overview for each idle mode state.

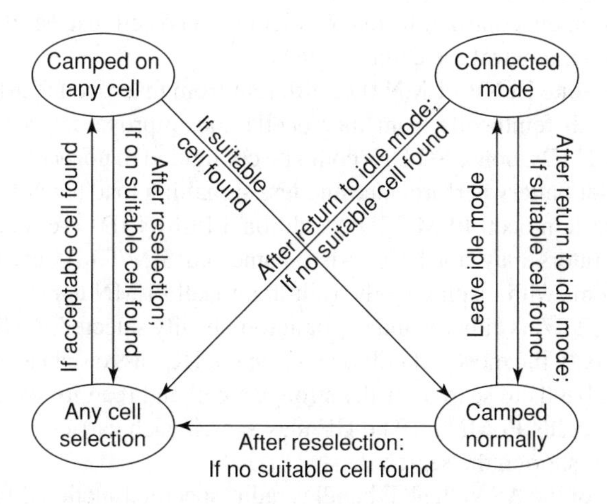

Figure 10.9 Idle mode states and state transitions.

Table 10.4 Overview for each idle mode state

State	Description
Camped normally	In the *camped normally state* the UE obtains normal service. When that UE finds a suitable cell after cell (re)selection, the UE selects this cell to camp on and reports this to the NAS, which then performs registration. After a successful registration, the UE enters the *camped normally state*. In this state, the UE will monitor relevant system information, and select and monitor accordingly indicated paging channels. Required measurements are performed for cell reselection by the UE to start the cell reselection process.
Any cell selection	In *any cell selection state* the UE attempts to find an acceptable cell of any PLMN to camp on. An acceptable cell is a cell that is not barred and that fulfils the cell selection criterion (see Section 10.4.3). The UE will search for a cell whose measured Reference Signal Received Power (RSRP) is at least −110 dB [6]. A UE leaves the *any cell selection state* when it finds an acceptable cell.
Camped on any cell	In the *camped on any cell state* the UE obtains limited service. In this state, similar to the *camped normally state*, the UE monitors relevant system information and selects and monitors indicated paging channels. Additionally, the UE performs measurements for cell reselection in order to execute cell reselection.
Connected mode	In the *connected mode state*, when the UE returns to the idle mode, the UE performs cell selection to find a suitable cell to camp on. This leads the UE to stay in the *camped normally state*. If a suitable cell is found, the registration procedure is performed by the NAS. If no suitable cell is found, cell selection by stored information is done (see Section 10.4.3).

10.4.2 PLMN Selection

The PLMN selection and reselection process is assumed to have higher priority than the cell selection and reselection process. This means that once coverage from the selected PLMN is obtained, the measurement control information elements and cell reselection parameters in the system information will control the choice of RAT.

In 3GPP LTE, separate HeNB PLMN IDs, different from macrocell PLMN IDs, are defined in order to distinguish femtocells from macrocells and improve cell selection procedures. Using separate PLMN IDs, non-CSG UEs could be configured to not access the HeNB PLMN, which results in better battery performance and less signalling load towards the CN. However, when using separate femtocell PLMN IDs, additional PLMN IDs are required, which some of the network operators may not have. Also, some old SIM/UICC cards may have some compatibility problems with displaying the right femtocell PLMN name.

The PLMN in E-UTRA is either manually or automatically selected. A UE's first cell search for a PLMN is normally the most difficult one, since the UE has to scan all RF channels in the E-UTRA frequency bands to search for the strongest cell and read its system information, in order to find out the cell's PLMN(s). The UE may search each carrier in turn or make use of stored information to shorten the search.

The control plane of the AS in the UE handles radio-specific functionality. The AS interacts with the NAS and will report available PLMNs to the NAS on request from the NAS, or autonomously. The NAS forms the highest stratum of the control plane between UE and MME at the radio interface. The NAS will provide a list of equivalent PLMNs, if available, which the AS will use for cell selection and cell reselection.

The PLMN identities are stored in a list by priority order. In the system information on the broadcast channel, the UE can receive one or multiple PLMN identities in a given cell. If the UE can read one or several PLMN identities in the strongest cell, each found PLMN will be reported to the NAS.

The search for PLMNs may be stopped by request of the NAS. Once the UE has selected a PLMN, the cell selection procedure will be performed in order to select a suitable cell of that PLMN for the UE to camp on.

If a CSG ID is provided by the NAS as part of PLMN selection, the UE will search for an acceptable or a suitable cell belonging to the provided CSG ID to camp on that cell. When the UE no longer camps on a cell with the provided CSG ID, the AS will inform NAS.

10.4.3 Cell Selection

Cell selection and cell reselection are the two basic mobility procedures in wireless mobile networks. In cell selection, the most suitable cell is chosen for a UE to camp on after it has selected a PLMN. Cell selection takes place when a UE is either powered on or has lost its previous coverage. While camping on this suitable cell the UE gets the system information that is broadcast by the Broadcast Channel (BCH) in 3GPP LTE. System information consists of cell- and network-specific parameters that are broadcast to allow UEs to connect successfully to the chosen suitable cell. System information includes for UEs in RRC_IDLE mode, for example, NAS common information and information about cell (re)selection parameters, neighbouring cell information. For UEs in RRC_CONNECTED mode the system information includes the common channel configuration information. The broadcast system information is divided into the following two categories [3]:

- **Master information block (MIB)**: The MIB consists of the most important and frequently transmitted parameters that are necessary for initial access to a cell. These parameters are carried on the BCH, and contact information such as the downlink system bandwidth, the number of resource blocks in downlink and the system frame number.
- ***SystemInformationBlock* (SIB)**: The system information is broadcast in SIBs, each of which contains the following set of functionally related parameters:
  - *SystemInformationBlockType1* contains parameters that are essential to decide whether a cell is suitable for cell selection or not, and also information about time domain scheduling of other SIBs,
  - *SystemInformationBlockType2* includes common and shared channel information, and
  - *SystemInformationBlockType3* to *SystemInformationBlockType8* contain parameters for intra-frequency, inter-frequency and inter-RAT cell reselection.

After the UE acquires system information it registers its presence in the tracking area. Subsequently the UE can receive paging information. The paging information is provided to upper layers, which may lead the UE to start an RRC connection to establish a call, for example.

When camping on a cell, the UE regularly searches for a better cell. This is known as performing cell reselection. The cell reselection procedure aims to find the best cell, which the UE can camp on and obtain services from. This procedure is always active in the idle mode, after the cell selection procedure has been completed and the first cell has been chosen. (Further information about cell reselection is given in Section 10.4.4) Depending on the UE's service level, the cells are categorized in 3GPP LTE as follows [6]:

- **Acceptable cell**: If the UE is not able to find a suitable cell, but manages to camp on a cell belonging to another PLMN, this cell is said to be an acceptable cell, and the UE obtains limited service, that is, it can only perform emergency calls – this is also the case when no USIM is present in the UE. The UE can always obtain emergency calls on an acceptable cell unless the cell is barred.
- **Barred cell**: With this cell the UE is not allowed to camp on it and obtain services. This cell indicates via its system information that it is barred.
- **Reserved cell**: As with the barred cell, the UE is not allowed to camp on it and obtain services. The indication is given via the system information. However, unlike barred cells, exceptions are defined for reserved cells. It is suitable when a UE camps on a cell that belongs to a registration area which is not allowed for regional provision of services. In this case, the UE is provided limited service.
- **Serving cell**: This is the cell on which the UE has camped.
- **Suitable cell**: If the UE has a valid USIM and the cell is part of either the selected or registered PLMN, or of a PLMN of the equivalent PLMN list, the UE may camp on it and obtain normal service.

Due to the large number and high density, the dynamic neighbour lists and variant access modes of femtocells, cell selection and reselection is a complex procedure in the femtocell environment. Within the femtocell environment, cell selection and reselection procedures should prioritize the UE to camp on allowed femtocells while at the same time avoiding UE registration on unauthorized femtocells. Therefore, 3GPP LTE introduced the support for manual CSG selection besides the two common cell selection procedures (see Figure 10.10). In the following, each of the procedures is described in detail.

There are two different cell selection procedures as illustrated in Figure 10.10, namely the *initial cell selection* and the *stored information cell selection*. In the initial cell selection procedure the UE scans all Radio Frequency (RF) channels in E-UTRA bands in order to find a suitable cell. On each supported carrier frequency the UE will search for the strongest cell and select the suitable cell that fulfils the cell selection criterion to reach the connected mode. This procedure does not require any prior knowledge of which RF channels are E-UTRA carriers. In contrast to this procedure, the stored information cell selection procedure requires stored information of carrier frequencies and also information about cell parameters from previously detected cells. If no suitable cell can be found in this case, the initial cell selection procedure has to be started. In neither of the cell selection processes are priorities between different frequencies and RATs considered.

As mentioned before, during cell selection the UE searches on all supported carrier frequencies of each RAT for the strongest cell, to find a suitable cell to camp on it. To speed up the search process, NAS can indicate the RATs of the selected PLMN, and the UE can use

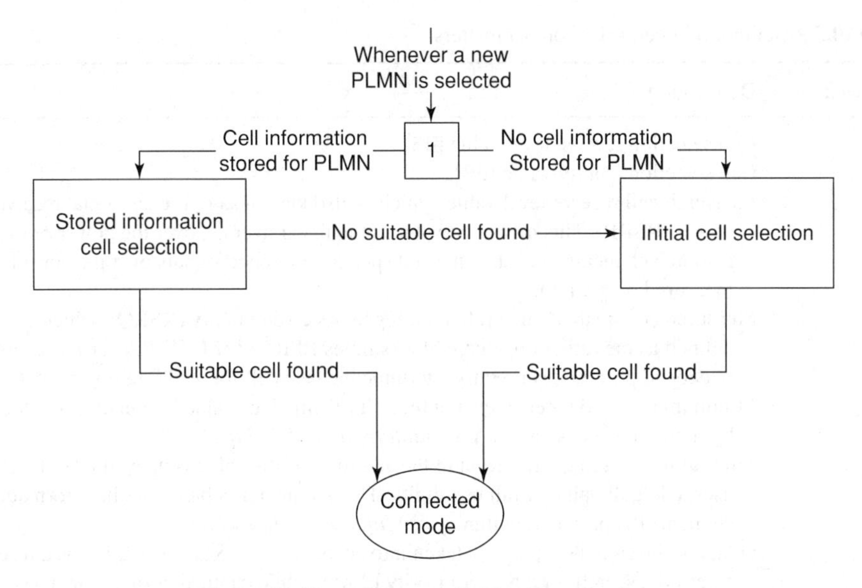

Figure 10.10 Different cell selection procedures.

information stored from previous accesses. Short cell selection duration is especially important with the steadily increasing number of frequencies and RATs that have to be searched.

In order to perform cell selection, the cell selection criterion has to be fulfilled. This is known as the *S-criterion* in 3GPP and is achieved when the cell selection receive level value S_{rxlev} and the cell selection quality value S_{qual} are larger than 0 dB, where:

$$S_{\mathrm{rxlev}} = Q_{\mathrm{rxlevmeas}} - (Q_{\mathrm{rxlevmin}} - Q_{\mathrm{rxlevminoffset}}) - P_{\mathrm{compensation}}, \quad (10.2)$$

$$S_{\mathrm{qual}} = Q_{\mathrm{qualmeas}} - (Q_{\mathrm{qualmin}} - Q_{\mathrm{qualminoffset}}). \quad (10.3)$$

The definition of the parameters is given in Table 10.5 [6]:

All parameters except Q_{qualmeas} and $Q_{\mathrm{rxlevmeas}}$ are provided via system information. In a real network, a UE will receive several cells from different network operators. After reading the *SystemInformationBlockType1* the UE knows if a cell belongs to its PLMN ID. First the UE will search for the strongest cell per frequency carrier, and then it will look for the PLMN ID by decoding *SystemInformationBlockType1* to decide if this PLMN ID is suitable. Finally, it will use the *S-criterion* to determine if the cell is suitable or not.

For cell selection, additional requirements are defined for specific cases. When a UE leaves the *connected mode*, it should typically select a cell to which it was connected. However, it is possible that the connection release message might include information that leads the UE to select a suitable cell on an E-UTRA carrier. If no suitable cell is found according to the cell selection criterion, the UE will perform cell selection by using the stored information cell selection procedure. In case of any cell selection mode, the UE looks for an acceptable cell of any PLMN to camp on by searching all supported RATs. However, if the UE finds a cell that meets the high quality criterion for that RAT, it may stop searching for another cell. The high quality criterion is fulfilled for an E-UTRAN cell if the RSRP is at least −110 dBm.

Table 10.5 Definition of cell selection parameters

Parameter	Description
S_{rxlev}	Cell selection receive level value [dB]
S_{qual}	Cell selection quality value [dB]
$Q_{rxlevmeas}$	Measured cell receive level value, which is also known as reference signal received power (RSRP). This measured value is the linear average over the power of the resource elements that carry the cell specific reference signals over the considered measured bandwidth.
$Q_{qualmeas}$	Measured cell quality value reference signal received quality (RSRQ), which is defined as the ratio of resource blocks times RSRP over E-UTRA carrier received wideband power. It is used to determine the best cell for a LTE radio connection.
$Q_{rxlevmin}$	Minimum required receive level in the cell [dBm]. This value is signalled as $Q_{rxlevmin}$ by higher layers as part of *SystemInformationBlockType1*.
$Q_{qualmin}$	Minimum required quality level in the cell [dB]. If the value is unavailable, the UE uses a default value of minus infinity. This parameter is based on the information element (IE) provided within *SystemInformationBlockType1*.
$Q_{rxlevminoffset}$	Offset to the signalled $Q_{rxlevmin}$ taken into account in the S_{rxlev} evaluation as a result of a periodic search for a higher priority PLMN while camped normally in a Visitor PLMN (VPLMN) [20]. It is based on the information element provided within *SystemInformationBlockType1*. The actual value of $Q_{rxlevminoffset}$ is calculated by multiplying IE with two. If the value is unavailable a default value of 0 dB is used.
$Q_{qualminoffset}$	Offset to the signalled $Q_{qualmin}$ taken into account in the S_{qual} evaluation as a result of a periodic search for a higher priority PLMN while camped normally in a VPLMN [20]. It is based on the information element provided within *SystemInformationBlockType1*. The actual value of $Q_{qualminoffset}$ is IE. If the value is unavailable the UE applies the default value of 0 dB. This parameter affects the minimum required quality level in the cell.
P_{EMAX}	Maximum transmission power level that a UE is allowed to use when transmitting on the uplink in the cell [dBm] defined as P_{EMAX} in [21]. It is based on the information element provided within *SystemInformationBlockType1*.
$P_{powerclass}$	Maximum radio frequency output power of the UE [dBm] according to the power class the UE belongs to as defined in [21].
$P_{compensation}$	$\max(P_{EMAX} - P_{powerclass}, 0)$ [dB]

Cell selection in the presence of CSG cells and hybrid cells can be performed either via the normal cell selection procedure like the initial cell selection procedure or the stored information based procedure or via manual selection of CSG cells. Manual CSG cell selection will be supported by the UE upon request from higher layers. If the NAS provides request to the UE to search for available CSGs, the AS scans all RF channels in the E-UTRA band to find cells with a CSG ID. For these cells the AS reads the HeNB name (if broadcast) from the Broadcast Control Channel (BCCH) on *SystemInformationBlockType9* and reports the CSG ID of the found cell, its HeNB name and PLMN to the NAS. The NAS then evaluates these reports from the AS for CSG selection and selects a CSG while requesting the AS to select a cell that belongs to this selected CSG. Subsequently, the AS selects a cell belonging to the selected CSG that is not barred or reserved and that fulfils the cell selection criteria. This procedure is known as manual CSG selection.

10.4.4 Cell Reselection

While camping on a suitable cell, the UE regularly searches for a better cell, that is the UE starts cell reselection. The cell reselection procedure aims to find the best cell in the selected PLMN ID and of its equivalent PLMN IDs, if there are any. A UE performs cell reselection only after having camped for at least one second on the current serving cell. For LTE, the radio link quality is the primary criterion for selecting a cell. Cell reselection between frequencies and RATs is primarily based on absolute priorities, where each frequency has an associated priority. Cell specific default values of the priorities are provided via the system information. However, for inter-frequency and inter-RAT cell reselection, other criteria will also be considered. In addition to cell specific values, E-UTRAN may assign UE specific values such as UE capability, subscriber type and call type. If equal priorities are assigned to multiple cells, cells are ranked based on radio link quality.

For cell reselection the UE first evaluates frequencies of all RATs based on their priorities in order to compare cells on relevant frequencies based on radio link quality by ranking the cells. Before reselecting the target cell, the UE verifies this cell's accessibility.

Priority handling

Absolute priorities of different E-UTRAN frequencies or inter-RAT frequencies are provided to the UE in the *RRCConnectionRelease* message via the system information, if they are not carried out by another RAT during inter-RAT cell (re)selection [3]. The *RRCConnectionRelease* message is used to command the release of an RRC connection and is sent from E-UTRAN to the UE. If system information is provided, an E-UTRAN frequency or inter-RAT frequency is listed without priority information, if the information element *cellReselectionPriority* is absent. This parameter contains the absolute priority of the concerned carrier frequency. The field of *cellReselectionPriority* has a higher priority than the priority information provided by the system information. Thus, if this field is not missing, the UE will ignore priorities provided by the system information and use the priorities given in *cellReselectionPriority*. For different idle mode states, the following different priority information has to be considered:

- **Camped on any state**: If the UE is in this state, it will only use priorities that are provided by system information.
- **Camped normally state**: If the UE camps on a suitable cell and obtains normal services and has only dedicated priorities other than for its current frequency, the UE considers its current frequency as lowest priority.
- **Camped on a suitable CSG cell state**: In this state the UE has to consider its current frequency as the highest priority frequency. The UE must not consider further priority information.

For any state, the UE performs only cell reselection for E-UTRAN frequencies and inter-RAT frequencies if the system information provides corresponding priorities.

Measurement rules

To avoid unnecessary measurements and to enable the UE to reduce battery consumption, measurements are limited for cell reselection. The UE may only perform intra-frequency measurements if the cell selection receive level value S_{rxlev} and the cell selection quality value S_{qual} of the serving cell are below or equal to a threshold.

If the current E-UTRA frequency is lower than an E-UTRAN inter-frequency and inter-RAT frequency the UE will perform measurements of higher priority frequencies. For an E-UTRAN inter-frequency with an equal or lower reselection priority and an inter-RAT frequency lower than the reselection priority of the current E-UTRA frequency, the UE will perform measurements on E-UTRAN inter-frequencies or inter-RAT frequency cells of equal or lower priority only under a certain condition: if the quality of the serving cell is below or equal to a threshold [6].

Inter-frequency and inter-RAT reselection

Absolute priorities of different E-UTRAN frequencies or inter-RAT frequencies are provided via system information. In addition to these cell specific priorities E-UTRAN may assign UE-specific priorities via dedicated signalling. For cell reselection, the UE has to consider only those frequencies that are indicated in the system information and for which it has priorities. *SystemInformationBlockType3* is an information element that contains cell reselection information that is common for intra-frequency, inter-frequency and inter-RAT cell reselection. In case of inter-frequency or inter-RAT cell reselection different parameters that are provided via *SystemInformationBlockType3* have to be checked, and several conditions have to be fulfilled. One of these conditions is the availability of the parameter *threshServingLowQ* which is provided by *SystemInformationBlockType3*. In addition to this parameter, it has to be checked whether the target cell has higher, lower or equal priority than the current serving cell.

In case of reselection to a cell on a higher priority frequency than the serving cell, and if a *threshServingLowQ* is provided, cell reselection is performed if the S-criterion of a higher priority E-UTRAN, UTRAN, GERAN or CDMA2000 frequency exceeds a high threshold $\mathrm{Thresh}_{\mathrm{X,High}}$. If *threshServingLowQ* is not provided, cell reselection is performed if the S-criterion of a higher priority RAT frequency exceeds $\mathrm{Thresh}_{\mathrm{X,High}}$.

In case of reselection to a cell on a lower priority frequency than the serving cell and if a *threshServingLowQ* is provided, cell reselection is performed if the S-criterion of a lower priority E-UTRAN, UTRAN, GERAN or CDMA2000 frequency exceeds a low threshold $\mathrm{Thresh}_{\mathrm{X,Low}}$ while the S-criterion of the serving cell is below the low threshold $\mathrm{Thresh}_{\mathrm{Serving,Low}}$. If *threshServingLowQ* is not provided, cell reselection is performed if the S-criterion of a lower priority RAT frequency exceeds $\mathrm{Thresh}_{\mathrm{X,Low}}$ while the S-criterion of the serving cell is below $\mathrm{Thresh}_{\mathrm{Serving,Low}}$.

In case of reselection to a cell with a priority frequency equal to the serving cell, cell reselection has to be performed based on ranking.

Intra-frequency and equal priority inter-frequency reselection

In case of intra-frequency and equal priority inter-frequency cells the UE ranks cells which fulfil the S-criterion for cell reselection by using the so-called R-criterion. The R-criterion generates ranking for the serving cell R_s and for neighbouring cells R_n as follows:

$$R_{\mathrm{s}} = Q_{\mathrm{meas,s}} + Q_{\mathrm{hyst,s}} \tag{10.4}$$

$$R_{\mathrm{n}} = Q_{\mathrm{meas,n}} + Q_{\mathrm{offset}} \tag{10.5}$$

where Q_{meas} is the measurement quantity used in cell reselections (RSRP) and Q_{hyst} is a parameter that specifies the degree of hysteresis for the ranking criteria. Q_{offset} is equal to the offset between the serving and neighbouring cells for intra-frequency. For inter-frequency cells Q_{offset} is equal to the offset between the serving and neighbouring cell plus the frequency specific offset. The indices 's' and 'n' stand for the serving cell and the neighbouring cell, respectively.

Cell ranking is performed for cells that fulfil the cell selection S-criterion. The UE reselects the highest-ranked candidate cell only if the cell is a suitable cell. If the highest-ranked candidate cell is not a suitable cell it will not be considered as a candidate cell for reselection for 300 seconds. Cell reselection is only performed if the new cell is better ranked than the serving cell and if the UE camped on its current serving cell for more than one second.

Accessibility verification

For the highest ranked cell or for the best cell according to absolute priority reselection criteria the UE has to check the cell's accessibility, that is if the cell is barred or reserved. If the cell is not accessible, the UE is not allowed to reselect this cell even for emergency calls. The UE is not allowed to consider this cell as a candidate cell for reselection for 300 seconds. This cell has to be excluded from the candidate cell list. However, if the cell is a CSG cell the UE selects another cell on the same frequency if the reselection criteria are fulfilled.

If the highest ranked cell or the best cell according to absolute priority reselection criteria is unsuitable because it belongs to a forbidden tracking area or to a non-equivalent PLMN, the UE is not allowed to consider this cell or cells on the same frequency as a candidate cell for reselection for 300 seconds.

In any case, if the UE is in any cell selection state, all limitations for reselection have to be removed.

10.4.5 Cell Reselection with Femtocells

Cell reselection is a more complex procedure in the femtocell environment than in homogeneous networks, in which only macrocells exist. Within the femtocell environment, reselection procedures should prioritize the UE to camp on allowed femtocells while at the same time avoiding UE registration on unauthorized femtocells. In 3GPP Release-8 the CSG ID and the UE autonomous search function have been introduced. Using these parameters and functionalities the cell reselection procedures for femtocells can be improved.

For cell reselection in the presence of CSG cells and hybrid cells the UE uses the autonomous search function which is intended to find at least previously visited allowed CSG cells on non-serving frequencies including inter-RAT frequencies. The autonomous cell search function is not specified and is left open to UE implementation. It is disabled if the UE's CSG whitelist is empty.

Different cases have to be considered if cell reselection is to be performed in a macrocellular network overlaid with femtocells. It has to be taken into account, whether a UE performs cell reselection from a CSG cell to a non-CSG cell or vice versa, or whether cell reselection takes place between CSG cells. In this section, the cell reselection procedures for these cases are described by distinguishing between intra-frequency, inter-frequency and inter-RAT frequency cell reselection.

Reselection from CSG cell to a non-CSG cell

During the self-configuration phase when an HeNB is powered on, the HeNB identifies the neighbouring macrocell by conducting measurements with its downlink receiver functionality – the so-called sniffer mode [22]. The macro neighbour cell list is defined in the HeNB based on these measurements and is indicated to HeNB-connected UEs via its system information as it is performed in the case of macrocells. Depending on this information, the UE can perform corresponding cell measurements for cell reselection purposes without performing an autonomous search function.

Here are some femtocell-related policies during cell reselection from an allowed CSG cell to a non-CSG cell/macrocell in intra-frequency, inter-frequency and inter-RAT frequency:

- *Intra-frequency cell reselection* – For intra-frequency cell reselection from an allowed CSG cell to a non-CSG cell/macrocell, the UE will apply the same cell ranking rules as defined for macrocells.
- *Inter-frequency cell reselection* – For inter-frequency cell reselection from an allowed CSG cell to a non-CSG cell/macrocell, the UE will follow the same cell ranking rules as defined for macrocells. While a UE camps on a suitable cell, the UE will consider the frequency of its serving cell as the highest priority frequency, as long as the serving cell is the highest ranked cell on that frequency. In this case the UE will not reselect a non-CSG cell/macrocell.
- *Inter-RAT cell reselection* – For inter-RAT cell reselection, inter-RAT procedures as defined for macrocells in that serving RAT shall be applied.

Reselection from a non-CSG cell to a CSG cell

Cell reselection to CSG cells is based on the UE's autonomous search function. The search function determines for itself when and where to search the allowed CSG cells. It does not need to be supported by the network with information about the dedicated frequencies to CSG cells. The UE will disable the autonomous search function for CSG cells if the UE's CSG whitelist is empty.

Cell reselection to CSG cells does not require the network to provide neighbour cell information to the UE. The neighbour cell information can be provided in order to help the UE if, for example, the network wants to trigger the UE to search for CSG cells.

Within *SystemInformationBlockType11bis* the information element 'dedicated CSG frequency' is provided. If this information element is present, the UE is allowed to use the autonomous search function only on these dedicated frequencies and on frequencies that are provided by system information.

Here are some femtocell-related policies during cell reselection from a non-CSG cell to a CSG cell in intra-frequency, inter-frequency and inter-RAT frequency reselection:

- *Intra-frequency cell reselection* – For intra-frequency cell reselection from a non-CSG cell to an allowed CSG cell the UE will follow the same cell ranking rules as those defined for macrocells. The UE will disable non-allowed CSG cells from the ranking. If a suitable cell is detected and if this cell is the highest ranked cell according to the R-criterion, the UE will reselect this concerned CSG cell.

  The most important issue that has to be considered in intra-frequency cell reselection to CSG cells is the design of a policy when the best ranked cell is a CSG cell whose

CSG ID is not in the UE's whitelist. In this case, it has to be clarified under what conditions the UE will perform cell reselection. Some proposals have been made, but these are still under discussion in 3GPP [23–25].

- *Inter-frequency cell reselection* – For intra-frequency cell reselection from a non-CSG cell to an allowed CSG cell the UE first distinguishes the CSG cells from non-CSG cells/macrocells instead of ranking all the measured cells as in the case of macrocells. If the selected cell is a CSG cell, whose CSG ID is in the UE's whitelist, the UE will perform cell ranking only on this CSG cell's frequency. The UE will reselect this cell if it is the highest ranked cell on that frequency, regardless of the network configured frequency priorities and irrespective of the cell reselection rules related to the cell the UE currently camps on.

  If multiple suitable CSG cells are detected on different frequencies and these cells have the strongest signal strength/quality on their frequencies, the UE can reselect any of them.
- *Inter-RAT cell reselection* – For inter-RAT cell reselection, reselection to an allowed CSG cell is supported when the UE camps on another RAT. UE requirements have to be considered as they are defined in the specifications of the concerned RAT, such as E-UTRAN, UTRAN, GERAN or CDMA2000.

Reselection from CSG cell to CSG cell

If a UE's whitelist is not empty, the UE is required to perform an autonomous search function in order to search for CSG cells that are not listed in the system information of its serving cell – that is if at least one CSG ID with associated PLMN ID is included in the UE's whitelist. For those CSG cells that are listed in the system information, normal cell measurements can be performed by the UE as for macrocells.

If the information element 'dedicated CSG frequency' is present within *SystemInformationBlockType11bis* the UE is allowed to use the autonomous search function only on these dedicated frequencies and on frequencies that are provided by the system information.

Here are some femtocell-related policies during cell reselection from a CSG cell to a CSG cell in intra-frequency, inter-frequency and inter-RAT frequency:

- *Intra-frequency cell reselection* – For reselection from a CSG cell to an allowed CSG cell in the same frequency, the UE will perform the same cell ranking rules as defined for macrocells. If the UE detects a suitable CSG cell on the same frequency and if the concerned cell is the highest ranked cell on that frequency, the UE will reselect to this cell.
- *Inter-frequency cell reselection* – For inter-frequency cell reselection from a CSG cell to an allowed CSG cell the UE will consider the frequency of its serving cell to be the highest priority frequency if this cell is the highest ranked cell.

  For reselection from a CSG cell to neighbour CSG cells that are listed in the system information of the serving CSG cell, the UE will apply cell measurement rules as defined for macrocells.

  If the UE detects a suitable CSG cell which is the strongest on that frequency and if this cell is listed in the system information of the serving cell on other frequencies, the UE will consider these frequencies to have equal priority to the current serving cell's frequency.
- *Inter-RAT cell reselection* – Where one or more suitable CSG cells are detected on another RAT the UE may reselect one of them.

Cell reselection with hybrid cells

For cell reselection with hybrid cells, the UE has to distinguish whether the considered cell is a hybrid cell or not. If the hybrid cell's CSG ID belongs to the UE's whitelist, the hybrid cell is considered as a CSG cell in the idle mode cell reselection procedures. In addition to normal cell reselection rules, the UE will use an autonomous search function in order to detect at least previously visited hybrid cells, whose CSG ID and associated PLMN ID are in the UE's whitelist. If the CSG ID and the associated PLMN ID of the detected hybrid cells are not in the UE's CSG whitelist, these cells have to be considered as normal cells. Otherwise the UE will treat detected hybrid cells as CSG cells and perform cell reselection according to CSG requirements [26].

References

1. 3GPP TS 23.401, 'General Packet Radio Service (GPRS) enhancements for Evolved Universal Terrestrial Radio Access Network (E-UTRAN) access', V9.9.0, June 2011.
2. 3GPP TR 25.820, '3G Home NodeB Study Item Technical Report', V8.2.0, September 2008.
3. 3GPP TS 36.331, 'Evolved Universal Terrestrial Radio Access (E-UTRA); Radio Resource Control (RRC); Protocol specification', V9.7.0, June 2011.
4. 3GPP TS 25.304, 'User Equipment (UE) procedures in idle mode and procedures for cell reselection in connected mode', V9.5.0, June 2011.
5. 3GPP TS 25.367, 'Mobility procedures for Home Node B (HNB); Overall description; Stage 2', V9.5.0, December 2010.
6. 3GPP TS 36.304, 'Evolved Universal Terrestrial Radio Access (E-UTRA); User Equipment (UE) procedures in idle mode', V9.7.0, June 2011.
7. 3GPP TS 23.003, 'Numbering, addressing and identification', V9.7.0, June 2011.
8. Qualcomm Europe, CATT, 'Renaming Allowed CSG List', 3GPP TSG-RAN R2-097120, November 2009.
9. 3GPP TS 24.285, 'Allowed Closed Subscriber Group (CSG) list; Management Object (MO)', V9.4.0, December 2010.
10. 3GPP TS 22.220, 'Service requirements for Home Node B (HNB) and Home eNode B (HeNB)', V9.7.0, June 2011.
11. 3GPP TR 23.830, 'Architecture aspects of Home NodeB and Home eNodeB', V9.0.0, September 2009.
12. 3GPP TS 25.467, 'UTRAN architecture for 3G Home Node B (HNB); Stage 2', V9.4.1, March 2011.
13. Qualcomm, 'LS on access control for CSG cells', 3GPP TSG-RAN R2-084948, September–October, 2008.
14. 3GPP TS 25.413, 'UTRAN Iu interface Radio Access Network Application Part (RANAP) signalling', V9.6.0, June 2011.
15. M. Simsek, T. Akbudak, B. Zhao and A. Czylwik, 'An LTE Femtocell Dynamic System Level Simulator', *ITG Workshop on Smart Antennas WSA 2010*, February 2010.
16. IEEE 802.16m-08/004r4
17. Femto Forum, 'Interference Management in UMTS Femtocells', December 2008.
18. 3GPP TR 36.814, 'Evolved Universal Terrestrial Radio Access (E-UTRA); Further advancements for E-UTRA physical layer aspects', V9.0.0, March 2010.
19. M. Simsek, H. Wu, B. Zhao, T. Akbudak and A. Czylwik, 'Performance of Different Cell Selection Modes in 3GPP-LTE Marco-/Femtocell Scenarios', *Wireless Advanced WiAd 2011*, June 2011.
20. Z. Bharucha, A. Saul, G. Auer and H. Haas, 'Dynamic Resource Partitioning for Downlink Femto-to-Macrocell Interference Avoidance', *EURASIP Journal on Wireless Communications and Networking*, vol. 2010, Article ID 143413.
21. Z. Fan and Y. Sun, 'Access and Handover Management for Femtocell Systems', *Vehicular Technology Conference (VTC 2010-Spring)*, 2010 IEEE 71st, 2010, pp. 15.
22. 3GPP TS 23.122, 'Non-Access-Stratum (NAS) functions related to Mobile Station (MS) in idle mode', V9.6.0, June 2011.

23. 3GPP TS 36.101, 'Evolved Universal Terrestrial Radio Access (E-UTRA); User Equipment (UE) radio transmission and reception', V9.8.0, June 2011.
24. 3GPP TR 25.967, 'Home Node B Radio Frequency (RF) Requirements (FDD)', V9.0.0, May 2009.
25. Nokia Corporation, 'LS on reselection handling towards non-allowed CSG cell', 3GPP TSG-RAN R2-084891, August 2008.
26. Qualcomm Europe, 'Intra-frequency reselection indicator for CSG cells', 3GPP TSG-RAN R2-085383, October 2008.
27. J. Zhang and G. de la Roche, *Femtocells: Technologies and Deployment*, John Wiley & Sons, December 2009

11

Distributed Location Management for Generalized HetNets. Case Study of All-wireless Networks of Femtocells

Josep Mangues-Bafalluy, Jaime Ferragut, and Manuel Requena-Esteso
Centre Tecnològic de Telecomunicacions de Catalunya, Spain

11.1 Introduction

The increase of capacity offered by wireless networks is closely associated with the reduction of cell radii. In fact, as pointed out in [1], out of a million-fold capacity increase since 1957, a 1600× factor of system capacity improvements is due to reduced cell sizes. Much less important are other factors, such as the availability of wider spectrum or better modulation schemes. This is highly relevant in a context in which the growth of mobile broadband traffic is starting to stress access and backhaul networks of mobile network operators (MNOs). This is one of the main driving forces behind heterogeneous cellular networks (HetNets). In the 3GPP vision, HetNets combine MNO-deployed cells (e.g., macrocells, microcells) with short-range, subscriber-deployed nodes (e.g., femtocells), but it is assumed that they all use 3GPP access technologies (e.g., LTE), and depending on the scenario, even the same technology [3].

In this chapter, we extend/generalize the HetNet vision along various lines. First, we consider diverse wireless technologies, including non-3GPP ones (e.g., WiFi). Second, we study the implications of HetNet deployments, not just at the radio resource management level, as usually

Heterogeneous Cellular Networks, First Edition. Edited by Rose Qingyang Hu and Yi Qian.

done, but also at the architectural level. Finally, we combine/integrate schemes coming from the data networking community with those defined in the 3GPP architecture. The following paragraphs further develop the motivation for such a HetNet vision as well as the approach followed for each of the above points.

11.1.1 Motivation

There are several factors contributing to a substantial increase in the complexity of mobile networks. Along with the heterogeneity of technologies used in the access and the backhaul, there is also the (partially) uncontrolled deployment of small cells. In fact, capacity-oriented deployments (as opposed to coverage-oriented ones) are expected to require a substantial number of small cells. Traditional pre-planning strategies are not applicable in this case owing to the complexity of the environment. Hence, more self-* functionalities should be put in place. Furthermore, the high volume of traffic that will be handled by these networks mandates for a better exploitation of all available technologies being in place. This should include 3GPP and non-3GPP technologies. For instance, WiFi is already being extensively used to offload traffic in the access, but it could also be exploited in alternative ways, as explained in the following section.

The complexity and dynamicity of the above networks advocate for network operations in which performance improvements are expected to mainly come from a coordinated operation of the network elements rather than from improvements of isolated elements (for example, better modulation). This has important architectural implications. In fact, the architecture should be flexible enough to integrate the above heterogeneity and should scale to unprecedented levels. This is particularly critical when wireless technologies are in use. All-wireless networks of femtocells take into account these constraints.

Furthermore, reliability of some of these deployments should be guaranteed through redundancy (that is, with the help of an adaptable network) rather than by deploying reliable isolated nodes, given the reduced control from the operator's side operator (e.g., small cell deployments in lampposts or at home). Data networking principles (already adopted by 3GPP) were designed to fulfil these needs. Therefore, all this potential should be fully exploited in these new environments.

In this chapter, we focus on all-wireless networks of femtocells, as they allow for fast, flexible and scalable deployments of femtocells. As already explained, these kinds of deployments have important architectural implications. Therefore, this also applies to one of the key building blocks of any mobile network, namely mobility management. Mobility management is traditionally divided into handoff management and location management. While the former in general receives much more attention, the latter is revisited in this chapter, in which an architectural solution complying with all the constraints and requirements introduced above is presented. New solutions are required in these kinds of deployments in which there is an all-wireless local backhaul between neighbouring femtocells. In this case, the consumption of the scarce wireless resources of such backhaul due to the transmission of control messaging associated to location management should be minimized. At the same time, large-scale deployments (e.g., outdoor deployments of small cells in lampposts) require schemes that also scale, for instance, by distributing the intelligence and the location databases, yet resulting in a reliable service. Furthermore, the integration of schemes coming from the data networking arena within a 3GPP architecture, which have very different philosophies, should also be done carefully. The following sections present our approach for tackling these issues.

11.1.2 Approach

Here, we briefly introduce the approach/solutions adopted to handle the issues identified above, which will be developed throughout the chapter in more detail.

First, and related to the heterogeneous access technologies point, there is an increasing trend to combine the use of WiFi networks and 3GPP cellular networks with various levels of integration. As a consequence, WiFi is being added and is becoming increasingly important in the HetNet vision. Moreover, WiFi is becoming a carrier-grade technology and, in addition, there are ongoing standardization efforts that would allow the offering of gigabit per second rates. Based on this, we not only consider WiFi for the access, but also for other types of links, for example, the local backhaul. Furthermore, the use of other access technologies is not precluded in these generic HetNet scenario.

Second, in conventional deployments of standalone femtocells, each femtocell has a wired connection (provided by an ISP) towards the core network of the mobile network operator (MNO). On the other hand, the EU project BeFEMTO [2] has defined alternative deployment scenarios, such as networks of femtocells (NoF). In this case, a local network (e.g., the enterprise LAN) is used to provide connectivity to the femtocells, and additional elements are introduced to confine local traffic (control and data) to the local network. This chapter argues for the idea of deploying all-wireless NoFs to reduce deployment and operation costs. In particular, it focuses on large-scale all-wireless NoFs that are connected to the core network of the MNO through LFGWs, appropriately distributed throughout the NoF. Sample application scenarios include shopping malls, airports, conferences and stadiums and, with the increasing importance of metro femtocells/small cells, it might also be of application in dense urban outdoor scenarios.

Such innovative HetNet deployments have several architectural implications. However, the HetNet scenario has been generally dealt with from a radio resource management perspective. On the other hand, in this chapter, we focus on the architectural implications of a particular instantiation of HetNet deployment, that is, all-wireless NoFs. In this direction, our architecture decentralizes some relevant functionality defined by 3GPP towards a more efficient operation of the network, particularly in local networks. This architecture is founded on two main components: (1) a new node, the LFGW, and (2) a geographic underlay network. The former allows scaling of the signalling towards the core by appropriately confining it to the local network, while the latter allows scaling in terms of signalling internal to the NoF by reducing the overhead sent over the air. Still, despite these modifications, the evolved packet core and user equipment remain unchanged. This is attained by building a geographically based underlay network (at layer 2.5 in the stack) in the NoF and by embedding into LFGWs functionality that is similar to that of home eNode-B gateways (HeNB GW).

Last but not least, with the move to an all-IP core network in the evolved packet system (EPS) [7, 8], there is also an increasing trend to integrate concepts traditionally linked to data networking in the networks of telecom operators. This chapter contributes to further extending this trend in the context of femtocell deployments. More specifically, our architecture integrates concepts from both wireless mesh networking and the 3GPP EPS architecture.

11.1.3 On Location Management in Generalized HetNets

One of the challenging issues to solve in generic femtocell deployments is mobility management. Nevertheless, most research and standardization efforts in general focus on handoff management (e.g., in the context of self-organized networks, SON [4]) and much less attention is paid to location management. In this respect, all-wireless NoFs are even more challenging, but in this case, the interest is shifted towards the edge of the network in the form of local mobility management. That is, the study of the procedures triggered by movements of user equipment (UE) internal to the NoF and the impact these may have on the core network. In this chapter, we focus on the optimization of those 3GPP procedures that handle mobility management inside the NoF. More specifically, we make optimizations to procedures related to location management inside a large-scale all-wireless NoF that exploits the availability of geographic information. Such procedures include paging and tracking area updates, as well as those handoff management procedures that also have implications for location management. Furthermore, we adapt the VIMLOC [6] distributed location management scheme (designed for large-scale wireless mesh networks) to the requirements of an all-wireless NoF. Our scheme is not only designed to scale inside the NoF, but also in a wider context. In fact, it is also designed to restrict mobility management signalling to the scope of NoFs as much as possible, and to only send selected signalling traffic to the core network of the MNO to inform the relevant nodes of significant changes of the mobility state of the UE. In this way, it also offloads the core network, which allows the whole network of the MNO to better scale. Finally, we revise the 3GPP procedures related to location management, namely paging, tracking area update and handoff management, and explain how relevant 3GPP messages are used to trigger VIMLOC messages so as to make these procedures better scale in an all-wireless NoF. Some message sequence charts of relevant procedures, which embed all the above concepts, are provided as example.

The rest of this chapter is structured as follows. First, we give an overview of geographic routing and location management and we explain the operation of VIMLOC (Section 11.2). After that, we explain the concept of all-wireless network of femtocells (Section 11.3). We also present an architecture for location management in the new geographic-based all-wireless NoF (Section 11.4). Section 11.5 describes in detail the message sequence charts for location management mechanisms. Finally, we conclude the chapter (Section 11.6).

11.2 Background on Geographic Routing and Geographic Location Management

The exploitation of the geographic information available at nodes is one of the two main pillars on which the work presented in this chapter is founded. Geographic routing relies on geographic position information to route packets towards a destination. In this case, geographic positions of nodes involved in the communication are carried in the header of the packet, which will be referred to as a geopacket. In this way, nodes decide the next-hop for geopackets by taking a look at the coordinates of neighbouring nodes stored in their neighbour table and by selecting, for instance, the closest one to the location of the destination.

The rationale behind exploiting geographic information is to benefit from the scalability of geographic routing protocols, since they store much less information at the nodes and they send much less control traffic through the network [5]. In this way, less radio resource is consumed, which makes these schemes interesting in multi-hop wireless networks, such as all-wireless NoFs.

On the other hand, location management focuses on managing databases that store the position/location of a node inside the network. In this way, new communications towards a node could eventually be established by previously requesting the location of the destination to such a database. In this chapter, we adapt a distributed location management scheme, called VIMLOC [6], which was originally designed for large-scale wireless mesh networks, to the requirements of an NoF. In VIMLOC, node_ID-to-location mappings are stored in wireless mesh routers (WMRs) inside a geographic region called the HGC (home geocluster), whose central coordinate is obtained by applying a well-known hash function to the node_ID. Each node may have multiple HGCs (e.g., two in Figure 11.1), depending on the reliability-vs.-control overhead tradeoff to be deployed. This mapping is also stored in the surroundings of each mobile node (MN). This region is called the VGC of that MN, and it is updated more often than the HGCs of that same MN. In this way, overhead is reduced at the cost of potentially having outdated information at HGCs. As the geopacket reaches the VGC, this outdated information (carried by geopackets after the sender requested it to HGCs) is corrected, and the geopacket is eventually diverted to the current location of the MN. Figure 11.1 presents the operation of VIMLOC. It also shows how location updates are geobroadcast towards the HGCs and location requests are geoanycasted towards HGCs.

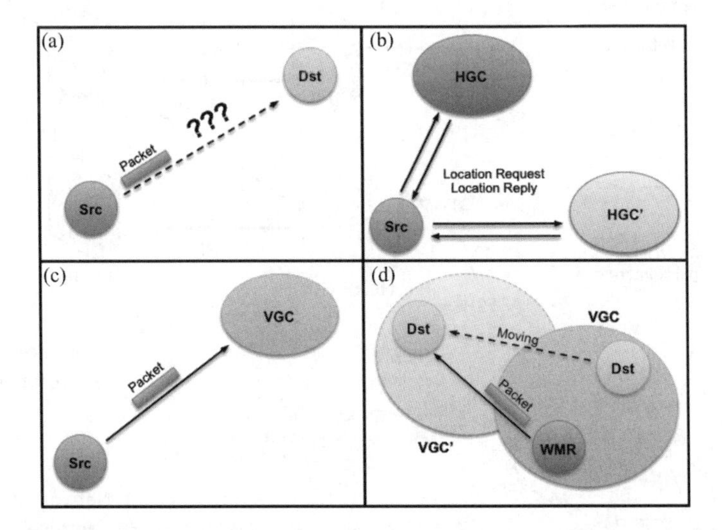

Figure 11.1 Operation of the distributed location management protocol VIMLOC. Location updates continually geobroadcast by nodes to their respective HGCs, obtained by applying geographic hash to their own nodeID. (a) Source node has traffic for destination nodeID. (b) Location requests/replies: source applies geographic hash to nodeID, obtains HGC regions and sends geoanycast request, node inside the HGC answers request with location of nodeID. (c) Source sends traffic to position of last location update (i.e., available in HGCs), which is possibly outdated. (d) Nodes inside VGC divert packets to actual destination of nodeID. Adapted from J. Mangues-Bafalluy, M. Requena-Esteso, J. Núñez-Martínez, A. Krendzel, "VIMLOC location management in wireless meshes: Experimental performance evaluation and comparison", Proceedings of IEEE ICC 2010.

11.3 All-wireless Networks of Femtocells

Figure 11.2 illustrates the concept of an all-wireless NoF integrated in the evolved packet system (EPS). The evolved packet core includes all standard network entities, namely the P-GW, S-GW, MME and HeNB GW. It also presents the most relevant 3GPP interfaces. In particular, the communication between the core and the NoF is done through the S1-MME (control) and S1-U (data) interfaces, and by traversing the network of an ISP. For more details on the functionality of each of these entities, the reader is referred to [7]. Inside the NoF, a special node – the local NoF gateway (LFGW) – is needed for interfacing the NoF and the core network, as explained in detail below. Furthermore, femtocells (or HeNBs) are connected to each other without the need for cable laying, hence cooperating to forward packets through multi-hop wireless paths towards/from LFGWs. It is in this generalized HetNet context that multi-radio femtocells can be fully exploited. In fact, it is already common to find femtocells that, in addition to a 3GPP access technology, also mount a WiFi radio. This WiFi radio could be further leveraged if it was used not only to provide access but also (or alternatively) as local backhaul.

Compared to the traditional deployment of (isolated) femtocells, an NoF allows a more cost-effective and flexible deployment by building a network between nearby femtocells, hence

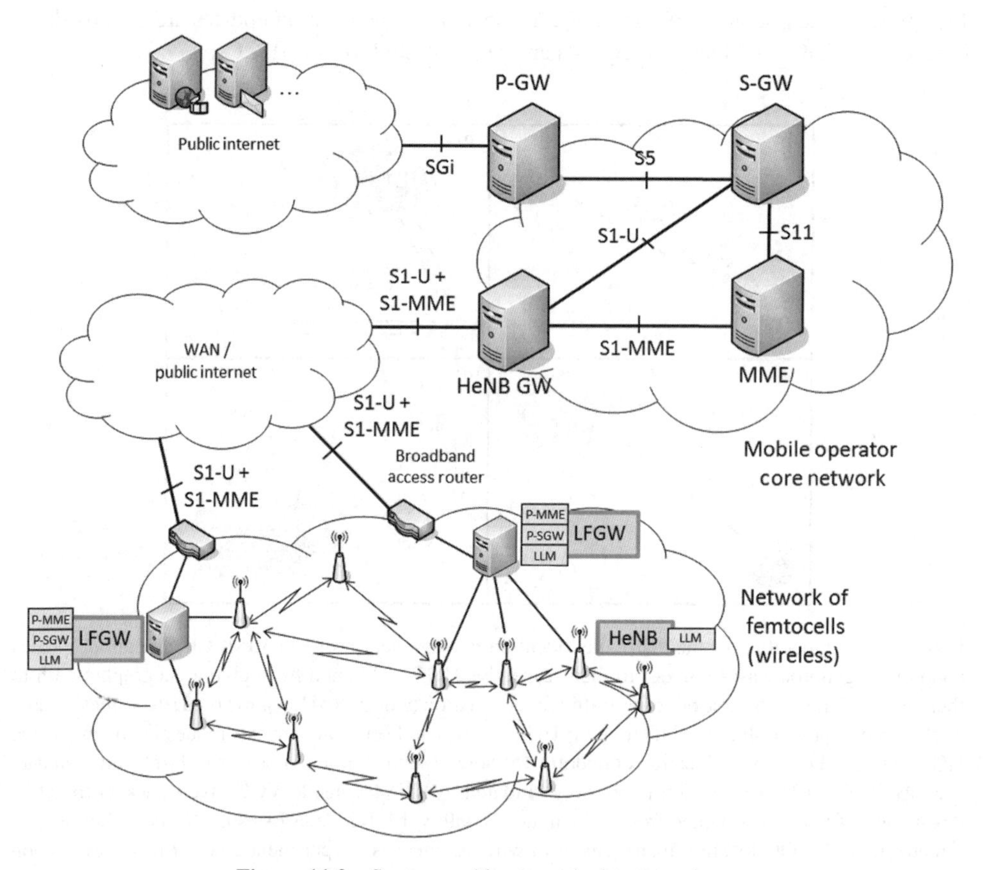

Figure 11.2 System architecture: deployment view.

allowing multiple femtocells to share the same connection to the core network of the MNO. This presents several advantages. First, it reduces deployment and operation costs for both the MNO and the local network operator (LNO). Second, it offloads signalling traffic from the core network of the MNO. Third, when combined with local IP access (LIPA) and selected IP traffic offload (SIPTO), it also helps offloading data traffic from the core network of the MNO. Furthermore, if the links between femtocells are wireless, the resulting network of femtocells is also heterogeneous (in a generalized way) in the sense that it combines data networking and 3GPP schemes to build a 3GPP-oriented wireless mesh network. In this way, the need for cable laying to reach each femtocell is eliminated.

The interest in wireless NoF deployments combined with the ever-increasing link rates offered by wireless technologies led us to make a case for large-scale all-wireless NoFs. The potential application scenarios may include airports, shopping malls and stadiums, and also, with the increasing importance of metro femtocells (or small cells), dense outdoor urban scenarios.

However, and due to the multi-hop wireless nature of such networks, regular 3GPP signalling procedures may consume too much radio resources. Here we further develops on the challenges of NoFs.

11.3.1 Challenges of All-wireless Networks of Femtocells

The integration of data networking concepts, such as a large-scale geographic-based all-wireless NoF, in a 3GPP context requires a great deal of architectural reasoning. This high-level architectural design has implications for relevant procedures of network operation, but the adaptation of such procedures to the new architecture is constrained by what has already been standardized by 3GPP. In practice, this translates into the constraint of not introducing modifications either to user equipment (UE) or to 3GPP procedures. Femtocells have also been thoroughly standardized in EPS but NoF-like scenarios are just starting to be considered.

Procedures related to mobility management are particularly relevant. In this chapter, we focus on location management within the NoF. More specifically, location management enables the network entities (NE) to track UE locations. In other words, location management solves the problem of mapping the UE identifier to its physical/logical location in the network. The UE identifier is permanent during the life of the UE and, depending on the context, it may be based on the international mobile subscriber identity (IMSI) and/or the serving temporary mobile subscriber identity (S-TMSI). On the other hand, the location address varies as the UE moves and it is used by the underlying transport network to route packets towards the UE.

In a large-scale NoF, a large number of UEs will be changing from one femtocell to another one in relatively short periods of time. Consequently, local location management must solve the scalability problem in terms of signalling towards the mobility management entity (MME) at the core network of the MNO. Therefore, signalling traffic must be kept local as much as possible. Another scalability problem is caused by the internal large-scale, all-wireless nature of the local network under consideration. In this case, the goal is to minimize over-the-air signalling traffic. Finally, another challenge for local location management is the distribution of the location database in the local network. This distribution affects the overall location update cost, as well as location request/paging procedures.

11.4 Architecture for Geographic-based All-wireless Networks of Femtocells

As explained above, the deployment of all-wireless NoFs as an instantiation of a generalized HetNet is constrained by 3GPP specifications. Therefore, the architectural high-level design principles are those of the evolved packet system (EPS) (e.g., all-IP architecture). Still, some adaptation is needed for the correct integration of NoFs in the EPS.

In this chapter we advocate the adequacy of concepts previously developed in the context of data networking over wireless mesh networks (e.g., geographic routing and distributed location management) to solve some of the issues that appear in large-scale all-wireless NoFs while leaving 3GPP procedures unchanged. In the following, we explain how we achieve this goal.

11.4.1 Overview of the Architecture

Bearing in mind the above scalability challenges, two main components were envisioned, namely a geographic network underlay that exploits geographic information for key network layer procedures (e.g., routing and mobility management) and an LFGW acting as an interface between the NoF and the core network of the MNO.

A key building block of our architecture for the NoF is a sublayer (referred to as the geo sublayer in Figure 11.3) inserted just below the IP layer in the stack. The geo sublayer is

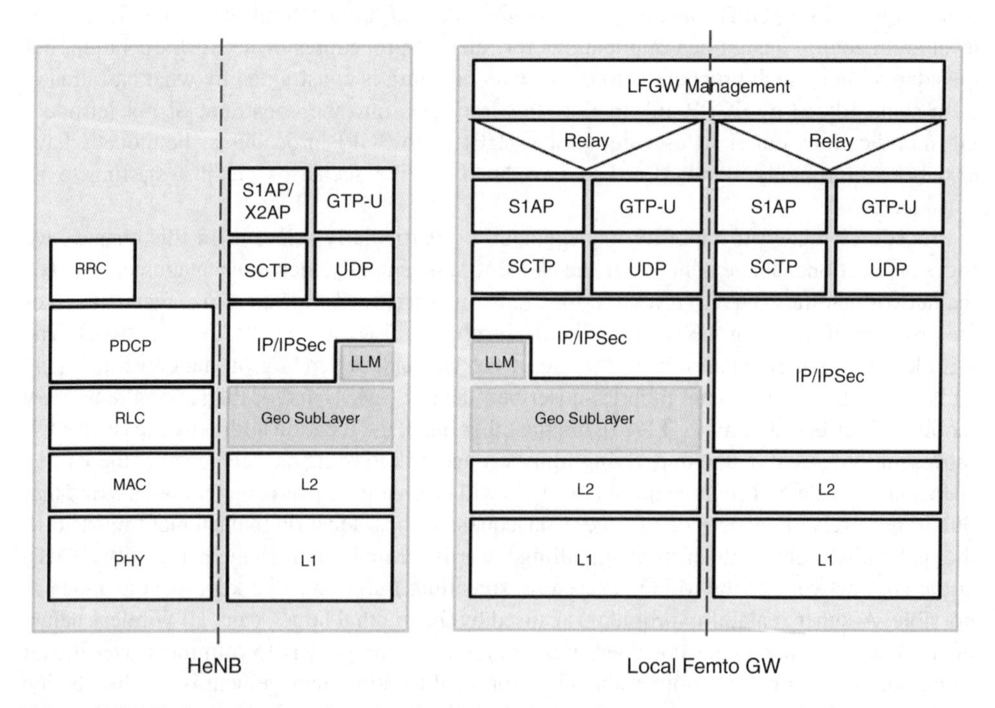

Figure 11.3 Node architecture for control and data planes: HeNB and local femto GW (LLM stands for local location management).

capable of handling geographic information and executing network layer functionality, such as routing or mobility management, based on this information. In fact, IP packets are transported as a payload of geopackets. The rationale behind exploiting geographic information is to benefit from the scalability of geographic routing protocols. In this way, less radio resource is consumed, which makes these schemes interesting in multi-hop wireless networks. This also requires maintaining an associated local location management service (LLM) in charge of mapping permanent (or long-lasting) UE identifiers (e.g., IMSI, S-TMSI) to the coordinates of the current position of the UE in the NoF. In fact, as explained in detail below, the coordinates used are those of the femtocell where the UE is currently camping (or was last seen).

Furthermore, the geographic underlay network is transparent to both IP and 3GPP messaging, hence not implying any modification to their specifications, and thus, to their regular operation. This component is the main facility responsible for providing a scalable solution inside the all-wireless NoF.

Additionally, scalability towards the core network is attained by introducing a new network node (the LFGW) acting as interface between the NoF and the core. There may be multiple LFGWs in an NoF, mainly depending on the network dimensions as well as its traffic patterns. These nodes will act as traffic aggregation points of the NoF (for both signalling and user data) towards the core network of the mobile network operator (MNO). In addition, they will have similar functionality to that of an HeNB GW in the sense that the network entities of the core network (MME and S-GW) and HeNBs send regular S1 messages, as if they were communicating with each other. However, these messages are being intercepted by the LFGW, which triggers the corresponding geographic procedures internal to the NoF without modifying the original traffic.

11.4.2 Network Entities Supporting Networks of Femtocells

Figure 11.3 illustrates the protocol stacks of LFGWs and femtocells (or HeNBs) of the NoF. Protocols for both the user plane and the control plane are represented. The following describes the functionality of these nodes in detail.

11.4.2.1 Local network of femtocells gateway (LFGW)

LFGWs are introduced in our architecture to be deployed in the NoF for fully benefitting from the reduction of signalling towards the core. They embed two network entities already defined for the core network, namely SGW and MME, but with their scope restricted to the NoF. They act as a proxy SGW (PSGW) for data traffic and proxy MME (PMME) for control traffic. In addition, they act on behalf of HeNBs in the NoF by making the whole NoF look like an HeNB from the point of view of the core. In this chapter, we assume that both network entities are packed in a single box to ease deployment. Therefore, this box exposes the regular S1-MME and S1-U interfaces towards both the core network and the HeNBs. The S11 interface (between the PSGW and the PMME) is kept internal to the LFGW.

There are two stacks represented in Figure 11.3, one facing the NoF and the other one facing the ISP that connects the NoF to the MNO. Nothing is modified in the latter. On the other hand, the geo sublayer is introduced for both the data and the control plane in the stack facing the NoF, which means that all IP packets sent through this interface will eventually be transferred through the network, encapsulated in geopackets. Furthermore, LLM (VIMLOC) messages are also transferred inside geopackets.

In terms of 3GPP message handling, the LFGW filters out those control messages that do not need to be sent through the NoF, because the LLM provides more precise information. For instance, when paging a UE, the LFGW does not transfer a paging message to each HeNB in the NoF. Instead, it just transfers those heading towards the cell where the UE is currently camped (and, potentially, neighbouring cells). In this way, less radio resource is consumed in the NoF. In brief, relevant messages of 3GPP procedures are used to trigger the appropriate LLM procedures, aiming at eventually reducing radio resource consumption inside the NoF.

Modifications to HeNBs

The functionality and 3GPP interfaces supported by HeNBs are not altered. The only difference in the protocol stack is the presence of the geo sublayer and the LLM in the stack facing the NoF. The regular relaying of messages from S1AP to RRC is not modified either. In this case, the main idea is the same as that of the LFGW, that is, only relevant messages are sent to the LFGW, and instead, certain messages (e.g., tracking area updates) are intercepted in order to trigger the appropriate VIMLOC messages. In this way, location information is distributed throughout the network, hence improving scalability.

Besides, and as part of the regular operation of VIMLOC, HeNBs may act as HGCs and/or VGCs of UEs in the NoF. This means that they store part of the distributed database built by VIMLOC, hence keeping information carried in location updates and answering location requests for those UEs. However, all these operations are transparent to 3GPP procedures.

Furthermore, the availability of WiFi radios in addition to 3GPP radios could also be leveraged to build the local backhaul of the all-wireless NoF among WiFi-enabled femtocells. Other wireless technologies could also be used for the same purpose.

11.4.3 Operation of the Network of Femtocells

Much in the same way of our implementation of VIMLOC in [10], the way in which the geo sublayer is exposed to the IP layer is in the form of a virtual interface through which the routing table in the kernel of the HeNB will send packets encapsulated in frames. The key aspect is that the geo sublayer answers the address resolution requests (ARP) by providing a fake MAC address. This allows the L2 frame carrying the IP packet to be built. Then, the IP packet is sent through the virtual interface down the stack to the geo sublayer, where the appropriate LLM-related processing takes place.

Furthermore, since the routing inside the NoF takes place at the geographic level, there is no need for pre-planning of IP addressing. This allows for a flat addressing scheme and eases the deployment and upgrade of NoFs.

The basic approach of the NoF architecture is to read relevant 3GPP messages related to location management as they traverse a given node (femtocell or LFGW) without modifying them, and to use them to trigger appropriate VIMLOC procedures in the geographic underlay network. In practice, this translates into: (1) using S1AP paging messages to trigger VIMLOC location requests, and (2) using tracking area update and handoff confirm messages to trigger VIMLOC location updates.

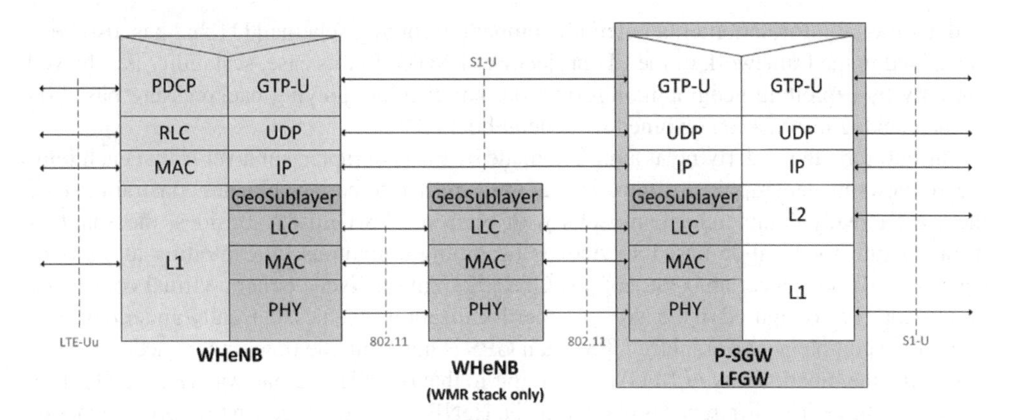

Figure 11.4 Sample WiFi-based all-wireless network of femtocells. User-plane protocol stacks for nodes inside the NoF.

Bearing in mind that this architecture is designed for large-scale all-wireless NoFs, its main advantage with respect to centralizing all location-related information at the PMME is that location servers, and so, location management signalling, is distributed throughout all the HeNBs of the network. This is precisely what makes this solution scale.

11.4.4 Sample Protocol Stacks for Wifi-based All-wireless NoFs

Figure 11.4 illustrates the user-plane protocol stack for nodes that compose the all-wireless NoF when WiFi is used as the technology for the local backhaul. In particular, it presents the stacks and interfaces exposed by the femtocells (referred to as wireless HeNB, or WHeNB, in the figure) and the LFGW (acting as P-SGW for the user plane) in which the physical and MAC layers are those of WiFi. Furthermore, each femtocell also contributes in building the mesh by forwarding traffic coming from other femtocells. The stack in the middle represents such transport-only functionality, which only requires processing geopackets up to the geo sublayer (that is, no 3GPP procedures involved). This behaviour is referred to in the figure as wireless mesh router (WMR). Besides, since only one stack is represented in the middle node of the figure, this WMR only mounts a single WiFi radio.

Notice that the modifications are added to the transport network (in fact, below IP), and are internal to the NoF. Therefore, there are no changes to either the 3GPP procedures or to the interfaces exposed by NoF nodes to the UEs and the EPC. A similar reasoning could be applied to control plane stacks.

11.4.5 Other Relevant Issues

The location management scheme described above will feed the associated geographic routing with the geographic information needed to forward/route geopackets. The same scalability challenges of location management constrain the design of the routing scheme. The description of the routing protocol is out of the scope of this chapter. However, the interested reader may find additional information on a potential companion routing scheme [11] that has also been conceived in the framework of the EU project BeFEMTO. In this case, scalability is achieved not only by exploiting geographic information, but also by applying backpressure based on queue backlog information obtained from neighbours.

Furthermore, the underlying assumption made by the geographic sublayer is that each femtocell knows its geographic position. This is not expected to be a significant constraint, since there are already commercial femtocells with built-in GPS chipsets. Besides, the enabling infrastructure for location-based services is becoming ubiquitous by exploiting all sorts of signals available at a certain spot, not just GPS [12]. Alternatively, either a virtual coordinate system may be configured in the NoF, or other localization schemes, currently under development, may complement or replace GPS when GPS is not available (e.g., indoor scenarios).

Finally, the functionality of LFGWs is similar to that of an HeNB gateway (HeNB GW), as explained above. The difference is that while the HeNB GW is deployed in the core network of the MNO, the LFGW is deployed in the NoF. This poses additional management and security constraints. However, these constraints are beyond the scope of this chapter.

11.5 Location Management Procedures

This section presents some relevant message sequence charts (MSCs) to illustrate our location management solution. The procedures described below are a modified version of the S1-based handoffs (HOs) explained in [7] and [8]. Additional messages have been included in order to implement our location management solution.

Figure 11.5 describes a taxonomy for the different mobility scenarios in an NoF. Mobility scenarios have been categorized according to the relocation of functional entities (namely P-MME/P-SGW and MME/S-GW) within the LNO and the MNO domains. For each mobility scenario, a summary of the most relevant implications on location management procedures has been included.

As an example of the details of operation of our architecture, the following describes message sequence charts for paging, and intra-local network, intra-MME S1-based HO with LFGW relocation.

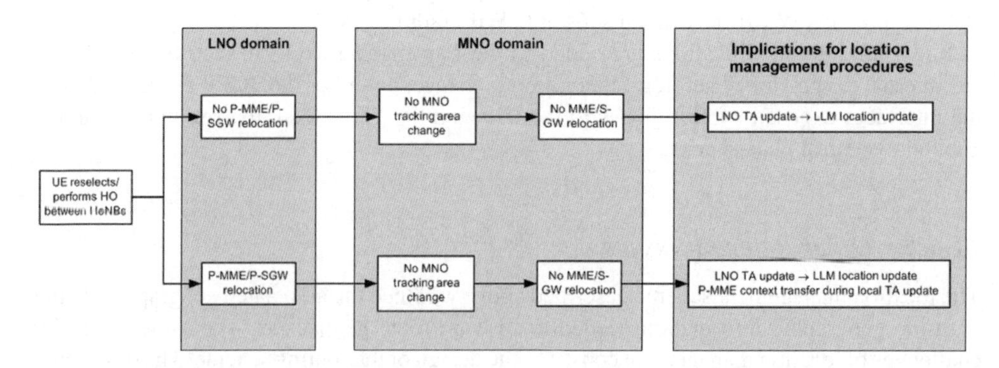

Figure 11.5 A taxonomy for mobility scenarios in a network of femtocells.

11.5.1 Paging

The purpose of the paging procedure is to enable the MME to page a UE in the specific eNB [9]. In an NoF, messages involved in the paging procedure need to be forwarded throughout the NoF in order to reach the appropriate HeNB. The MSC for this location management procedure is described in detail in Figure 11.6.

When a UE is in idle mode (more specifically, ECM-IDLE state) and a downlink IP packet from an external network is received at the P-GW, the EPC needs to establish a connection with the UE in order to deliver the incoming packet. Once this packet has reached the S-GW, it is forwarded on the specific S1 bearer to the P-SGW of the LFGW that manages the HeNB where the UE was last seen. After reaching the P-SGW, a GTP-C downlink data notification message is sent to the corresponding P-MME. The reception of this message triggers the paging procedure in the P-MME.

As explained in [9], the paging procedure requires the MME to send S1-AP paging request messages to all eNBs in the tracking area (TA) where the UE was last seen. In our solution, the LLM/VIMLOC sublayer located in the P-MME intercepts and stores all MME-originated S1-AP paging request messages and, in turn, sends a location request to the HGC indicating the S-TMSI of the UE that needs to be paged. Once the request is received and processed at the HGC, a location reply message indicating the IP and geographical addresses of the HeNB where the UE is currently camped on is sent back to the P-MME. This allows the LLM sublayer in the P-MME to discard all stored S1-AP paging request messages, except for that intended for the HeNB where the UE is currently camped. Finally, the P-MME sends a single

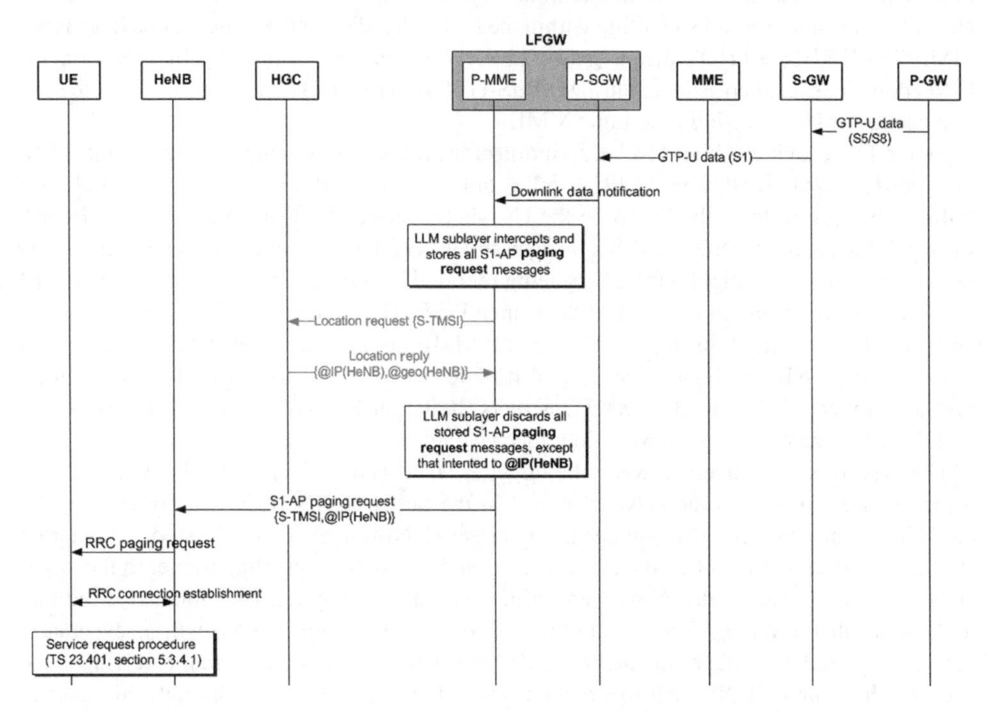

Figure 11.6 Message sequence chart for a paging procedure in a network of femtocells.

S1-AP paging request message to the HeNB where the UE is currently camped and the UE is effectively paged.

As a final remark, it must be noted that for geographic forwarding, the only information needed is the geographic position of the femtocell and not its IP address. However, this IP address is still needed for the correct operation of the all-IP transport required by the EPS.

11.5.2 Handoff

A handoff procedure takes place when a UE in active mode (more specifically, in ECM-CONNECTED state) moves from a source to a target HeNB. HOs are required in order to keep UE reachability within the EPC, and to ensure a seamless transmission and reception of user data [7]. 3GPP TS define X2- and S1-based HOs, depending on the logical interface that is used to deliver control- and user-plane data during the handoff procedure. In our scenario (Figure 11.7), we have considered S1-based HOs between HeNBs in the NoF that involve LFGW relocation (hence P-MME and P-SGW relocation), but no MME/S-GW relocation or TA updates.

From a location management point of view, HOs in an NoF entail two procedures that need to be addressed: LLM location updates in the HGC when a UE moves between HeNBs, and control- and user-plane path switches in the MME/P-MME, and the S-GW/P-SGW during LFGW relocations.

Our solution for intra-local network (intra-LN), intra-MME S1-based HOs with LFGW relocation combines aspects from traditional 3GPP intra- and inter-MME/S-GW S1-based HOs [7,8]. In summary, all signalling within the NoF is handled in the same way as in an inter-MME/S-GW S1-based HO (since it involves LFGW relocation), while signalling reaching the MNO domain is handled as in an intra-MME/S-GW S1-based HO (since the EPC sees the HO as a change of HeNBs within the same MME).

Figure 11.7 provides a high-level description of the messages exchanged during an intra-LN, intra-MME S1-based HOs with LFGW relocation in an NoF. Prior to step 1, downlink user traffic is being sent from the S-GW to the UE via the source P-SGW and the source HeNB. In step 1, the source HeNB decides to trigger an HO via the S1 interface based on the power levels for serving and neighbour HeNBs being reported by the UE. In step 2, the source HeNB sends a handover required message to the source P-MME in order to notify that an HO to the target HeNB is required. In step 3, the source P-MME notifies the target P-MME that an HO to one of its HeNBs is about to occur, and it indicates so with a forward relocation request message. In step 4 the target P-MME instructs the target P-SGW to allocate resources for uplink user-plane traffic on the S1-U interface.

The messages exchanged between the target P-MME and the target HeNB in steps 5 and 6 are required to perform admission control in the target HeNB. This is a critical step, as the HO can only proceed if resources in the target HeNB are available. In step 7, the target P-MME instructs the target P-SGW to create an indirect data forwarding tunnel to the target HeNB in order to ensure downlink data traffic continuity during HO execution. It is indirect in the sense that traffic originally sent to the UE through the source HeNB will be eventually delivered to the UE through the target HeNB after handoff. Therefore, when receiving these packets, the source HeNB will divert them towards the target HeNB through this interim indirect forwarding tunnel. In step 8, the target P-MME notifies the source P-MME that the

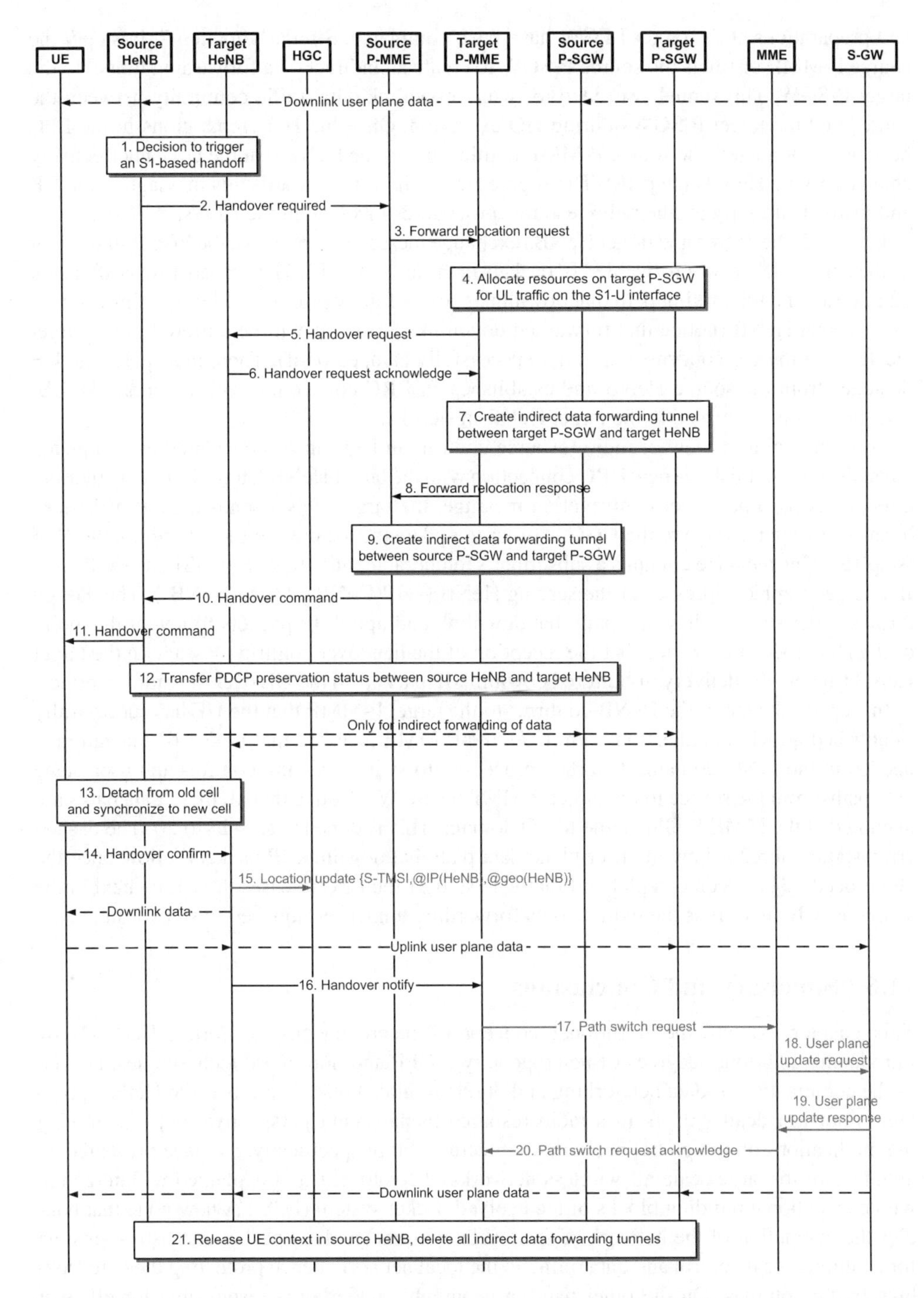

Figure 11.7 Message sequence chart for an intra-local network, inter-LFGW S1-based handoff without MME/S-GW relocation or tracking area update in a network of femtocells.

HO preparations at the target LFGW have been completed. Similarly to step 7, in step 9 the source P-MME instructs the source P-SGW to create an indirect data forwarding tunnel to the target P-SGW. This tunnel is needed to ensure downlink data traffic continuity between the source and the target P-SGWs during HO execution. Once the HO preparations in the EPC have been completed, the source P-MME notifies the source HeNB that the UE can effectively change serving HeNBs (step 10). The source HeNB, in turn, forwards this message to the UE and instructs it to trigger the radio resource procedures involved in the HO (step 11).

In step 12, the source and target HeNBs exchange messages to preserve the PDCP status, that is, the correct delivery of forwarded IP packets in the target HeNB. The dashed arrows after step 12 describe the forwarding path that downlink packets follow prior to the UE synchronization to the target HeNB (notice that forwarded downlink IP packets will be buffered in the target HeNB until the UE confirms that it has successfully camped on it). Then, in step 13, the UE detaches from the source HeNB and establishes an RRC connection with the target HeNB. This process is confirmed to the target HeNB in step 14.

From the point of view of radio resource control, an HO can be considered as completed when the UE establishes a new RRC connection with the target HeNB (step 14). In our solution, after receiving a handover confirmation message, the target HeNB sends a LLM/VIMLOC location update message to the HGC in order to update the location of the UE within the NoF (step 15). This message contains a subscriber's temporal identity (e.g., S-TMSI), as well as the IP and geographic addresses of the serving HeNB (@IP(HeNB), @geo(HeNB)). The dashed arrows under step 15 show the path that downlink and uplink IP packets follow at this point of the HO procedure. Notice that the reception of the handover confirm message in the target HeNB triggers the delivery of buffered downlink IP packets to the UE over the radio interface.

In step 16, the target the HeNB confirms to the target P-MME that the UE has successfully established an RRC connection with it. As the P-MME is aware that no MME relocation is needed in the MNO domain, it starts a procedure to switch existing control- and user-plane data paths from the source to the target P-MMEs/P-SGWs. Notice that all these data paths are anchored at the MME/S-GW in the MNO domain. This is done in steps 17 to 20. The dashed arrows after step 20 show the user-plane data path that downlink IP packets follow once the HO procedure has been completed. Finally, in step 21 the EPC releases the UE context in the source HeNB, as well as the indirect data forwarding tunnels established in steps 7 and 9.

11.6 Summary and Conclusions

This chapter extends the traditional concept of heterogeneous networks (HetNets) by introducing additional degrees of heterogeneity (3GPP and non-3GPP technologies, as well as the combination of data networking and 3GPP architectures). Therefore, the HetNet problem is not just dealt with from a radio resource management perspective, but by studying the implications on the global network architecture. More specifically, we have presented an architecture for large-scale all-wireless networks of femtocells and explained its interaction with conventional building blocks of the evolved packet system (EPS). A new node that handles the interaction of the network of femtocells with the evolved packet core is responsible for confining local control and data traffic in the local network, hence protecting the core from high traffic volumes. On the other hand, a geographic underlay is responsible for efficient transport and location management by distributing and reducing the overhead sent over the air. Furthermore, user equipment and the evolved packet core remain unchanged.

The main conclusion of this chapter is that innovative HetNet deployments are feasible if the traditional HetNet vision is generalized. Furthermore, efficient operation can only be achieved by means of an architectural view of the whole network, which also entails a generalization of the traditional radio resource management viewpoint generally found in 3GPP-only HetNets. Such an architectural view has been applied to the design of the location management scheme presented in this chapter.

Acknowledgements

This work has been partially supported by the Generalitat de Catalunya under grant 2009-SGR-940, the Spanish Ministry of Science and Innovation under grant TEC2008-06826 and the Spanish Ministry of Education under grant FPU AP2009-5000, and performed in the framework of the ICT project ICT-4-248523 BeFEMTO, which is partly funded by the European Union. The authors would like to acknowledge the contributions of their colleagues from the BeFEMTO consortium.

References

1. V. Chandrasekhar, J. G. Andrews and A. Gatherer, Femtocell networks: a survey, *IEEE Communications Magazine*, 46(9), pp. 59–67, (2008). doi:10.1109/MCOM.2008.4623708.
2. BeFEMTO project website (2011), http://www.ict-befemto.eu. Accessed 11 October 2011.
3. A. Khandekar, N. Bhushan, J. Tingfang and V. Vanghi. LTE-Advanced: Heterogeneous networks, in Proceedings of the *European Wireless Conference 2010*, Lucca (Italy), 12–15 April 2010. doi:10.1109/EW.2010.5483516.
4. 3GPP TR 36.902, Evolved Universal Terrestrial Radio Access Network (E-UTRAN); Self-configuring and self-optimizing network (SON) use cases and solutions.
5. M. Mauve, J. Widmer and H. Hartenstein, A survey on position-based routing in mobile ad hoc networks, *IEEE Network*, 15(6), pp. 30–39, (2001). doi:10.1109/65.967595.
6. J. Mangues-Bafalluy, M. Requena-Esteso, J. Núñez-Martínez and A. Krendzel, VIMLOC location management in wireless meshes: Experimental performance evaluation and comparison, in *Proceedings of the IEEE International Conference on Communications (ICC2010)*, 23–27 May 2010. doi:10.1109/ICC.2010.5502669.
7. 3GPP TS 23.401, General Packet Radio Service (GPRS) enhancements for Evolved Universal Terrestrial Radio Access Network (E-UTRAN) access.
8. 3GPP TS 36.300, Evolved Universal Terrestrial Radio Access (E-UTRA) and Evolved Universal Terrestrial Radio Access (E-UTRAN); Overall description; Stage 2.
9. 3GPP TS 36.413, Evolved Universal Terrestrial Radio Access Network (E-UTRAN); S1 Application Protocol (S1AP).
10. A. Krendzel, J. Mangues-Bafalluy, M. Requena-Esteso and J. Núñez, VIMLOC: Virtual Home Region Multi-Hash Location Service in Wireless Mesh Networks, in Proceedings of the 1st *IFIP Wireless Days 2008*, Dubai (United Arab Emirates), 24–27 November 2008. doi:10.1109/WD.2008.4812852.
11. J. Núñez-Martínez, J. Mangues-Bafalluy and M. Portoles-Comeras, Studying Practical Any-to-Any Backpressure Routing in WiFi Mesh Networks from a Lyapunov Optimization Perspective, in *Proceedings of the Fifth IEEE International Workshop on Enabling Technologies and Standards for Wireless Mesh Networking (MeshTech'11)*, Valencia (Spain), 17 October 2011.
12. Mobile Europe, Softbank to add Skyhook location into femtocells, http://www.mobileeurope.co.uk/news/press-wire/8952-softbank-to-add-skyhook-location-into-femtocells. Accessed 12 October 2011.

12

Vertical Handover in Heterogeneous Networks: a Comparative Experimental and Simulation-based Investigation

Giovanni Spigoni,[1] Stefano Busanelli,[2] Marco Martalò,[3]
Gianluigi Ferrari,[1] and Nicola Iotti[2]

[1] *University of Parma, Italy*
[2] *Guglielmo Srl, Italy*
[3] *University of Parma, Italy, E-Campus University, Novedrate (CO), Italy*

12.1 Introduction

The continuously growing traffic generated by Mobile Terminals (MTs) – such as smartphones, tablets, netbooks and other mobile Internet devices – is nowadays one of the biggest challenges for mobile network operators, especially because this process is not likely to vanish, at least from a short-time perspective. Therefore, in order to prevent network saturation phenomena, the operators are forced to increase their network capacity more quickly than the customers' demand increases. This goal will probably be achieved by a combination of methods: (1) increasing the available bandwidth in cooperation with the public communications agency, for example by exploiting the spectrum holes; (2) increasing the cell spectral efficiency through technology upgrades, such as by switching from 3G to the upcoming Long Term Evolution (LTE) [1] or WiMAX [2] technologies; (3) reducing the number of users per macrocell by either reducing the cell size or offloading data traffic through WiFi access points or femtocells [3, 4]. An example of the last solution, which is of interest in this chapter, can be found in 3G networks in customers' homes, where the deployment of the so-called home-evolved

Heterogeneous Cellular Networks, First Edition. Edited by Rose Qingyang Hu and Yi Qian.

Node-B (HeNB) allows to overlap reduced-size cells (femtocells) on top of the macrocell of the corresponding Node-B base station [5]. The femtocell configuration is an example of a heterogeneous network (HetNet), since the involved devices (HeNB and Node-B) have different capabilities (e.g., the coverage range), even though they share the same technology [5]. This configuration is viable if both HeNB and Node-B belong to the same network operator and this could be a serious limitation for its commercial deployment, since the customer has to agree to keep an HeNB in their home, feeding it with their self-funded internet connection.

Considering the diffusion of WiFi access points (e.g., IEEE 802.11a/b/g/n [6]), from the customer viewpoint a more attractive solution would consist in jointly using their 3G cellular and WiFi connections. In this case, the devices are different not only in terms of capabilities, as in the femtocell case, but also in terms of technology. Moreover, the UMTS and WiFi networks may belong to two different non-related operators. It is interesting to observe that unlike in a femtocell configuration, the WiFi connection can be used both to replace the 3G connection (the so-called WiFi offloading [7]) and to increase the bandwidth of the 3G connection [8]. From the point of view of a mobile network operator, the hybrid UMTS-WiFi solution is more appealing, since its helps reducing the traffic load on the 3G network. On the other hand, the user has a real advantage only if the WiFi connection can offer an economic saving or a throughput benefit, with respect to the 3G connection. For this reason, the choice cannot be taken by the operator alone, but the customer has to be, to some extent, involved in the decision process. This is particularly true when the UMTS and WiFi networks belong to two different operators without a specific commercial agreement.

In a classic cellular network, the switch between two different network base stations is governed by a relatively simple *Horizontal HandOver* (HHO) mechanism [9] and, therefore, it is seamless from the user perspective. In a HetNet, with a single involved technology (e.g., UMTS network with femtocell) HHO is still possible but it is more complex, especially in the transition from a macrocell to a femtocell. In the case of a hybrid HetNet with two involved technologies, as considered in this chapter, the complexity is even higher, since it is necessary to use the so-called *Vertical HandOver* (VHO) mechanisms. The VHO will therefore play a key role in future hybrid HetNets. For this reason, in this chapter we discuss the potential and the limitations of VHO in HetNets, on the basis of both experimental and simulation results obtained with two novel low-complexity VHO algorithms. In particular, the experiments are conducted through a small testbed composed of a single MT and a few base stations, whereas the simulation analysis is carried out to investigate large-scale scenarios involving several MTs and base stations.

This chapter is structured as follows. In Section 12.2, we provide the reader with some preliminaries on VHO. In Section 12.3, we present experimental results, with two recently proposed low-complexity VHO algorithms, in a realistic mixed outdoor/indoor scenario. In Section 12.4, an OPNET-based simulator for performance analysis of the VHO algorithm is presented: this allows to investigate more complex scenarios with a greater number of MTs. On the basis of the experimental and simulation results, in Section 12.5 we reflect on the practical role of VHO in HetNets. Finally, concluding remarks are given in Section 12.6.

12.2 Preliminaries on VHO

A VHO procedure is composed of three main phases: initiation, decision and execution [10]. During the initiation phase, the MT (or the network controller) triggers the handover procedure, according to the specific networks' conditions. In the second phase, the VHO algorithm chooses the new access point according to a predetermined set of metrics, such as the received signal strength indicator (RSSI), the network connection time, the available bandwidth, the power consumption, the monetary cost, the security level and, obviously, the user preferences [11]. During the final execution phase, all signalling operations for communication reestablishment and data transfer are carried out. The most relevant international standardization effort regarding VHO and continuous communications, the IEEE 802.21 standard, only refers to the first two phases (initiation and decision) that are relatively technology-independent, but it deliberately ignores the execution phase [12]. Similarly, the access network discovery and selection function (ANDSF), defined by the 3GPP [1] consortium, assists user equipment (UE) to discover non-3GPP access networks [13]. The execution phase is the most delicate task of the handover, since it directly impacts on the behavior of the applications running on the MTs. As of today, most of the VHO approaches, for example that considered in [14], leverage on some flavours of mobile IP [15], a level-3 solution that is based on the idea of maintaining the same IP address in every network visited by the MT. For example, the 3GPP consortium leverages on three mobile IP-based protocols: dual-stack mobile IPv6 (DS-MIPv6) [16], proxy mobile IPv6 with dual-stack extensions [17], proxy mobile IPv6 and mobile IPv4 (PMIPv6) [18]. However, there is still no universal and definitive solution, and there are several works based on different approaches, such as UPMT [19], based on the 'IP in UDP' tunnelling that provides per-application flow management or based on the application-level session initiation protocol (SIP) [20], mostly because it can better support voice over IP (VoIP) applications [21].

There are several possible classifications of the VHO algorithms. In particular, they can be distinguished into *no-coupling* and *coupling*. The first group of VHO algorithms refers to scenarios without any form of cooperation between the involved players (users and network operators) [10]. This situation offers the highest degree of freedom to the user, at the price of an increased complexity of the whole handover procedure and of a degraded performance. Clearly, with a higher level of coupling (namely, loose or tight [22]), a better performance can be achieved.

In a no-coupling scenario (the one of interest here and described in detail in Section 12.3), handover times are typically long. Therefore, in order to avoid any lack of connectivity during the handover *execution* phase, it is necessary to adopt a *make-before-break* approach. In other words, the old connection is torn down only after the new connection has been established, thus yielding to a period of coexistence of the two connections, during which the MT becomes a temporary multi-homed host. The management of a multi-homed host during the execution phase is an open problem, without a universal solution and, currently, every operating system (OS) has its own solution for this problem [23].

12.3 Experimental Investigation

12.3.1 VHO Decision Algorithms

In [24], two novel low-complexity VHO decision algorithms are proposed. The first decision algorithm is RSSI-based, while the second is a hybrid RSSI/goodput-based decision algorithm. The dataflow of the latter is shown in Figure 12.1, where the portion inside the dotted circle is the only difference with respect to the dataflow of the RSSI-based VHO algorithm. This

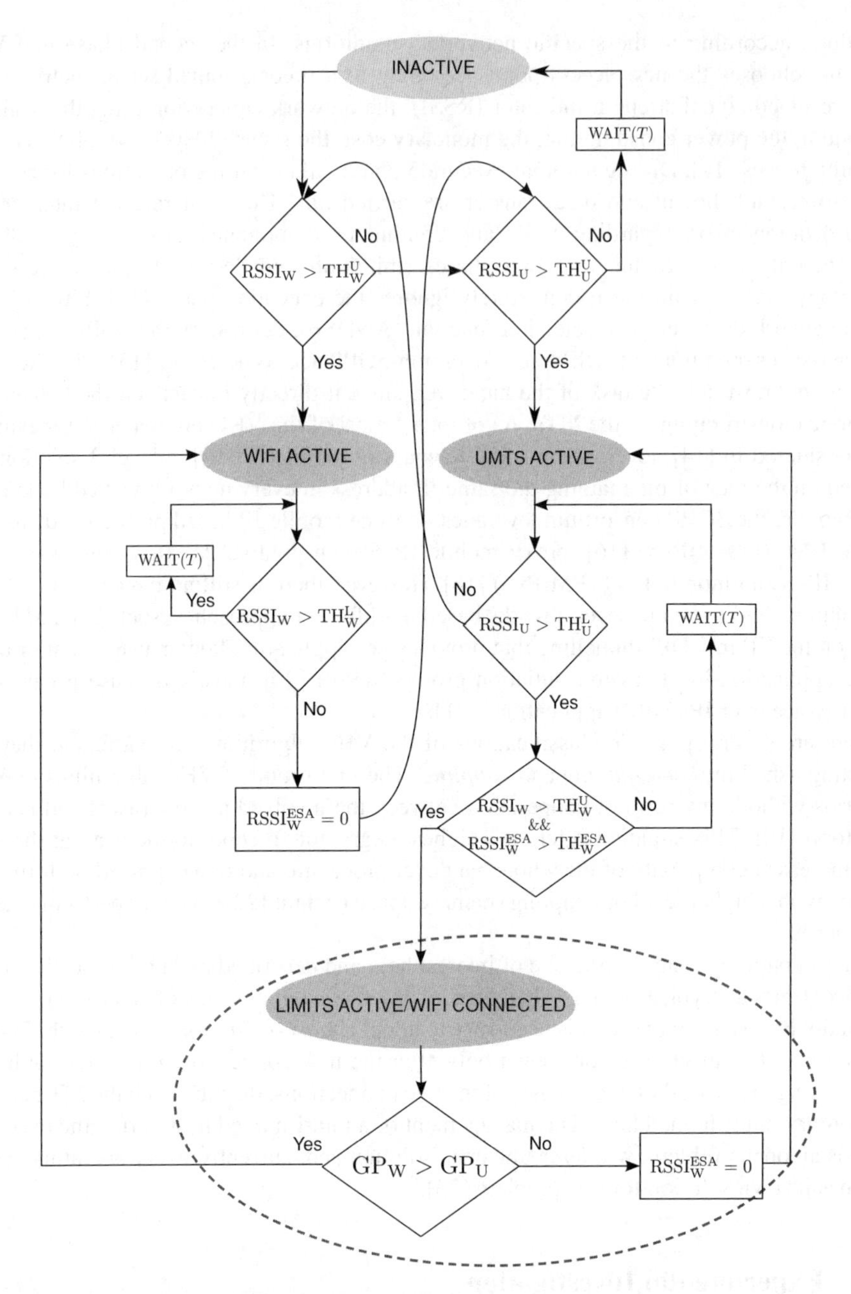

Figure 12.1 Dataflow of the hybrid RSSI/goodput-based VHO algorithm. *Source:* S. Busanelli, M. Martalò, G. Ferrari and G. Spigoni, 'Vertical handover between WiFi and UMTS networks: experimental performance analysis', International Journal of Energy, Information and Communications, Vol. 2, Issue 1, pp. 75–96, February 2011. Reproduced with permission from SERSC.

difference is due to the fact that, while the RSSI-based VHO decision algorithm is based on the implicit assumption that, whenever available, an IEEE 802.11 network always guarantees a better service than a 3G network, the hybrid RSSI/goodput-based VHO algorithm weighs both RSSI and goodput before making a decision. This extension is motivated to avoid switching from the UMTS network to the WiFi network when the latter offers a smaller effective bandwidth.

In the RSSI-based VHO algorithm, the instantaneous RSSI value of interface x, denoted by $RSSI_x$, is compared with two thresholds, TH_x^U and TH_x^L. The *lower* threshold TH_x^L is used to determine when the RSSI is not sufficient to guarantee a stable connection, so, it is slightly higher than the corresponding interface sensitivity. Clearly, when $RSSI_x < TH_x^L$ the connection on the interface x is torn down. On the other hand, the *upper* threshold TH_x^U is used to determine if the measured RSSI is sufficient to establish a stable connection. To this end, we assume that $TH_x^U > TH_x^L$. The use of two thresholds (per network interface) is the first countermeasure against the ping-pong effect, that is, the continuous switching between the two networks when the MT moves on the border of the WiFi network coverage region. This is considered as one of the crucial problems in the design of VHO algorithms.

According to Figure 12.1, when in the INACTIVE state, the MT measures, with a period T (dimension: [s]), the RSSI level at each network interface. As soon as the first (of the two) RSSI level overcomes its upper threshold, the corresponding interface notifies the event to the VHO manager, triggering the execution of the authentication, authorization, and accounting (AAA) procedure to join the selected network. We observe that if both networks are available, the priority is always given to the WiFi network. If the AAA procedure in the selected network x succeeds, the state of the MT switches from INACTIVE to 'X ACTIVE.' Duc to the asymmetric nature of the algorithm, the WIFI ACTIVE and the UMTS ACTIVE states have to be treated separately. More details can be found in [24].

In the hybrid RSSI/goodput algorithm, there is an additional state, the WiFi CONNECTED/UMTS ACTIVE state (highlighted at the bottom), where the MT is authorized in both networks. The presence of this state is expedient to estimate the bandwidths of both networks. The bandwidth is estimated by measuring the time necessary to download a 400 kilobyte file from a remote host (for the ease of simplicity, the file is hosted by a Google server). Another bandwidth estimation technique, well suited to the LTE technology, is based on a more complex algorithm called WBest [25]. The advantage of this algorithm is in the smaller amount of data downloaded from the server and, thus, on the lower cost of the bandwidth test. This algorithm, in fact, uses a train of packets, sent by the server, to calculate the average delay between the packets introduced by the multi-hop network path. On the basis of this delay and of the size of the packets, it is possible to estimate with sufficient reliability the available bandwidth of a network path with a significantly fewer exchanged bytes with respect to the file download approach – approximately 100 kB against 400 kB.

Owing to the asymmetric nature of the algorithm, the MT can move towards this new state only from the UMTS ACTIVE state. In particular, during this transition the MT performs the AAA in the IEEE 802.11 network. Then, the MT remains in the WIFI CONNECTED/UMTS ACTIVE state for all the time needed for estimating the goodput of both networks. As soon as the new measurements, denoted respectively as GP_W and GP_U, are available, the VHO algorithm decides to switch to the WIFI ACTIVE or to come back to the UMTS CONNECTED state. In the latter, the MT disconnects from the WiFi network and resets its RSSI, in order to reduce the waste of resources. From a practical point of view, when $RSSI_W^{ESA} > TH_W^{ESA}$ the goodput is periodically estimated at a variable, but low, rate given by the inverse of the sum of the time necessary to complete the AAA procedure and the time necessary to rise again the value of $\text{RSSI}_{\text{W}}^{\text{ESA}}$.

Finally, due to the long time needed by the WiFi AAA procedure, during the UMTS ACTIVE → WIFI CONNECTED/UMTS ACTIVE transition there are some hidden transitional states, not shown here for simplicity. In particular, when the AAA procedure fails, the transition to the WIFI CONNECTED/UMTS ACTIVE state cannot be carried out and it is necessary to come back to the UMTS ACTIVE state.

12.3.2 Experimental Setup and Results

In [24], the performance of the VHO algorithms described in Section 12.3.1 is evaluated in a realistic environment, using commercially available connectivity service providers and using standard mobile terminal devices. In particular, the experimental test is performed using a notebook running the Windows 7 OS, equipped with a Broadcom IEEE 802.11g compliant network interface and integrated by a UMTS USB Huawei dongle, using the UMTS standard.

The IEEE 802.11 connectivity was offered by a hot-spot owned by one of the biggest Italian wireless internet service providers (WISPs), namely Guglielmo Srl [26]. The hotspot is given by a Browan IEEE 802.11 access point integrated with a captive portal, while the authentication server (AS) is remotely located, as in the standard WISP roaming (WISPr) configuration [27]. The proprietary AAA procedure foresees two additional message exchanges with respect to the WISPr directives [27], thus increasing the time needed to complete the AAA procedure [24]. The UMTS connectivity was instead offered by the public land mobile network (PLMN) of Telecom Italia, one of the most important Italian mobile operators. The sequence of messages needed to complete the AAA procedure is the same as used in a typical 3G network. We observe that using either WiFi or 3G connection we have no direct control on the traffic generated by other users.

The two VHO mechanisms described in Section 12.3.1 are implemented on top of a so-called smart client (SC) software.[1] According to the WISPs directives, an SC is an application studied for enhancing the user experience, making the AAA procedure automatic. Basically, the goal of the SC is that of constantly monitoring the status of the available connections and executing the VHO. Additionally, once the MT initiates a VHO, the SC automatically has to take care of the appropriate AAA procedure. The SC controls both the network interfaces, working with every IEEE 802.11 device able to provide a real-time RSSI information, and with every 3G device (e.g., modem 3G, dongle USB) that supports the Microsoft remote access service (RAS) API [28]. Due to the make-before-break approach, the SC also has to manage the routing functionalities of the OS, in order to make non-critical the multi-homed situation that appears after the authentication on the second network interface [23, 29].

In Figure 12.2, the experimentally estimated (upon time discretization in 0.312 s bins) probability mass function (PMF) of the handover time is shown. From the results in Figure 12.2, it emerges that the handover times towards WiFi and UMTS networks have very different behaviours. In particular, the PMF of the handover time towards the UMTS network

[1]This version of the SC also supports the Microsoft Vista OS. A version for Android-based platforms has also been implemented.

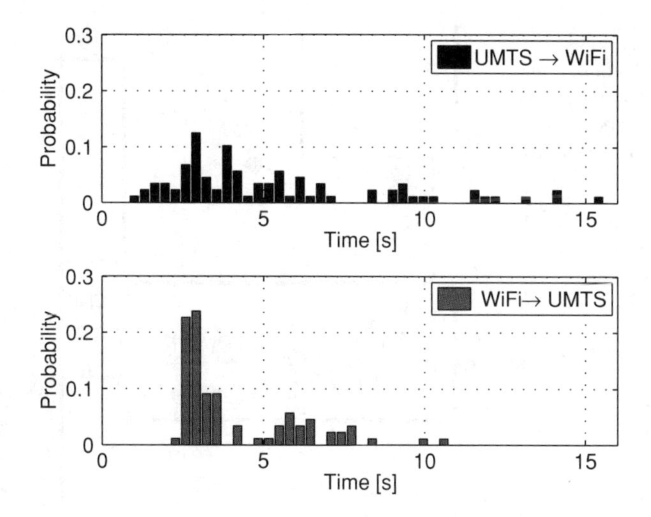

Figure 12.2 PMFs of the handover time of the VHO algorithm presented in Section 12.3.1: from UMTS to WiFi (upper) and from WiFi to UMTS (lower). *Source:* S. Busanelli, M. Martalò, G. Ferrari and G. Spigoni, 'Vertical handover between WiFi and UMTS networks: experimental performance analysis', International Journal of Energy, Information and Communications, Vol. 2, Issue 1, pp. 75–96, February 2011. Reproduced with permission from SERSC.

concentrates around its average value 4.13 s – the standard deviation is 1.76 s. Note that the few values above 10 s can be considered as outliers. On the other hand, the handover time towards the WiFi network is more 'unpredictable' than that towards the UMTS network, as the PMF of the handover time is characterized by a higher average value (5.43 s) and a much higher standard deviation (3.30 s). At the same time, one should observe that the minimum value is very small (1.22 s).

From the results at the top of Figure 12.2, it can be observed that the handover time from the UMTS network to the WiFi network spreads between 1 s and 10 s. This relatively high variability has several causes. First of all, in order to save energy, the MT is supposed to log out from a given network once the VHO manager has selected the other network. Sometimes (more often in the WiFi network), the logout fails and the remote authentication server keeps the authentication state for a certain timeout (roughly 60 s), before automatically logging the user out. In these cases, frequent UMTS → WiFi transitions (in a region at the border of both UMTS and WiFi networks) can experience a short handover time since the MT is de facto already authenticated to the network. Moreover, while the authentication procedure at MAC layer has, in practice, no impact, the release of an IP address by a DHCP (WiFi network) introduces significant randomness. Finally, when the MT is close to some furniture, the RSSI experiences large oscillations that can delay the AAA procedure. Conversely, the RSSI of the UMTS network is more stable and the probability of experiencing such large variations is very small.

From the results shown at the bottom of Figure 12.2, it can be observed that the handover time from the WiFi network to the UMTS network is generally shorter and more predictable (i.e., its PMF is more concentrated) than that in the opposite direction.

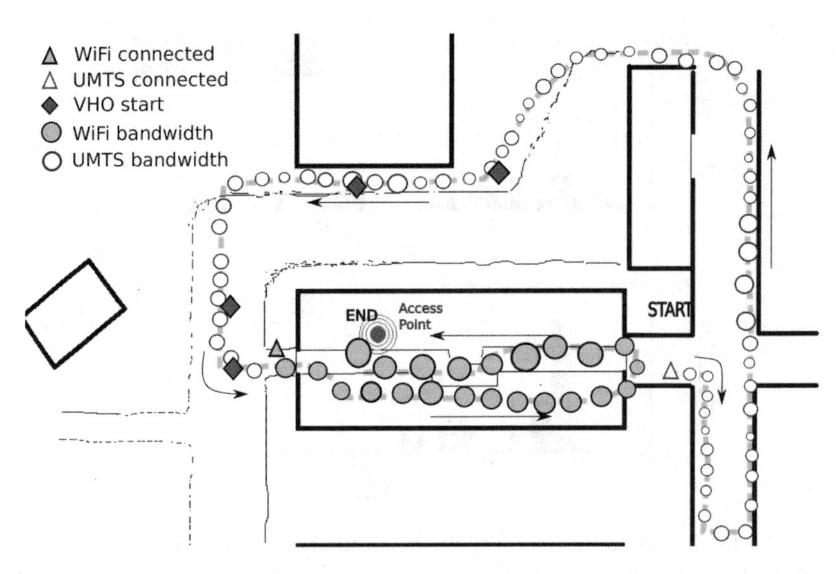

Figure 12.3 Throughput improvement using the hybrid RSSI/goodput-based algorithm. *Source:* S. Busanelli, M. Martalò, G. Ferrari and G. Spigoni, 'Vertical handover between WiFi and UMTS networks: experimental performance analysis', International Journal of Energy, Information and Communications, Vol. 2, Issue 1, pp. 75–96, February 2011. Reproduced with permission from SERSC.

However, due to the no-coupling and the lack of any optimization, the handover time may be long also in the transition towards the UMTS network. This result is somewhat to be expected, since the proposed VHO algorithm is designed to be used for slowly moving mobile MT, such as people moving from a place to another.

In order to measure the goodput, we focus on a single walking path, chosen from the experimental data set. The selected path is shown in Figure 12.3, along with the layout of the environment where the tests were performed. Our experiments were conducted in a building within the Department of Information Engineering of the University of Parma. The nearest UMTS base station is roughly 1 km away from the building, and it offers a 7.2 Mbit/s downlink (384 kbit/s uplink) bandwidth, being compliant with the UMTS specifications. We placed the hotspot in the WASN Lab, at 1 m above groundlevel. The hotspot is fed by an optical fibre network with 100 Mbit/s of symmetric bandwidth, but the hotspot imposes a symmetric limit on the available bandwidth equal to 7.2 Mbit/s, similar to that provided to typical customers. Despite the identical nominal downlink UMTS bandwidth (7.2 Mbit/s), the WiFi network has often outperformed, in our tests, the UMTS network.

The test were performed by walking through the building, keeping the notebook in hand and measuring (1) the time needed to perform the handover and (2) the goodput variations. The tests were always performed during working hours, in order to obtain results associated with realistic daytime situations.

The bold solid lines represent reinforced concrete walls, that cause strong signal attenuation. Near to a glass window or a door (where the bold solid lines are interrupted), the signal attenuation is clearly much less. The path followed by the user is represented by a dashed line and is delimited by the words 'START' and 'END.' The circles drawn along the path represent the measured available goodput; in particular, the diameter of the circle is proportional to the available goodput. The filled circles are where data was sent via the IEEE 802.11 interface, while at the to empty circles the UMTS interface was used.

A (filled) diamond denotes the beginning of a VHO procedure, while a triangle indicates when the procedure was successfully completed. The filled triangles indicate that the VHO procedure has established a WiFi connection, while empty triangles denote the establishment of a UMTS connection. We stress the fact that between diamonds and triangles the MT is still connected with the old network, in order to avoid partial loss of connectivity before finalizing the VHO. Finally, the distance between the circles is directly proportional to the duration of the bandwidth test and, hence, inversely proportional to the available bandwidth. In Figure 12.4, the RSSI and goodput relative to the hybrid RSSI/goodput VHO algorithm, along a sample path of about 200 m (partly indoor and partly outdoor), are shown as functions of time. For the sake of clarity, a direct comparison with the RSSI-based VHO algorithm is also considered. In the top graph, $RSSI_W$ is shown together with the corresponding upper and lower thresholds;

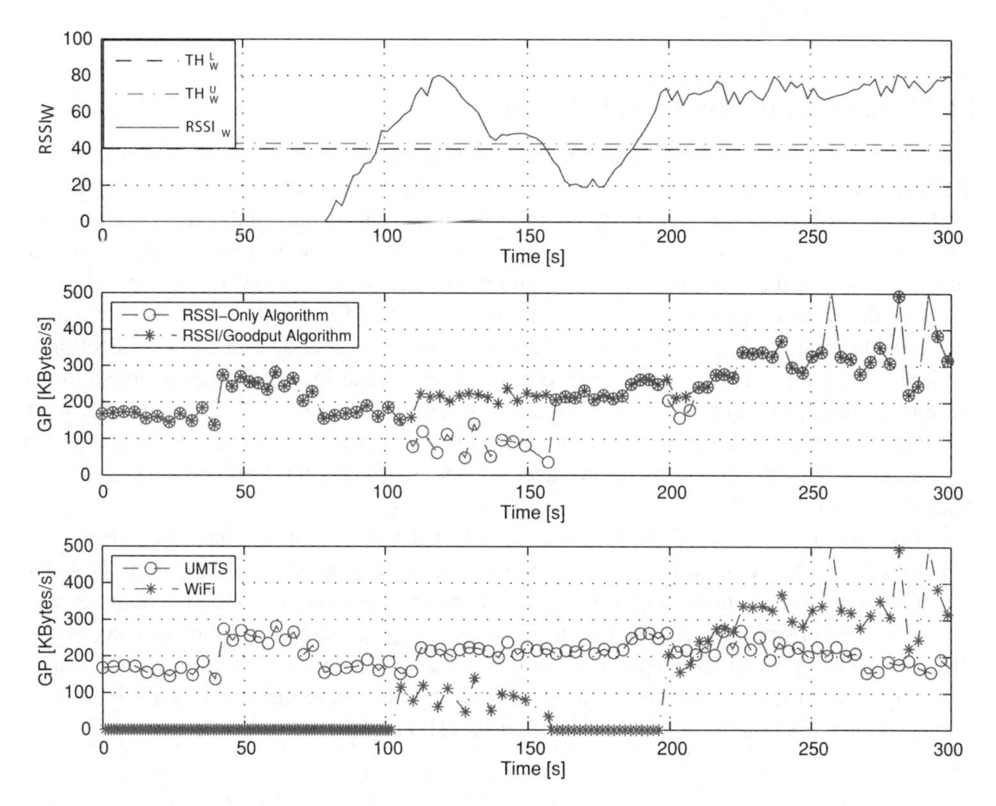

Figure 12.4 RSSI and goodput experienced by the MT following a sample path using the hybrid RSSI/goodput VHO algorithm. *Source:* S. Busanelli, M. Martalò, G. Ferrari and G. Spigoni, 'Vertical handover between WiFi and UMTS networks: experimental performance analysis', International Journal of Energy, Information and Communications, Vol. 2, Issue 1, pp. 75–96, February 2011. Reproduced with permission from SERSC.

in the bottom graph, the estimated goodputs GP_W and GP_U are directly compared; finally, in the middle graph the overall goodput guaranteed by the RSSI-based VHO algorithm (denoted as GP) and the hybrid RSSI/goodput-based VHO algorithm (denoted ad GP^H) are directly compared. According to the results in Figure 12.4, in the initial phase the MT is disconnected from the WiFi network, because of the too low received power. In the top graph, it can be observed that at a given point (a particular position along the path), $RSSI_W$ starts to quickly increase and soon tops the threshold TH_W^U. At this moment, the RSSI-based algorithm begins the VHO towards the WiFi network, ignoring the fact that the effective goodput available in the WiFi network is lower (as can be seen in the bottom graph). On the other hand, in the case of the hybrid VHO algorithm, the MT starts the bandwidth estimation process, after which it decides to keep the UMTS connection because it becomes aware of its higher available goodput. It can be seen from the middle graph that the hybrid RSSI/goodput-based VHO algorithm guarantees a better (over time) goodput performance than that of the RSSI-based algorithm, the only 'penalty' being slightly higher handover time and complexity. These prices to be paid are due to the presence of a double connectivity situation, which requires to properly configure the OS routing table, in order to perform the bandwidth test on both networks, without penalizing the user.

12.4 Simulation-based Investigation

In this section, we present a simulation-based investigation of the proposed VHO algorithms in HetNet scenarios by relying on the Opnet simulator [30]. We first examine a scenario where a single node is moving across the network, in order to reproduce and validate the experimental results presented in Section 12.3.2. However, as we will describe in more detail below, the current version of the simulated VHO algorithm between WiFi and UMTS networks is only based on the received power of the WiFi interface. The UMTS signal, in fact, is assumed to be always present, since in realistic urban scenarios one can assume that a 3G connection is available all the time. Then, we will extend the analyzed scenario, considering a large number of nodes randomly moving in the HetNet and possibly carrying out, VHO procedures. Note also that the realistic AAA procedures for 3G and WiFi networks are not exactly replicated in the simulator, but their effects are reproduced by adding, during the VHO process, a random delay drawn from the PMF shown in Figure 12.2.

The performance of the hybrid RSSI/goodput-based VHO algorithm has also been investigated in the presence of an LTE cellular network. Considering an LTE system, in the place of a 3G system, can be interesting as LTE systems are being deployed in many countries and, therefore, the performance analysis with this technology is timely. In order to analyze the VHO between WiFi and LTE networks, we created a scenario where 40 mobile nodes move randomly across the coverage area of an LTE base station, where there is also a WiFi access point. All the mobile nodes implement the hybrid VHO algorithm.

12.4.1 The OPNET Simulator

OPNET is a modular discrete-event simulator providing support for several technologies: among many others, WiFi, UMTS and LTE networks. The WiFi implementation adheres to the IEEE 802.11g standard, whereas the UMTS implementation, supposed to be compliant with the 3GPP Release-5, does not support HSDPA and it offers a maximum downlink throughput of 384 kbps (the HSDPA maximum nominal throughput by contrast is 14.4 Mbit). The LTE implementation is compliant with the 3GPP Releases-8/9: for instance, it is possible to select the number of antennas of the devices in the multiple-input multiple-output (MIMO) setting and preset the spectrum bandwidth. Note that the LTE system provides a high data rate downlink connection which can be compared with that of an HSPA+ equipped 3G system.[2] Note that a comparison between WiFi and UMTS (without HSPA) available bandwidths wouldn't make sense, since the WiFi data rate is (almost always) much higher than the UMTS data rate.

In all the networks of interest, the MTs implement the entire protocol stack, including the application layer on which desired applications can be run. In particular, we consider three downlink scenarios. In the first two scenarios, there is a file transfer protocol (FTP) application running and each node downloads, every second from an FTP server, a file of a given size, which depends on the scenario of interest. In the second scenario, the MTs launch their applications one at a time every 10 s. The first MT waits 100 s before running its application, in order to let every MT associate to the UMTS network. In the last scenario, instead, each node executes, every 60 s on average, a hypertext transfer protocol (HTTP) application downloading a web page of a given size. The LTE MIMO configuration is set to 2×1 (two transmitter antennas and one receiver antenna) for the downlink and to 1×2 for the uplink. The spectrum bandwidth is set to 20 MHz. Wireless communications between MTs and APs (or base stations) follow a signal propagation model where fading with a Ricean distribution is also included. However, in the considered settings, the impairments due to fading are relatively small and therefore the signal attenuation is very similar to a free-space model.

The main challenge in simulating a VHO algorithm is the implementation of a node module able to jointly control both a WiFi and a cellular (UMTS or LTE) radio interface. In our simulator, for the couple WiFi/UMTS this goal has been pragmatically achieved by coupling two independents nodes, equipped, respectively, with a WiFi and a UMTS network interface and the corresponding protocol stack, as shown in Figure 12.5. However, since the nodes are forced to move together through the network at a fixed distance of 10 cm from each other, they appear, from a network perspective, as a single node.

The VHO algorithm is implemented at the MAC layer of the WiFi node, that is in the module denoted as *wireless_lan_mac* in Figure 12.5. As already anticipated at the beginning of this section, the power level of the UMTS signal is assumed to be always greater than $\mathrm{TH}_{\mathrm{U}}^{\mathrm{U}}$. Moreover, the power of a beacon frame sent by the WiFi access point is computed and filtered as described in Section 12.3.1. If the conditions triggering the VHO procedure represented in Figure 12.1 are verified, the MAC layer module sends an interrupt to the application layer of both WiFi and UMTS nodes. In our simulator, the threshold is set to -83 dBm. If the VHO is towards the WiFi network (i.e., if $\mathrm{RSSI}_{\mathrm{W}} > \mathrm{TH}_{\mathrm{W}}^{\mathrm{U}}$ and $\mathrm{RSSI}_{\mathrm{W}}^{\mathrm{ESA}} > \mathrm{TH}_{\mathrm{W}}^{\mathrm{ESA}}$), the FTP application on the WiFi node starts receiving data. On the other hand, the application on the UMTS node stops the data generation, in order to simulate a hard disconnection at this interface. The opposite operations are performed if the VHO is towards the UMTS network.

[2]Downlink connection data rates for HSPA+ and LTE are not the same, although both technologies provide several tens of Mbps depending on the physical configuration (multiple-antenna techniques, spectrum bandwidth, multi-cell technique, terminal category, etc.)

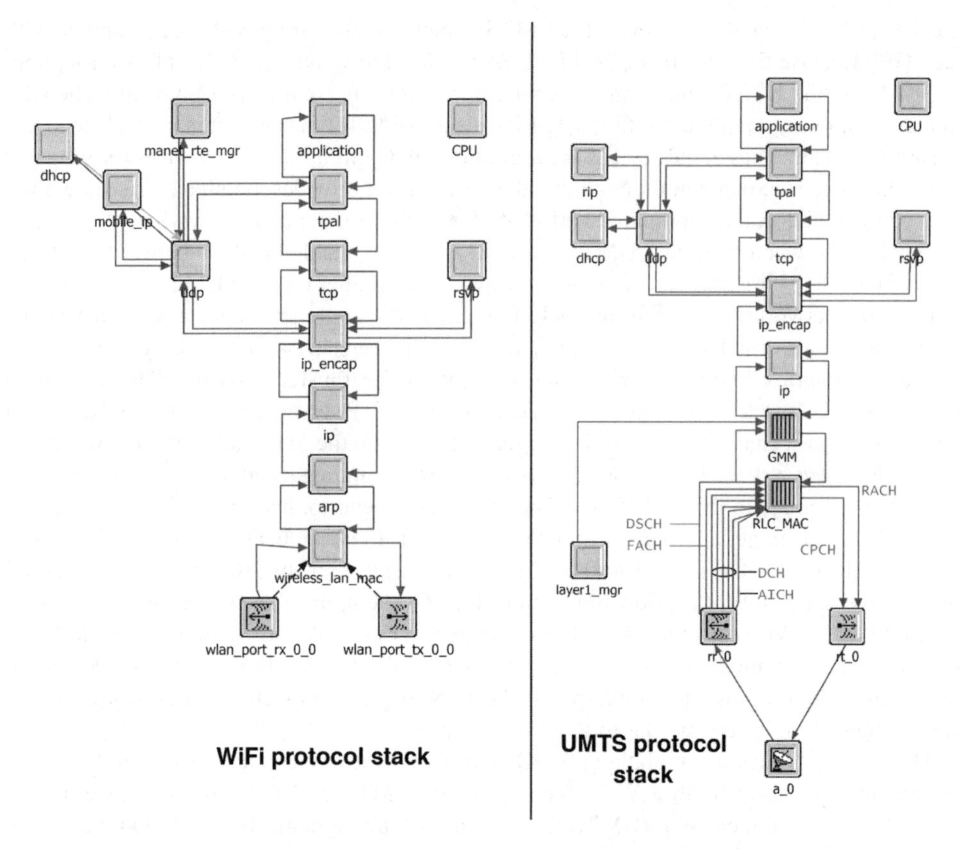

Figure 12.5 Protocol stack of a WiFi node and a UMTS node in the OPNET simulator.

In the case of WiFi/LTE VHO, we have developed a new Opnet node having both the WiFi and LTE interfaces as shown in Figure 12.6. In this node, the *dg_change* module is responsible for the commutation of the traffic between WiFi and LTE interfaces. The *wireless_lan_mac* module computes the WiFi power level as explained in the previous paragraph, while the *bw_test* module executes the available bandwidth (in both WiFi and LTE networks) estimation. Both the *bw_test* and the *wireless_lan_mac* modules communicate their metrics to the *dg_change* module which, if necessary, triggers a VHO. In this case, the WBest-based bandwidth estimation algorithm is applied.

12.4.2 Performance Results

Here, we analyze the performance predicted by the OPNET simulator for three possible scenarios of interest. In the first case, we consider the same scenario used to experimentally validate the VHO algorithm and shown in Figure 12.3: a single node moves, several times, from the WiFi coverage area to the UMTS cell and vice versa. This scenario is expedient to verify, trend-wise, the experimental results presented in Section 12.3 and, therefore, to

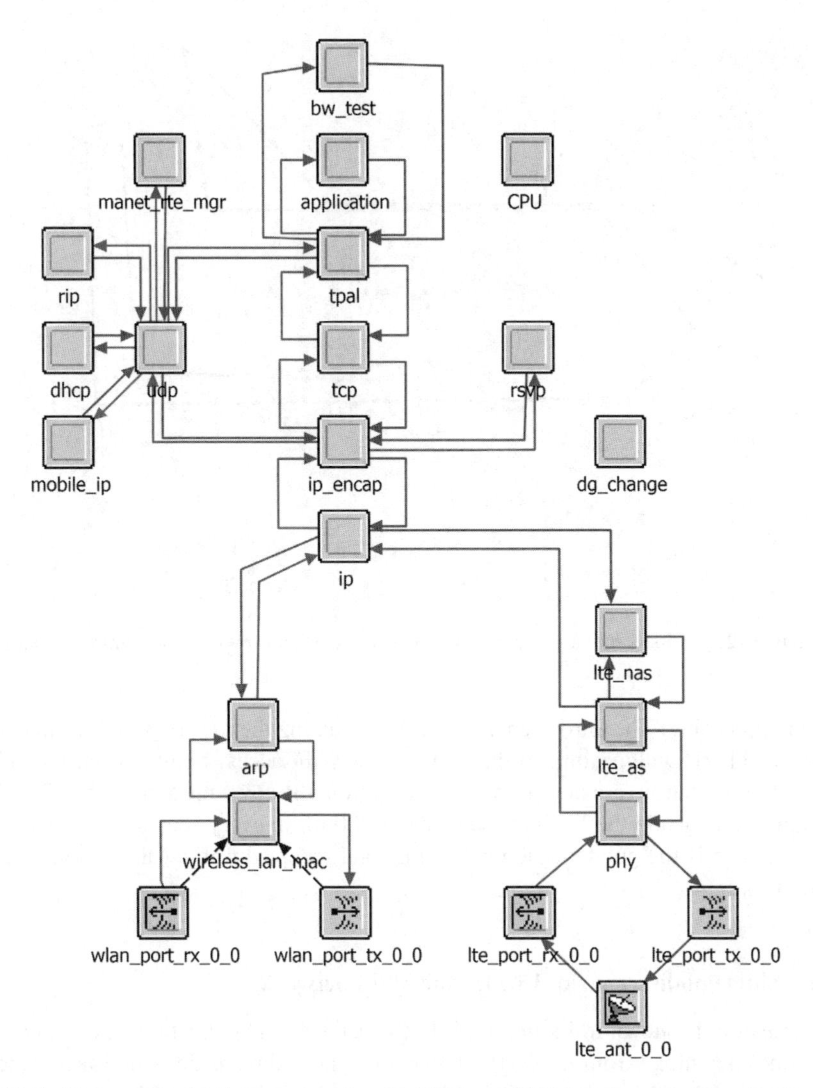

Figure 12.6 Protocol stack of the new node with WiFi and LTE interfaces developed with OPNET Modeler. The module *bw_test* is responsible for the computation of the available bandwidth estimation, while the *dg_change* module manipulates the IP routing table in order to change the default gateway and to redirect specific traffic over a given network.

obtain a sanity-check of our simulator. In the second case, 40 nodes randomly move inside the same area of interest. A unique WiFi access point is placed in this area, whereas four UMTS antennas are placed to ensure total cellular coverage over the entire area. This is representative of a realistic scenario where many users may be in the same city area close to an access point (e.g., a crowded square) and may want to connect to the internet (e.g., through their smartphones). In all simulations, the metrics are measured starting from the end of the initial transient, of duration equal to 100 s, in order to allow the UMTS MTs to perform the network

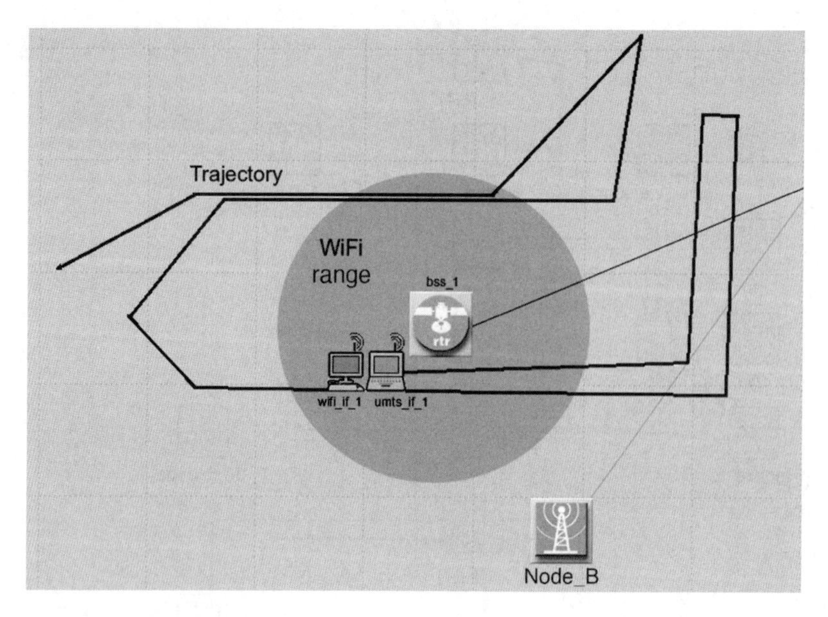

Figure 12.7 First simulation scenario with a single MT following a deterministic path.

association procedure. The third scenario aims at validating the experimental results obtained with the hybrid VHO algorithm. In this scenario, a WiFi access point and an LTE eNodeB (i.e., an LTE base station) are placed in the same region: the coverage area of the LTE eNodeB has a 1 km radius, while the WiFi range is about 400 m. In every scenario, the performance metric of interest is the goodput, defined as the total received traffic (dimension: bytes/s) at application layer.

12.4.2.1 Single node scenario: UMTS and WiFi networks

The first scenario is shown in Figure 12.7. In this case, both applications try to periodically (each second) download from the FTP server a file whose dimension is interface-dependent: 35 kB from the UMTS interface and 50 kB from the WiFi interface. This difference in file size has the sole purpose to differentiate in the graph the UMTS traffic from the WiFi traffic. The node moves on the deterministic path, highlighted in Figure 12.7 and representative of the realistic one considered in Figure 12.3, at a constant speed $v = 2$ m/s. The total length of the path is approximately 1.4 km.

In Figure 12.8 (a), the received power (in dBm) of the beacon frames is shown. Beacon frames are sent by the WiFi AP every 0.5 seconds. In this simulated scenario, the curve is smoother than the corresponding curve in Figure 12.4. This is due to the fact that, unlike the experimental scenario where the signal is impaired by the reflections of the buildings, in the simulated scenario only reflections from the ground are considered, so the effects on the signal are very similar to that of a free-space scenario. In Figure 12.8 (b), the goodput, due to either WiFi or UMTS connections, of the node moving along the path described in Figure 12.7 is shown as a function of time. As one can see, every time a beacon frame is received from

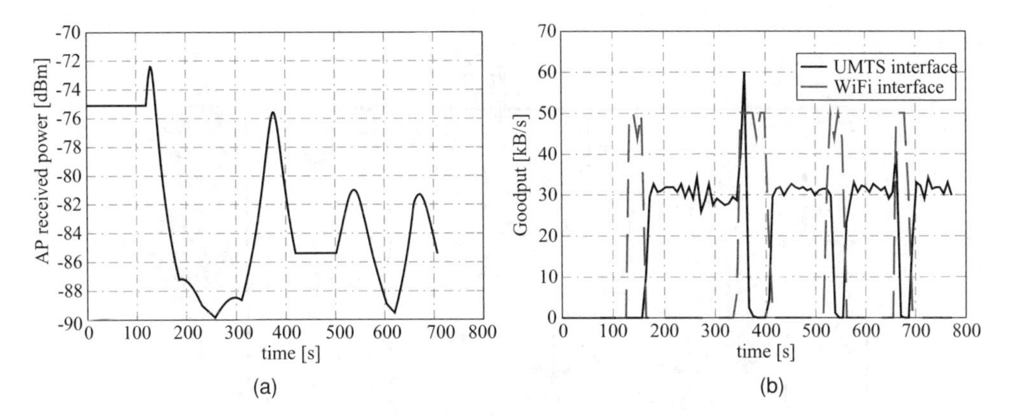

Figure 12.8 (a) Received power on the WiFi interface and (b) goodput on both the interfaces for the single node scenario.

the WiFi access point, as shown in Figure 12.8 (a), the VHO algorithm is triggered and the goodput on the WiFi interface increases, whereas the UMTS interface is triggered down. We recall that the threshold on the received power is set, in our simulator, to −83 dBm. This confirms the experimentally acquired results shown in Section 12.3.

12.4.2.2 Multiple node scenario: UMTS and WiFi networks

In Figure 12.9, the second simulated scenario of interest is shown. In this case, there are 40 MTs randomly moving in the depicted square area of interest. The MTs move according to the random waypoint mobility model [31] with a speed uniformly distributed in the interval between 0 and 5 m/s. The WiFi AP is placed inside the coverage UMTS area; when MTs get sufficiently close to the AP, they stop the UMTS connection and start sending and receiving data through the WiFi interface. The small filled circle represents the WiFi signal range, corresponding to a transmission power equal to 7 dBm (approximately 5 mW). Every MT tries to download, every second, a file of 30 kB so that the total goodput received by the 40 MTs should be 1200 kB/s. The number of UMTS base stations is set to four to ensure UMTS connectivity for all MTs moving inside the perimeter denoted as 'UMTS domain.' The radio network controllers (RNCs) and the core network nodes (SGSN and GGSN) of the UMTS system are also shown, as well as the two FTP servers.

In Figure 12.10, we show (a) the goodput and (b) the corresponding number of connected MTs as functions of time. Sixty independent simulation runs were performed to eliminate statistical fluctuations of the results. In all simulations, the metrics are measured starting from the end of the initial transient, of duration equal to 100 s, in order to allow the UMTS MTs to perform the network association procedure. One can first observe that the goodput is an increasing function of the time. This is due to the fact that, as time passes, the number of MTs in the network increases and, therefore, the total received traffic on each (or both) the interfaces becomes higher. Moreover, from Figure 12.10 (a), one can see that when both the MTs' interfaces are active, the total received goodput approaches 1000 kB/s, which is close to

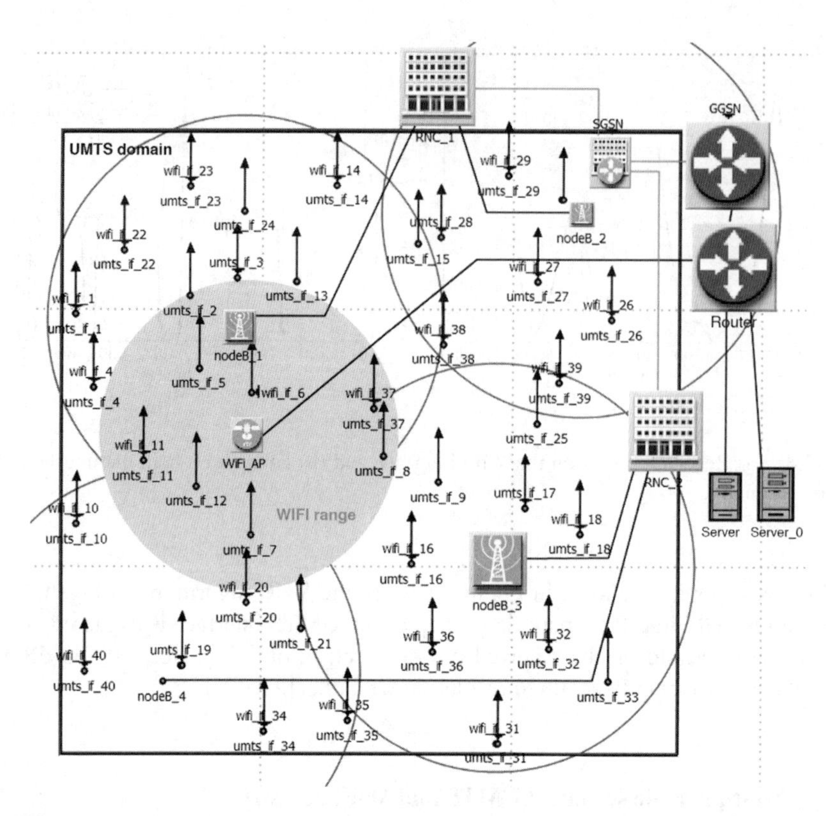

Figure 12.9 Second simulation scenario with multiple MTs moving according to the random waypoint mobility model. MTs have both WiFi and UMTS interfaces.

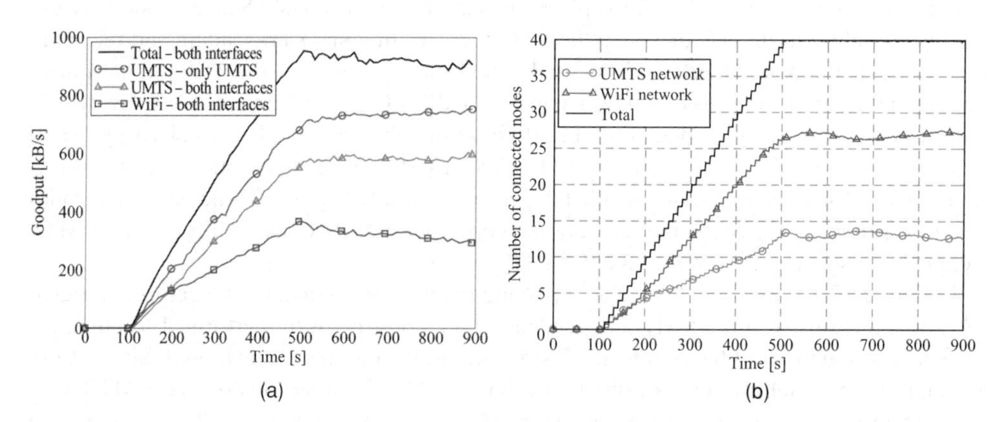

Figure 12.10 (a) Goodput and (b) number of the transmitting MTs in each network for the multiple node scenario.

the maximum possible value of 1200 kB/s. On the other hand, when only the UMTS interface is active, the traffic saturates to approximately 800 kB/s. A similar result holds for the scenario where both interfaces are active. Therefore, the presence of a WiFi connection supplies further connectivity needed to reach the (theoretical) highest possible goodput.

12.4.2.3 Multiple node scenario: LTE and WiFi networks

In Figure 12.11, the third simulated scenario of interest is shown. In this case as well, there are 40 MTs moving across the area depicted by a quadrangle, the filled circle being the coverage area of the WiFi access point. The mobility model and the maximum speed of the nodes are the same as those in the second scenario. Both WiFi access point and LTE system (eNodeB and evolved packet core, EPC) are connected, as well as four HTTP servers, to the internet cloud.

In Figure 12.12, the estimated bandwidth is shown, as a function of time, for a single node in the network. For the first part of the simulation, both WiFi and LTE network have no traffic to be delivered and, therefore, the estimated bandwidth corresponds to the maximum capacity offered by the networks. After approximately nine minutes, some traffic is generated on the LTE network and, after a few seconds, the MT detects the strong decrease of available bandwidth over the LTE network. Consequently, the MT executes a VHO procedure towards the WiFi network.

In Figure 12.13, the performance of the hybrid VHO algorithm is compared, in terms of total aggregated goodput (dimension: bytes/s), with that of the RSSI-based algorithm. The

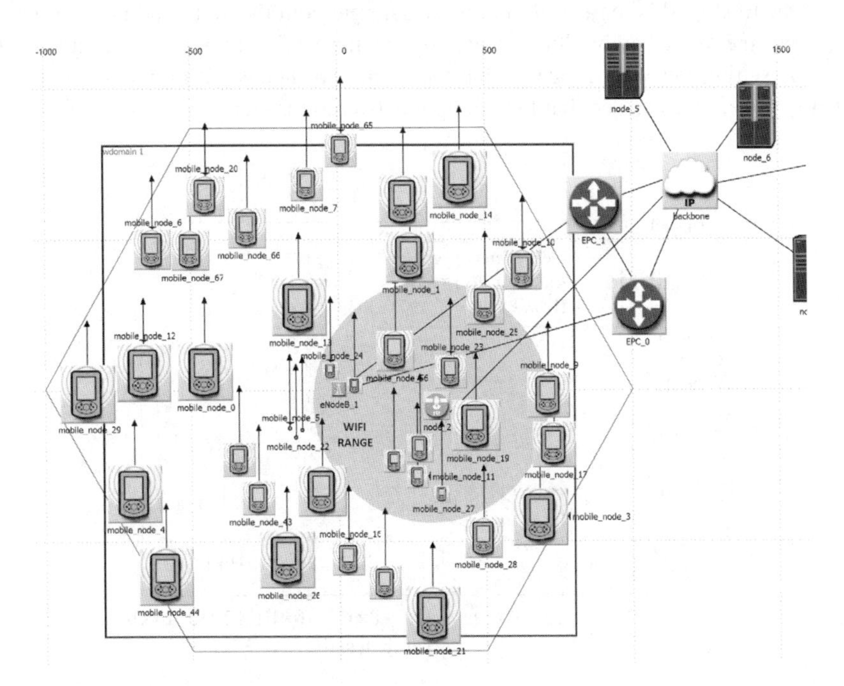

Figure 12.11 Third simulation scenario with multiple MTs moving according to the random waypoint mobility model. MTs have both WiFi and LTE interfaces.

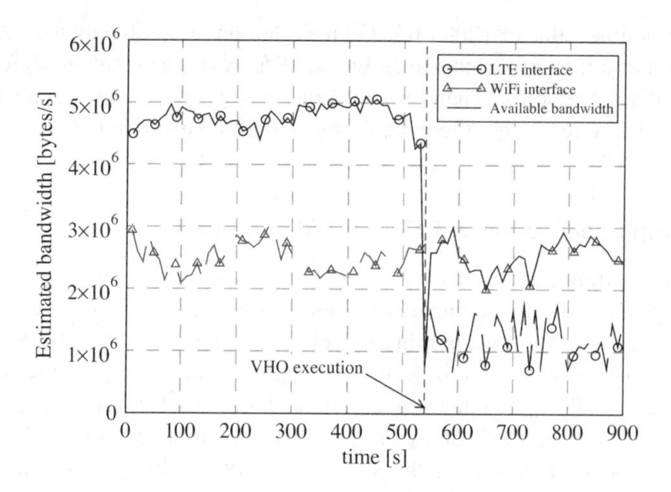

Figure 12.12 Estimated bandwidth, as a function of time, for a single node in the third scenario.

results are obtained executing 16 15-minute long simulation runs – eight runs for the RSSI-based algorithm and eight runs for the hybrid algorithm – of the third scenario. The network increases at each simulation run, due to the fact that the average size of the downloaded HTTP pages increases. For every run, we compute the average aggregated goodput of the networks. Note that the total load is increased in order to saturate both the WiFi and the LTE networks. In particular, the WiFi bandwidth saturates before the LTE network is overloaded – this is obtained by setting the WiFi range so that the number of nodes within the WiFi coverage is sufficiently large. One can see that the goodput delivered through the WiFi interface does not

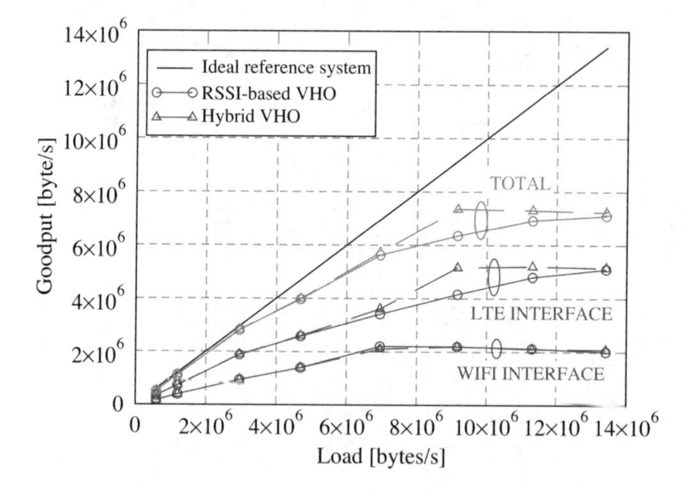

Figure 12.13 Total aggregated goodput of the networks for both the hybrid and the RSSI-based VHO algorithms.

change considerably between the hybrid and the RSSI-based algorithms.[3] The most noticeable difference between the two algorithms can be observed through the LTE interface. Obviously, the goodput reached by the LTE network in saturation conditions does not change. However, with the hybrid algorithm the goodput increases faster, due to the fact that, when the WiFi can no longer transport further data and the LTE band is not overloaded, the nodes under WiFi coverage will switch their connections towards LTE. This fact could be seen as a load balancing between LTE and WiFi implicitly executed by the VHO hybrid algorithm. In fact, thanks to the bandwidth estimation metric, the mobile nodes are somehow aware of the saturation of the WiFi network and can select the LTE connection.

For the sake of completeness, we point out that the total goodput is never equal to the total load requested by the mobile nodes. This is due to the implementation of the HTTP application: the time interval between two HTTP requests is not keep fixed to 60 s, but it has an exponential distribution with mean value equal to 60 s. Therefore, it may happen that an HTTP request arrives when the previous request (for the same MT) has not been served. At this point, the HTTP client interrupts the download in progress to start the new one. The total offered load does not take into account this reduction in the requested data.

12.5 Discussion on the VHO in HetNets

On the basis of the experimental and simulation results presented and discussed in Section 12.3 and Section 12.4, respectively, the following reflections on the role of VHO in HetNets can be carried out.

12.5.1 Role of the (Internal) Decision Algorithm

From the obtained experimental results, it turns out that the VHO decision algorithm is relevant to mitigating the ping-pong effect. However, the RSSI/goodput-based decision algorithms guarantee a very good performance, provided that the received QoS indicator (either the RSSI or a running estimate of the available goodput) is properly filtered, in order to avoid sudden (and temporary) handover decisions.

12.5.2 Role of the Authentication Procedures

The authentication procedure seems to be the real bottleneck of the VHO procedure, especially in a no-coupling scenario, where the MT needs to switch between different operators. In this case, the only solution is a radical simplification of this procedure, possibly demanding some security mechanism to lower levels of the protocol stack. On the other hand, in the case of tight coupling, where the two networks involved in the VHO are operated by the same provider, it might be possible to significantly simplify the authentication procedure, thus reducing the VHO time.

[3] A slight decrease can be observed in the case of the RSSI-based algorithm due to the increasing number of packet collisions at the WiFi physical layer.

12.5.3 Impact of VHO on HetNet Coverage

The impact of the VHO on network coverage depends on the networks between which it is carried out. More precisely, in the case of two heterogeneous networks with radically different coverage ranges (e.g., UMTS and WiFi), the VHO procedure has basically no impact on the coverage extension, but it is mainly expedient to select, between two simultaneously available networks, the one which guarantees the highest QoS – according to the chosen performance indicator. On the other hand, VHO might play a key role in efficiently extending the coverage when carried out between networks with more similar coverage ranges, such as UMTS and WiMAX networks. The same experimental investigation described in this chapter can also be applied to different pairs of networks, provided that proper connection procedures and VHO algorithms are implemented in the SC.

12.5.4 Impact of VHO on HetNet Capacity

The impact of VHO on the capacity of a HetNet is not easy to evaluate. In fact, an efficient VHO mechanism allows each MT to connect to the best currently available network. In particular, if the chosen QoS indicator is the available goodput, the selection, by each node, of the network which guarantees the highest goodput implies, from a single-user perspective, maximization of the capacity. Roughly speaking, Figure 12.4 shows that the choice of the network with the best available goodput allows a doubling of the bandwidth experienced by a single MT. In general, VHO will likely be a key ingredient in performing efficient cellular offloading [8] and will thus have a crucial role in 4G systems [11].

However, as known in the realm of game theory, the maximization of each user's utility does not necessarily imply the maximization of the entire HetNet utility [32]. In fact, there may be a large number of MTs that would like to connect to the network with the best goodput (e.g., the WiFi network), thus leading to a violation of the minimum QoS on this network and, therefore, to a congestion. This opens several interesting research perspectives, as it is expected that a centralized control of a HetNet will really allow us to exploit its potential. On the other hand, efficient decentralized control strategies will likely play a key role in future systems.

12.6 Conclusions

In this chapter, we have reflected on the role of VHO in future HetNets. In particular, on the basis of internetworking experimental results obtained with low-complexity novel VHO algorithms (relying on RSSI and goodput measurements) [24], we have drawn some simple conclusions on the potential and limitations of VHO in HetNets. The main conclusion is that the VHO procedure in loosely coupled heterogeneous networks experiences a high handover time, mostly due to the latency induced by the AAA procedures. Therefore, the design of effective VHO mechanisms requires us to consider a top-down interaction from the high layers of the protocol stack to the bottom layers of the same. Moreover, the simulation-based investigation has shown that the use of VHO has the potential to perform cellular offloading, thus increasing network capacity.

Acknowledgment

This work is sponsored in part by Guglielmo Srl and in part by the project 'Cross-Network Effective Traffic Alerts Dissemination' (X-NETAD, Eureka Label E! 6252 [33]), sponsored by the Ministry of Foreign Affairs (Italy) and the Israeli Industry Center for R&D (Israel) under the 'Israel-Italy Joint Innovation Program for Industrial, Scientific and Technological Cooperation in R&D'.

We would like to thank Ing. G. Guerri (Guglielmo Srl) for his continuous support and help.

References

1. 3rd Generation Partnership Project, '3GPP', Website: http://www.3gpp.org.
2. Institute of Electrical and Electronics Engineers, 'IEEE Std 802.16TM-2009. Part 16: Air Interface for Broadband Wireless Access Systems', 2009.
3. J. Andrews, H. Claussen, M. Dohler, S. Rangan and M. Reed, 'Femtocells: Past, present, and future', *IEEE J. Select. Areas Commun.*, vol. 30, no. 3, pp. 497–508, April 2012.
4. V. Chandrasekhar, J. Andrews and A. Gatherer, 'Femtocell networks: a survey', *IEEE Commun. Mag.*, vol. 46, no. 9, pp. 59–67, September 2008.
5. A. Ghosh, N. Mangalvedhe, R. Ratasuk, B. Mondal, M. Cudak, E. Visotsky, T. Thomas, J. Andrews, P. Xia, H. Jo, H. Dhillon and T. Novlan, 'Heterogeneous cellular networks: From theory to practice', *IEEE Commun. Mag.*, vol. 50, no. 6, pp. 54–64, June 2012.
6. Institute of Electrical and Electronics Engineers, 'IEEE Std 802.11TM-2007. Part 11: Wireless LAN Medium Access Control (MAC) and Physical Layer (PHY) specifications', 2007.
7. K. Lee, I. Rhee, J. Lee, Y. Yi and S. Chong, 'Mobile data offloading: how much can WiFi deliver?' *SIGCOMM Comput. Commun. Rev.*, vol. 40, no. 4, pp. 425–426, October 2010.
8. A. Balasubramanian, R. Mahajan and A. Venkataramani, 'Augmenting mobile 3G using WiFi', in *Proc. of the Int. Conf. on Mobile Systems, Applications, and Services* (MobiSys), San Francisco, CA, USA, June 2010, pp. 209–222.
9. J. Manner, M. Kojo, T. Suihko, P. Eardley and D. Wisely, 'IETF RFC 3753, Mobility related terminology', 2004.
10. G. Lampropoulos, N. Passas, L. Merakos and A. Kaloxylos, 'Handover management architectures in integrated WLAN/cellular networks', *IEEE Communications Surveys & Tutorials*, vol. 7, no. 4, pp. 30–44, October 2005.
11. X. Yan, Y. A. Sekercioglu and S.Narayanan, 'A survey of vertical handover decision algorithms in 4G heterogeneous wireless networks', *Elsevier Computer Networks*, vol. 54, no. 11, pp. 1848–1863, August 2010.
12. Institute of Electrical and Electronics Engineers, 'IEEE Std 802.12TM-2008. Part 21: Media Independent Handover Services', 2008.
13. J. Sachs and M. Olsson, 'Access network discovery and selection in the evolved 3GPP multi-access system architecture', *European Transactions on Telecommunications*, vol. 21, no. 6, pp. 544–557, April 2010.
14. S. Sharma, I. Baek and T. Chiueh, 'OmniCon: A mobile IP-based vertical handoff system for wireless LAN and GPRS links', *Software: Practice and Experience*, vol. 37, no. 7, pp. 779–798, 2007.
15. C. Perkins, 'IETF RFC 3344, IP Mobility Support for IPv4', 2002.
16. H. Soliman, 'IETF RFC 5555, Mobile IPv6 Support for Dual Stack Hosts and Routers (DS-MIPv6)', June 2009.
17. R. Wakikawa and S. Gundavelli, 'IETF RFC 5844, IPv4 Support for Proxy Mobile IPv6 (DS-MIPv6)', May 2010.
18. S. Gundavelli, K. Leun, V. Devarapalli, K. Chowdhury and B. Patil, 'IETF RFC 5213, Proxy Mobile IPv6 (PMIPv6)', August 2008.
19. M. Bonola, S. Salsano and A. Polidoro, 'UPMT: universal per-application mobility management using tunnels', in *Proc. IEEE Global Telecommun. Conf.* (GLOBECOM), Honolulu, HI, USA, November 2009, pp. 2811–2818.
20. J. Rosenberg, H. Schulzrinne, G. Camarillo, A. Johnston, J. Peterson, R. Sparks, M. Handley and E. Schooler, 'IETF RFC 3261, SIP: session initiation protocol', 2002.

21. S. Salsano, A. Polidoro, C. Mingardi, S. Niccolini and L. Veltri, 'SIP-based mobility management in next generation networks', *IEEE Wireless Communications*, vol. 15, no. 2, pp. 92–99, April 2008.
22. J.-Y. Song, S.-W. Lee and D.-H. Cho, 'Hybrid coupling scheme for UMTS and wireless LAN interworking', in *Proc. IEEE Vehicular Tech. Conf.* (VTC-Fall), vol. 4, Orlando, FL, USA, October 2003, pp. 2247–2251.
23. M. Wasserman and P. Seite, 'IETF Draft (Work in progress) – Current Practices for Multiple Interface Hosts', July 2011, Available at: http://tools.ietf.org/html/draft-ietf-mif-current-practices-12.
24. S. Busanelli, M. Martalò, G. Ferrari, G. Spigoni and N. Iotti, 'Vertical handover between WiFi and UMTS networks: experimental performance analysis', *International Journal of Energy, Information and Communications*, vol. 2, no. 1, pp. 75–96, February 2011.
25. M. Li, M. Claypool and R. Kinicki, 'Wbest: a bandwidth estimation tool for IEEE 802.11 wireless networks', in *IEEE Conference on Local Computer Networks* (LCN), Montreal, Canada, October 2008, pp. 374–381.
26. Guglielmo S.r.l., Website: www.guglielmo.biz.
27. B. Anton, B. Bullock and J. Short, 'Best current practices for wireless internet service provider (WISP) roaming', Wi-Fi Alliance, Tech. Rep., February 2003.
28. 'Remote Access Service (RAS) Windows API', Available: http://msdn.microsoft.com/en-us/library/bb545687(VS.85).aspx.
29. R. Braden, 'IETF RFC 1122, Requirements for Internet Hosts – Communication Layers', 1989.
30. Opnet Website: http://www.opnet.com.
31. D. B. Johnson and D. A. Maltz, 'Dynamic source routing in ad hoc wireless networks', in *Mobile Computing*, T. Imielinski and H. Korth, Eds. Kluwer Academic Publishers, 1996, pp. 153–181.
32. T. Jie, A. Klein and D. Brown, 'Natural cooperation in wireless networks', *IEEE Signal Processing Mag.*, vol. 26, no. 5, pp. 98–106, September 2009.
33. 'Eureka project 6252 X-NETAD', Website: http://www.eurekanetwork.org/project/-/id/6252.

Part III

Deployment, Standardization and Field Trials

13

Evolution of HetNet Technologies in LTE-advanced Standards

Young-Han Nam, Boon Loong Ng, Jinkyu Han, and Jianzhong (Charlie) Zhang
Samsung Telecommunications America, USA

13.1 Introduction

In the 3GPP LTE-Advanced workshop held in Shenzhen, China, in 2008, introduction of heterogeneous networks was proposed as a method of increasing the overall network capacity by means of further cell splitting (e.g., [1]). Heterogeneous networks include base stations (or enhanced NodeBs, eNBs) with different orders of transmission powers, which is the main distinguishing feature from the homogeneous networks where base stations transmit with the same order of transmission power.

Depending on the cell coverage and the backhaul assumptions, base stations are classified into macro, pico, femto, remote radio head (RRH) and relays. Macro base stations are those base stations considered in the traditional homogenous networks, transmitting signals with a relatively high power, such as 20 W for 5 MHz and 40 W for 10 MHz bandwidth. On the other hand, the new types of base stations transmitting signals with a relatively low power are characterized in the 3GPP technical report 36.814 [2] (see Table 13.1).

During the 3GPP LTE-Advanced Rel-10 work item phase, two deployment scenarios – macro–femto and macro–pico scenarios – were primarily investigated and new interference scenarios were identified. In the macro–femto case, downlink transmission from a home eNB (HeNB) interferes with non-member user equipment's (UE) signal reception from a macro eNB (MeNB). Meanwhile, uplink transmission from non-member UE to the MeNB interferes with HeNB's signal reception from the HeNB's member UEs. On the other hand, in the macro–pico case, interference scenarios associated with a new operation option, namely cell range expansion (CRE), are considered. CRE expands the cell range of pico eNBs (PeNBs), so that

Heterogeneous Cellular Networks, First Edition. Edited by Rose Qingyang Hu and Yi Qian.

Table 13.1 Categorization of the new nodes

	Deployment	Access	Backhaul
Remote radio head	Placed indoors or outdoors.	Open to all UEs	Optical fibre/microwave. Several μs latency to macro
Pico eNB	Node for hotzone cells. Placed indoors or outdoors. Typically planned deployment.	Open to all UEs	X2[1]
Home eNB	Node for femtocells. Placed indoors. Consumer deployed.	Open only to member UEs, i.e., closed subscriber group	No X2
Relay nodes	Placed indoors or outdoors.	Open to all UEs	Through air-interface with a macrocell (for in-band RN case)

PeNBs can serve more UEs. This facilitates load balancing between the MeNB and the PeNBs, as well as enabling better uplink cell association. One challenge to the CRE implementation is that pico UEs in the range-expanded region experience significant downlink interference from the MeNB.

3GPP LTE-Advanced standards provide necessary specifications to handle these interference scenarios. The corresponding work item is called enhanced inter-cell interference-coordination (eICIC) [3]. In this chapter, the LTE-A specifications related to the eICIC work item are explained and discussed in detail.

13.2 Deployment Scenarios for LTE-advanced HetNet

As the number of subscribers and the demand for bandwidth grow rapidly, homogeneous network topology based on the typical MeNB deployment can barely meet the subscribers' requirements. In general, traffic demands are concentrated in the small and densely populated areas, or so-called hotspots. In these areas, the throughput achievable from a typical macrocell would not be sufficient to guarantee a certain level of user experience. By placing low power nodes (LPNs) in such hotspots, we can offload a significant amount of traffic from MeNB to LPNs. To enhance further offloading capability, CRE technology was proposed in Rel-10 and it allows a UE to be attached and served by a cell with weaker downlink signals.

In developing the LTE-Advanced HetNet technology, 3GPP radio-access-network workgroup 1 (or RAN1) took into account the deployment scenario where macrocells and LPNs utilize the same frequency band. This inherently leads to the problem of intra-frequency interference, which is difficult to handle in the overlapped coverage area between macrocells

[1] X2 is the standardized interface between eNBs in 3GPP LTE.

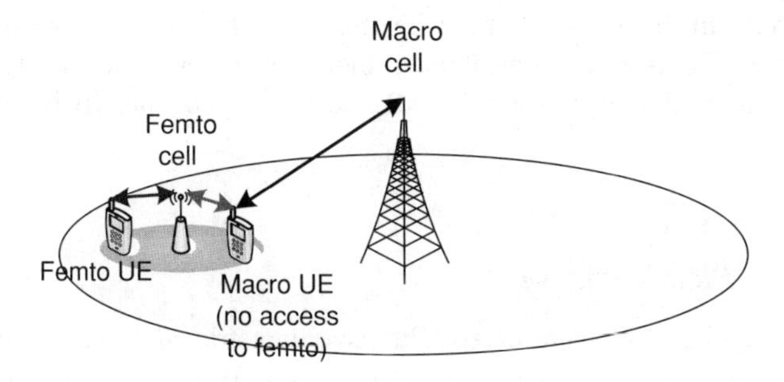

Figure 13.1 Macro–femto scenario.

and LPNs. Moreover, since CRE allows a UE to access a weaker cell, the signal to interference plus noise ratio (SINR), sometimes known as the geometry of a UE, becomes even lower. To understand the interference problem clearly, the two interference scenarios illustrated in Figure 13.1 and Figure 13.2 are considered.

13.2.1 Macro–Femto Scenario

A femtocell serves a closed subscriber group (CSG) where access to the cell is restricted only to member UEs. When a non-member UE stays in the femtocell coverage area, it experiences strong interference from the femtocell. Figure 13.1 illustrates the macro–femto scenario, where a femto UE and a macro UE are served by a femtocell and a macrocell, respectively. Even though the macro UE is closer to the femtocell, it cannot access the femtocell and can only get service from the distant MeNB.

In the downlink, the macro UE is essentially being jammed by the femtocell. An appropriate femtocell power control [11] can reduce the probability of severe interference, but frequent outage is still likely.

In the uplink, the femtocell is being jammed by the macro UE. Since the macro UE is power controlled by the macrocell, it will cause strong and bursty interference to the femtocell. Noise padding – the scheme of injecting artificial noise into the received uplink transmission – can smooth out interference, but it also decreases capacity.

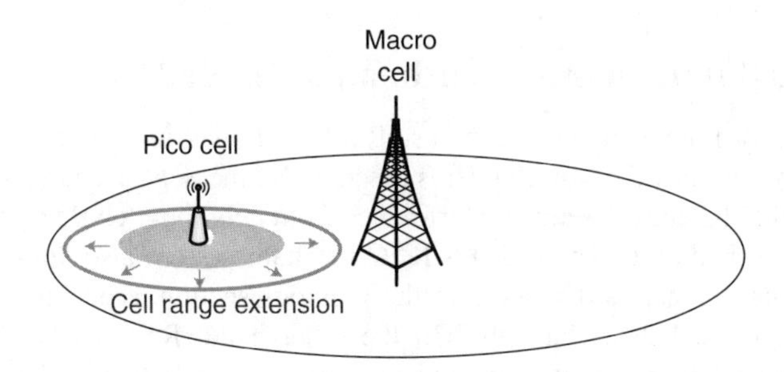

Figure 13.2 Macro–pico scenario.

Although both the macrocell and femtocell suffer from the inter-cell interference problem, as outage to macro UE is often considered a more severe problem, the femtocell/eNB is typically referred to as the aggressor cell/eNB and the macrocell/eNB is referred to as the victim cell/eNB.

13.2.2 Macro–Pico Scenario

A picocell is a small cell to which macro UEs have unrestricted access. In this macro–pico scenario, even when the received signal strength from a picocell is weaker than that from a macrocell, it can sometimes be beneficial for a UE to be served by the picocell when the picocell can assign the UE more bandwidth than the macrocell. Due to unrestricted access, in theory, UE should always be associated with a cell with good SINR. However, in practice, UE association with a cell with the highest received SINR may not always be desirable, and in some cases it may result in no pico UE accessing to the picocell.

The system bandwidth needs to be shared by the UEs belonging to a cell. Since the macrocell coverage is wider than the picocell coverage, it can be assumed in general that more UEs are accessing the macrocell. The network is supposed to serve all the UEs fairly and as a result the expected bandwidth assigned to an individual UE is inversely proportional to the number of UEs in a cell. Therefore, a UE belonging to the picocell can have more chance to be assigned a larger bandwidth. Furthermore, offloading to picocells achieves load balancing between the macro and picocells and improves the overall system capacity.

CRE has been introduced to enable this offloading operation as shown in Figure 13.2. The UEs should be able to discover picocells in expanded range and handover to the picocells.

However, the legacy signals and channels from macrocells still remain as strong interference to the pico UEs. The first release of LTE (Rel-8) defined downlink cell-specific reference signal (CRS) for measurement and demodulation reference, primary synchronization signal (PSS) and secondary synchronization signal (SSS) for frequency/time synchronization and cell ID indication, and physical broadcast channel (PBCH) for delivery of the master system information. While the other signals and channels can be controlled by the macrocell to avoid severe interference to a pico UE, these legacy signals and channels have to be transmitted to maintain backward compatibility.

In the macro–pico scenario, the macrocell/eNB is typically referred to as the aggressor cell/eNB and the picocell/eNB is referred to as the victim cell/eNB.

13.3 Inter-cell Interference Coordination for HetNet

Considering mainly homogeneous networks, Rel-8 LTE has introduced various inter-cell interference coordination (ICIC) mechanisms based on X2 message exchanges between neighbour eNBs. Rel-8 ICIC aims to reduce interference from neighbour eNBs by coordinating the eNBs' frequency scheduling. These Rel-8 ICIC mechanisms are also applicable to HetNet specific interference scenarios discussed in the previous section, especially for interference management of physical data channels. On the other hand, Rel-10 LTE-Advanced introduced mainly two additional ICIC mechanisms, carrier-aggregation (CA) based ICIC and non-CA based ICIC, which aim to mitigate interference of control channels, which are physical

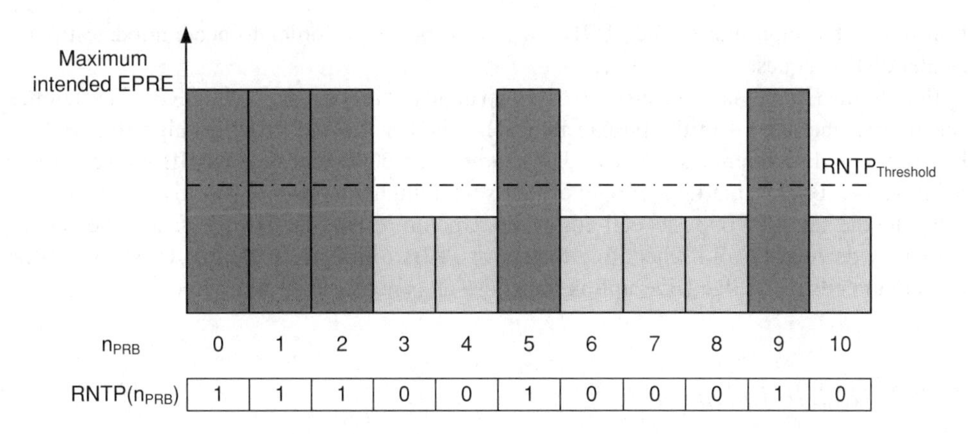

Figure 13.3 RNTP bitmap.

downlink control channel (PDCCH), physical hybrid-ARQ indicator channel (PHICH) and physical control format indicator channel (PCFICH) as well as the physical data channels.

13.3.1 Rel-8/9 ICIC

Rel-8/9 LTE ICIC provides useful mechanisms for coordinating interference of data channels, which are physical downlink shared channel (PDSCH) and physical uplink shared channel (PUSCH). The Rel-8/9 ICIC mechanism relies on message exchanges between eNBs involved in the ICIC. The ICIC message exchanges are through the X2 interface, whose latency is in the order of 10s of milliseconds (or 10s of subframes[2]). Hence, it may take at least in the order of 100s of milliseconds for an eNB to update scheduling based on the exchanged ICIC messages. Considering that eNBs' frequency-domain scheduling may change every subframe, the Rel-8/9 ICIC has a limitation in that it can only coordinate inter-cell interference in relatively long timescale.

Rel-8/9 introduced three X2 messages for ICIC: relative narrow-band transmission power (RNTP) for downlink ICIC, and overload indicator (OI) and high-interference indicator (HII) for uplink ICIC [5].

The RNTP message [5] includes an RNTP bitmap (an information element (IE) named RNTP per physical resource block (PRB)) and other information, such as RNTP threshold. The RNTP bitmap is a proactive indicator on DL power restriction per PRB in a cell, where each bit indicates whether a corresponding PRB's maximum intended energy per resource element (EPRE) exceeds the RNTP threshold or not [4], as illustrated in Figure 13.3. Neighbour eNBs may utilize the RNTP for interference-aware PDSCH scheduling in the subsequent subframes.

The OI message includes the UL interference OI list, where the i-th element in the list indicates the interference level of the i-th PRB pair. The interference level of each PRB can be chosen among high, medium and low, based on eNB's uplink interference measurement results. On receiving the OI message, an eNB may instruct its UEs who are scheduled to

[2]In 3GPP LTE, a subframe spans 1 msec, and it is the minimum time unit for scheduling a transport block, or a medium-access (MAC) layer packet.

transmit in the high-interference PRBs to reduce power, in order to accommodate the OI sender eNB's request.

Finally, the HII message is a proactive bitmap indicator reporting on each PRB's interference sensitivity. The i-th bit in the bitmap indicates whether the i-th PRB has high (bit $= 0$) or low (bit $= 1$) interference sensitivity. For example, an eNB may send an HII message to its neighbour eNBs with marking of a set of PRBs with 'high interference sensitivity', in order to schedule the set of PRBs to its cell-edge UEs. Upon receiving the HII message, a neighbour eNB may decide to avoid scheduling the set of PRBs to its own cell-edge UEs, so that the neighbour cells' cell-edge UEs' uplink transmission is not severely interfered.

13.3.2 Rel-10 Enhanced ICIC

Rel-10 enhanced ICIC (eICIC) is introduced to coordinate downlink physical control channel interference in the HetNet interference scenarios as well as to further improve the data channel reception quality.

In LTE Rel-8/9, interference management of downlink control channels is achieved by cell-specific resource offsets and interference randomization techniques. The randomization techniques can in general maintain a low interference level in lightly loaded traffic conditions because the number of control channel REs colliding with neighbour cells' control channels is small. However, the solution cannot manage control channel interference in heavily loaded conditions, especially when the interference on individual REs is high as in the HetNet interference scenarios.

Rel-10 introduced two new eICIC methods to resolve the control channel interference issue, which are *CA-based eICIC* and *time-domain eICIC*. The basic method of CA-based eICIC is cross-carrier scheduling, which enables an aggressor eNB to transmit PDCCH scheduling PDSCH of one cell (say cell A) from another cell (say cell B) on different carrier frequency. This allows a victim eNB operating at the same frequency as cell B to transmit PDCCH without being interfered by the aggressor eNB in cell B. Similarly, if the victim eNB controls multiple cells of different carrier frequencies, cross-carrier scheduling also enables the PDCCH of a cell suffering from high interference in its control region to be transmitted from another cell of different carrier frequency which is relatively interference free. In the case of time-domain eICIC, an aggressor eNB is configured to transmit only a small number of physical channels (or none at all) in designated subframes called almost blank subframes (ABS), so that the victim eNBs' downlink transmission is less interfered by the aggressor. In other words, the aggressor eNB's downlink subframe is lightly loaded in ABSs, and hence it can maintain low interference level to the neighbour victim eNBs' downlink subframe.

The Rel-10 eICIC methods provide a greater design flexibility to the network, in that they allow the network to actively coordinate inter-cell interference for the downlink control/data region by configuring eICIC parameters, for example, number of ABS.

CA-based eICIC – CA-based eICIC is supported in Rel-10 LTE as a way to coordinate inter-cell interference of the downlink control channels for heterogeneous network deployment. It is suitable for a deployment scenario where either the aggressor eNB or the victim eNB, or both, are deploying carrier aggregation with the same carrier frequencies.

In CA-based eICIC, the downlink carriers of an eNB are partitioned into two types: one is used for transmission of the downlink data as well as the downlink control signalling; the other

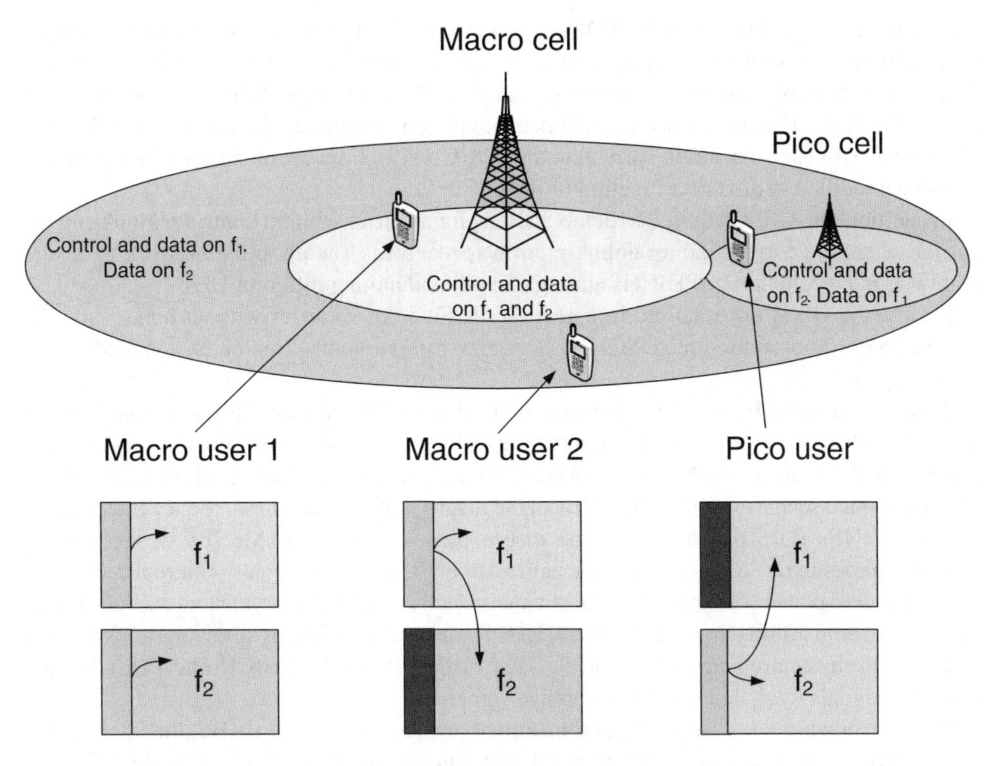

Figure 13.4 CA-based ICIC.

is used for transmission of the downlink data and no downlink control signalling (or downlink control signalling with reduced transmission power). By ensuring that the carriers with control region do not overlap in frequencies among neighbouring cells, inter-cell interference of the downlink control channels can be avoided/mitigated. An example deployment scenario is shown in Figure 13.4, where the MeNB configures carrier f_1 as the carrier with control region and carrier f_2 as the carrier without (or with reduced powered) control region; meanwhile, the pico eNB configures carrier f_2 as the carrier with control region and carrier f_1 as the carrier without (or with reduced powered) control region.

It is noted that the configuration of carrier without control region is UE-specific, that is, a carrier that is configured *without* control region may be configured as a carrier *with* control region for another UE – one that is close to the cell centre. Carriers aggregated for a UE consist of one primary component carrier and one or more secondary component carriers. The primary component carrier cannot be configured as a carrier without control region; that is, only a secondary component carrier can be configured as a carrier without control region.

From a UE's perspective, a carrier that is configured without control region will be scheduled by *cross-carrier scheduling*. When cross-scheduling is configured, the UE will obtain the PDCCH for the carrier without control region from another carrier with control region. A carrier with control region may be used to transmit the downlink control channels of one or more carriers, and the UE determines the target carrier of the PDCCH by reading the 3-bit carrier indicator field (CIF) in the PDCCH. Therefore, the UE configured with cross-carrier scheduling is also said to be configured with the CIF. Note that the CIF is the same as the serving cell index of a carrier. In order to mitigate PDCCH blocking on the carrier with control region, the PDCCH search space is expanded and there is a separate definition of the PDCCH search space for each carrier. In particular, the PDCCH search space for carrier without control region is modified to also be a function of the CIF.

In addition, the UE is required to detect PHICH for a carrier without control region from the carrier where the corresponding uplink grant was received. It is up to the network to ensure that there is no collision of PHICHs among carriers and among different UEs.

Finally, the UE is not required to detect the PCFICH of a carrier without control region; the starting symbol index for PDSCH for a carrier without control region is signalled by the higher layers.

Time-domain eICIC – In time-domain eICIC, inter-cell interference is coordinated among multiple cells in time domain, via exchange of X2 signalling, or operation and maintenance (OAM) configuration of ABS patterns. ABSs are configured at aggressor eNBs (i.e., MeNBs in the macro–pico scenario, and femto eNBs in the macro–femto scenario) to reduce interference to victim eNBs (i.e., pico eNBs in the macro–pico scenario and MeNBs in the macro–femto scenario) in the ABS. In ABS, the aggressor eNB transmits signals with reduced power (including zero power) on some physical channels, such as PDSCH carrying unicast data traffic. Backward compatibility to earlier release LTE is ensured in ABS, by allowing transmission of essential physical channels (e.g., PSS, SSS, PBCH, PCFICH, PHICH and PDCCH) and reference signals (e.g., cell-specific reference signals).

When no downlink unicast traffic is transmitted in an aggressor cell's ABS, the interference to the victim cells in the data region is reduced. On the other hand, it is also clear that the PDCCH is lightly loaded in an aggressor cell's ABS because of the lack of the downlink unicast data scheduling in the downlink control region, and hence the interference to the victim cells is reduced in the downlink control region as well. The interference scenarios related to the time-domain eICIC is illustrated in Figure 13.5. In non-ABS, an aggressor cell's downlink transmissions highly interfere with a victim cell's downlink transmissions. On the other hand, when ABS is configured in the aggressor cell, the interference to the victim cell is reduced, and hence the victim cell may schedule victim UEs in the ABS.

There are two ways to configure ABS in the LTE-Advanced [6] – OAM configuration and X2 message exchange. For the macro–femto scenario where CSG cells are aggressor cells, OAM configuration can be used to configure ABS to the CSG cells. OAM can configure an ABS pattern to the CSG cells in case the CSG cells' downlink transmissions severely interfere with MeNB's downlink transmission to non-member macro UEs. At the same time, OAM can also inform the MeNB of the ABS pattern configured to the CSG cells, so that the MeNB schedules the non-member macro UEs in an ABS. For the macro–pico scenario where macrocells are aggressor cells, an X2 message is transmitted from the aggressor cell to the victim cell, informing the victim cell of the ABS pattern configuration of the aggressor cell. For this operation, the OAM may configure association between eNBs to use the time-domain ICIC, and/or may ensure that a common subset of the ABS patterns of the interfering cells exists when it is desirable.

The load and interference coordination information may be transferred over the X2 interface between eNBs controlling intra-frequency neighbouring cells using the load indication procedure. The X2 message (LOAD INFORMATION message) conveying the ABS information

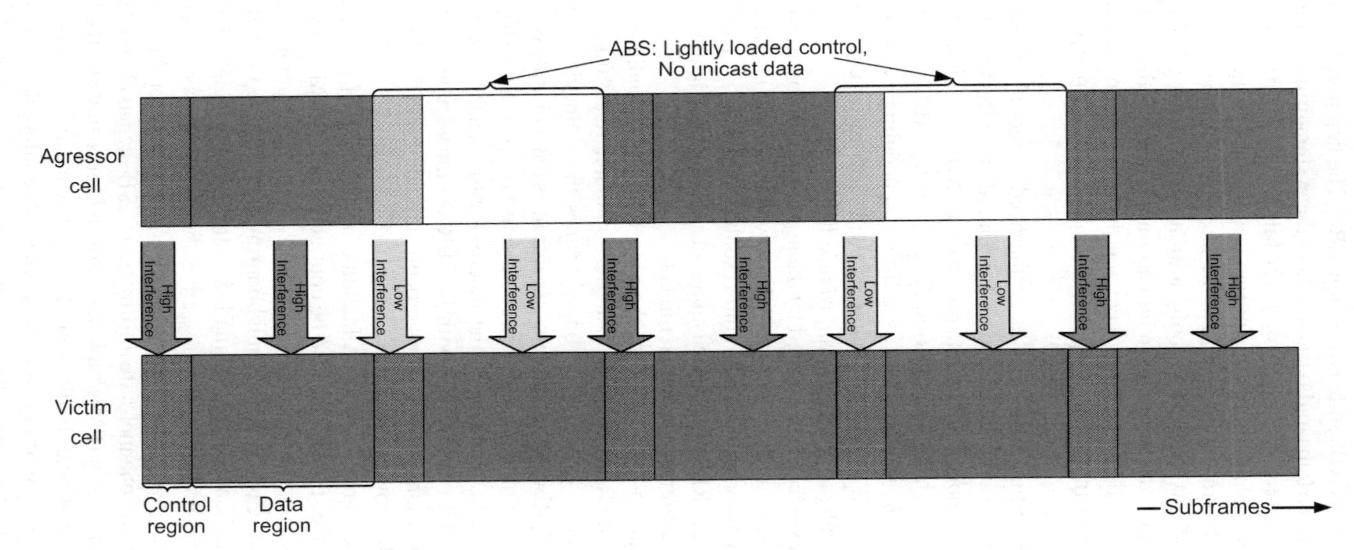

Figure 13.5 Time-domain eICIC with configuration of ABS in aggressor cells.

of the aggressor cell controlled by the sending eNB (defined in Section 9.2.54 in [5]) includes a bitmap IE, *ABS pattern info*, where *i*-th bit represents a DL subframe type. When the bit is 1, it implies the *i*-th subframe is ABS; otherwise, non-ABS. The first position of the ABS pattern info corresponds to subframe 0 in a radio frame where the system frame number (SFN) is zero. The ABS pattern is restarted each time SFN = 0. The bitmap size is 40 bits for FDD and 70 bits for TDD. For TDD, the actual maximum of subframes addressed by the 70-bit bitmap depends on the TDD UL-DL configuration, for example the maximum number of subframes is 20 for UL-DL configuration 1–5; 60 for UL-DL configuration 6, and 70 for UL-DL configuration 0. The receiving eNB may take the information into account when scheduling UEs on the victim cell controlled by the receiving eNB. Upon receiving the information, the receiving eNB will consider the information as immediately applicable and valid until reception of a new LOAD INFORMATION message carrying an update.

To allow for the eNB setting ABS to evaluate the need to modify its ABS configuration, the eNB can request for resource status update from another eNB using the ABS to protect its UEs. For this purpose, upon receiving the request, the eNB controlling the victim cell can send the *ABS status* IE which includes information about the ABS among the set of ABSs indicated by *ABS pattern info* that has been used for DL scheduling by the victim cell (*usable ABS pattern Info* IE) as well as the percentage of used ABS resources (*DL ABS status* IE) among the ABS indicated by *usable ABS pattern info* IE.

When the protected resources are provided by the aggressor cell for the victim cell through configuration of ABS by the aggressor cell, the UEs served by the victim cell can be configured to perform radio link monitoring (RLM)/ radio resource management (RRM) measurement on the protected resources (RLM/RRM measurement resource restriction) in order to obtain accurate RLM/RRM measurement of the victim cell.

For instance, in the macro–pico scenario, the pico UE can be configured by the picocell with CRE (victim cell) with RLM/RRM measurement resource restriction so that RLM/RRM measurement of the picocell is performed on subframes that coincide with the ABS of the macrocell (aggressor cell). RLM measurement on subframes that coincide with the non-ABS of the macrocell which contain high interference could cause the pico UE to trigger unnecessary radio link failure procedure. Similarly, in the macro–femto scenario, the macro UE can be configured by the macrocell (victim cell) with RLM/RRM measurement resource restriction so that the RLM/RRM measurement of the macrocell coincides with the ABS of the femtocell (aggressor cell), which has low interference.

For the macro–pico scenario, in order to facilitate accurate RRM measurement of neighbouring picocells with CRE, the UE can also be configured with RRM measurement resource restriction used for measuring those neighbouring cells. Since measurement restriction is only needed for picocells with CRE, a list of neighbouring cells where the RRM measurement resource restriction is applicable is also signalled to the UE. For neighbouring cells that do not belong to the list, the UE is still free to use any subframe for measurement.

Since the desired signal levels as well as the interference levels for ABS and for non-ABS can be significantly different, channel state information (CSI) such as channel quality indicator (CQI)/precoding matrix indicator (PMI)/rank indicator (RI), measured by the UE for each type of subframe can also be significantly different. If the network has the knowledge of the CSI from the UE for each type of subframe, the appropriate scheduling decision and transmission format such as the transmitted rank, the transmitted PMI and the modulation and coding scheme (MCS) level for each subframe can be determined. For this reason, resource restriction for CSI measurement can also be configured to the UE. In particular, two subframe subsets can be configured, where one subframe subset can correspond to the non-ABS and another can correspond to the ABS. CSI measurement resource restriction is supported for both aperiodic and periodic CSI reporting.

In Rel-10 LTE, RLM/RRM/CSI measurement resource restriction is only applicable to the primary component carrier. Moreover, the RLM/RRM/CSI measurement resource restriction is only supported when the UE is in RRC-connected mode and the configuration is provided by dedicated RRC signalling. In addition, for neighbouring cell measurement, only neighbouring cells operating at the same carrier frequency as the primary component carrier may be configured with RRM measurement resource restriction.

If the X2 interface is available, a subset of ABS recommended for configuring measurement towards the UEs can be provided by an eNB controlling the aggressor cell to another eNB controlling the victim cell as part of the LOAD INFORMATION message through the load indication procedure. The X2 message concerned is the bitmap IE *measurement subset* which indicates a subset of ABS indicated by the IE *ABS pattern info*.

13.3.3 System-level Performance of HetNet with Time-domain eICIC

Here, the system-level performance results of the HetNet with the time-domain eICIC are presented for macro–pico scenario and macro–femto scenario.

Macro–pico scenario – For the macro–pico scenario, the system-level performance is evaluated for the HetNet with the time-domain eICIC with CRE [8]. For the simulation, 57-cell hexagonal three-sector layout is generated according to [2], and two, four or ten picocells (or hotzones) are dropped in each macrocell (or sector) according to Figure 13.6.

Four different CRE bias values – 0 dB, 5 dB, 10 dB and 20 dB – are used in the simulations. The UE's cell selection is conducted as follows:

$$N_{Cell}^{ID} = \arg\max_{i \in A} (RSRP_i + \beta_i),$$

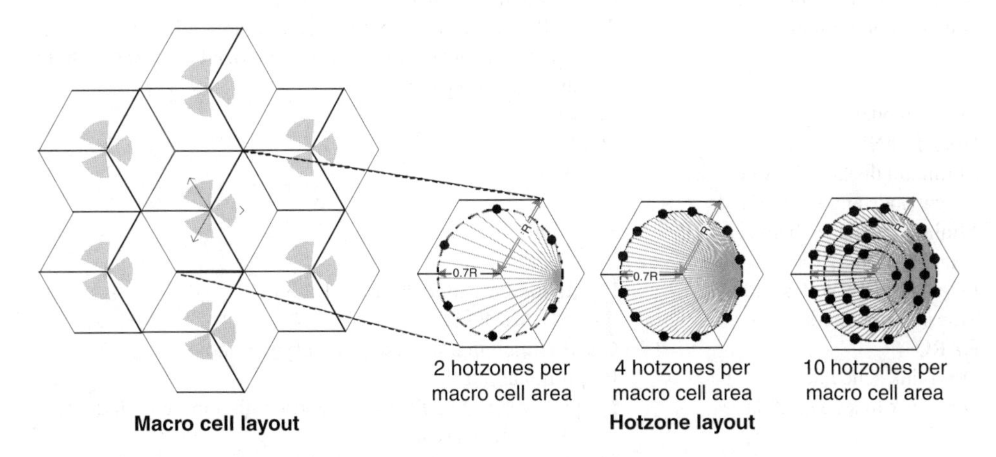

Figure 13.6 Cell Layout for the macro–pico simulations.

where $\beta_i = \beta$ if cell i is a picocell, $\beta_i = 0$ if cell i is a macrocell; $RSRP_i$ is the reference signal received power (RSRP) of cell i and N_{Cell}^{ID} is the physical cell ID of the associated cell of the UE. Furthermore, the CRE bias for the picocells is β dB and A is the set of physical cell IDs of the neighbour cells for which a UE is able to measure RSRP.

It is assumed that the ABSs in all the macrocells are synchronized, and the number of ABS in each radio frame (10 subframes) is selected such that the ratio of the number of ABSs to the number of non-ABSs is the same as the ratio of the UEs associated with picos to the UEs associated with macros.

Details of the simulation assumptions are listed in Table 13.2.

Figure 13.7 shows the system-level simulation results for the no-CRE case, that is, when $\beta =$ 0 dB. Without CRE, most of the UEs in the cellular network are associated with macrocells, and the performance gain from adding picos is limited. Only 7%, 9% and 36% gains are observed for the cell average throughput with two, four and ten picocells, respectively.

Table 13.2 Simulation assumptions for the macro–pico scenario

Parameter	Assumption
Homogeneous deployments	3GPP case 1 set based on [7] (ISD 500 m, 2.0 GHz CF and 3 km/h)
Macrocell antenna pattern (horizontal, vertical, 3D)	Table A.2.1.1-2 in [2]
Channel model	3GPP spatial channel model (SCM)
Total eNB TX power	46 dBm – 10 MHz carrier
Nodes per macrocell	2, 4, 10 hotzones per macrocell
Path loss model	Macro to UE: L = 128.1 + 37.6log10(R), R in km Hotzone to UE: L = 140.7 + 36.7log10(R), R in km (model 1)
Shadowing standard deviation	10 dB
Penetration loss	20 dB
Hotzone cell antenna pattern	0 dB (omnidirectional)
Antenna configuration	TX: 2, RX: 2 (eNB: uncorrelated co-polarized four wavelengths between antennas UE: vertically polarized antennas with 0.5 wavelengths separation)
Traffic model	Full buffer
Hotzone eNB TX power	30 dBm
Minimum distance between UE/ new node and regular node	>= 35 m
Minimum distance between UE and hotzone	>= 10 m
Placing of new nodes and UEs	Configuration 1 in table A.2.1.1.2-3 in [2]
Transmission mode	2×2 SU MIMO
HARQ	Chase combining, non-adaptive, asynchronous
Downlink scheduler	Proportional fair
Downlink link adaptation	Frequency selective PMI/CQI report with 5 msec periodicity and 6 msec delay
Downlink receiver type	MMSE with ideal channel estimation and no feedback errors

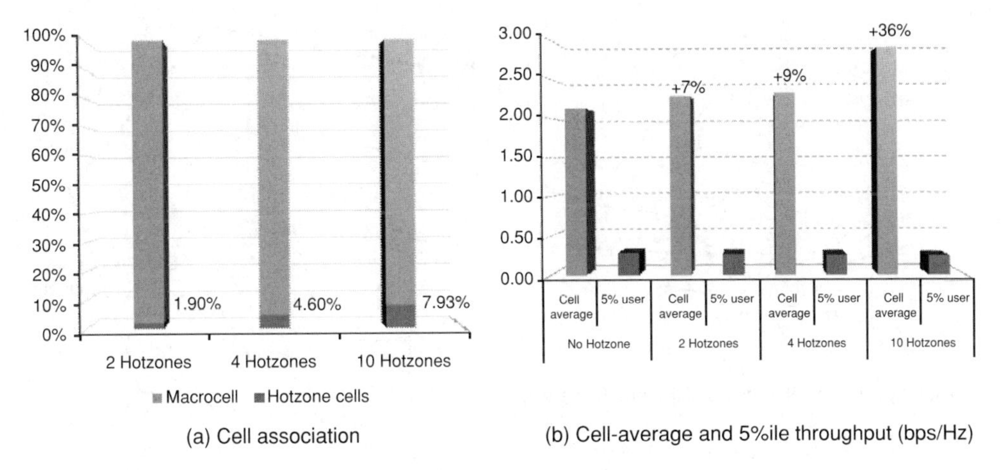

Figure 13.7 Simulation results for the no-CRE ($\beta = 0$ dB) case in the macro–pico scenario.

On the other hand, with CRE, the picocell association ratio increases as the CRE bias β increases. Figure 13.8(a) shows that with $\beta = 5$, 10, 20 dB CRE bias, the picocell association ratios for the two-pico, four-pico and ten-pico cases become 25.40%, 36.34% and 62.30%, respectively. As picos serve more UEs, the performance also improves, as shown in Figure 13.8(b). In terms of cell-average throughput, two-pico, four pico and ten-pico cases achieves 39%, 91% and 231% gain against no-pico case, with 20 dB CRE bias. On the other hand, 5%ile throughput improves a little bit against no pico case, 3.7% and 17% in four-pico and ten-pico cases, respectively.

Macro–femto scenario – For the macro–femto scenario, the system-level performance is evaluated for the HetNet with and without the time-domain eICIC [9, 10]. For the simulation, 57-cell hexagonal three-sector layout is generated according to [2], and a cluster of femto

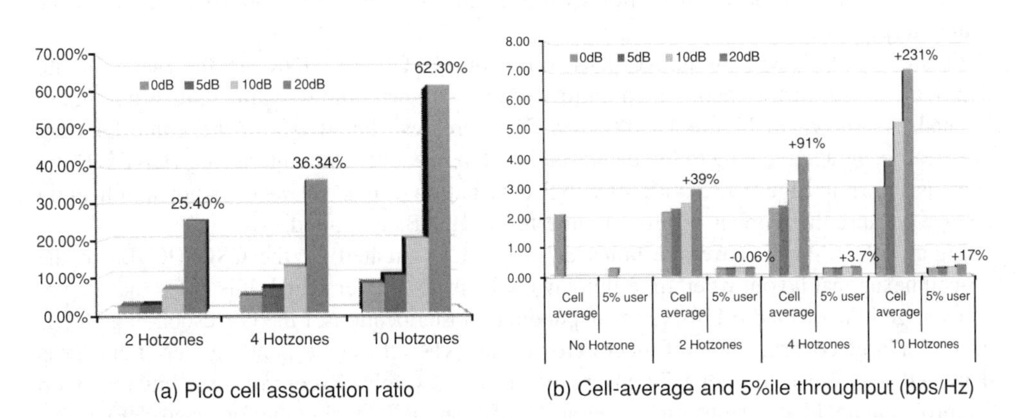

Figure 13.8 Simulation results with CRE ($\beta = 5$, 10, 20 dB) in the macro–pico scenario.

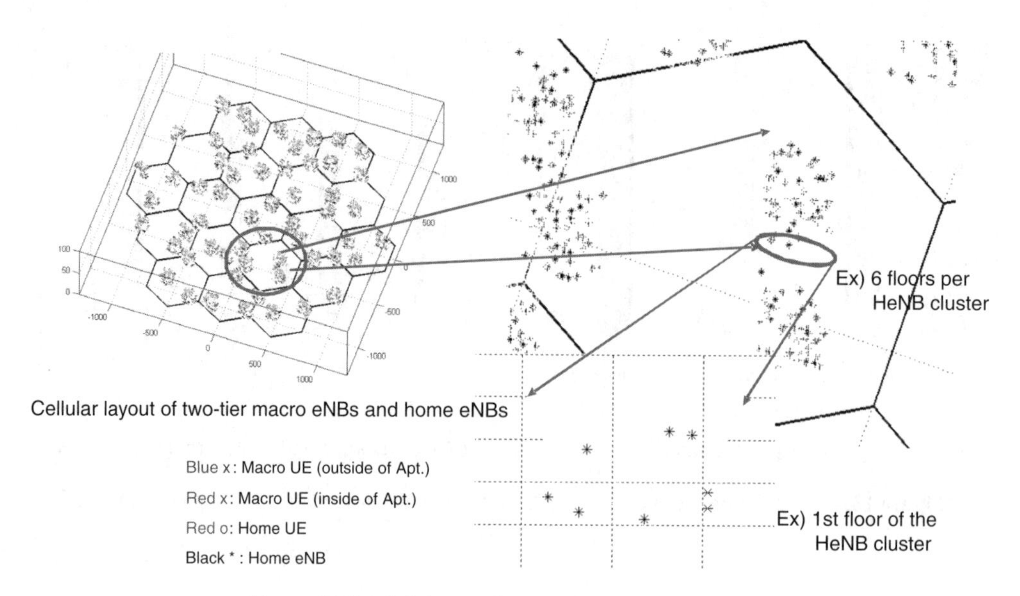

Figure 13.9 Cell layout for macro–femto simulations.

eNBs (or HeNBs) is dropped in each macrocell (or sector) according to Figure 13.9. Each cluster of HeNBs is located according to the *dual strip model* [10], where the cluster of HeNBs is placed in two apartment buildings spaced apart by a street of width 10 m, each of which has N_f floors, and each floor has 10×2 apartment units of size 10 m × 10 m each; in the example shown in Figure 13.9, the number of floors $N_f = 6$. The average number of femtos in each cluster is determined by a parameter called *deployment ratio* (DR), or r_D, where $0 \leq r_D \leq 1$; given the parameter r_D, the average number of HeNBs in each floor is $20r_D$. An apartment unit can have up to one HeNB which serves one home UE (or a CSG UE), where both the HeNBs and home UE are dropped randomly in the apartment unit. Furthermore, ten macro UEs (or non-CSG UEs) are randomly dropped in the macro geographical area according to the uniform distribution. More details for the simulation assumption can be found in [2].

Figure 13.10 shows cumulative density function (CDF) of SINRs in the macro–femto deployment with various deployment ratios and $N_f = 1$ when time-domain eICIC is not implemented. As shown in Figure 13.10(b), as DR increases, the SINRs of the home UEs are drastically degraded, owing to the increased interference level from interfering HeNBs. However, as shown in Figure 13.10(a), the SINR distribution of macro UEs does not significantly change because the number of severely interfering HeNBs is limited.

Figure 13.11 shows coverage holes of macro UEs created by the CSG HeNBs in the 57-cell hexagonal layout when time-domain eICIC is not implemented. It is noted that size of a coverage hole of macro UEs varies depending on the location of the corresponding HeNB cluster; a large coverage hole of macro UEs occurs when the corresponding HeNB cluster is deployed at a macrocell's edge. To ensure the macro UEs' reliable control signalling reception (control channel block error ratio $< 1\%$ at -8 dB), an eICIC technique for macro UEs in the dead-zone – that is, victim UEs – is necessary.

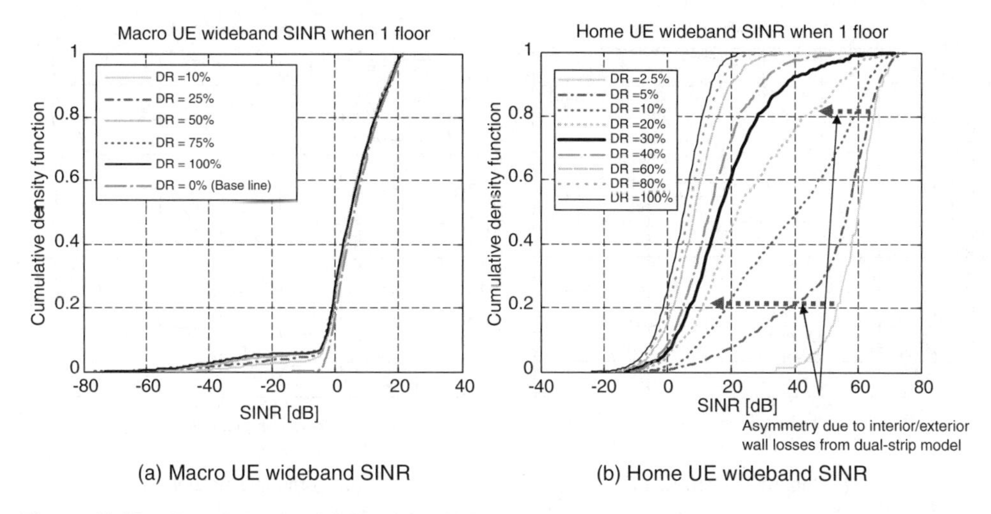

Figure 13.10 Cumulative density function of SINRs in the macro–femto deployment with various DRs, when time-domain eICIC is not implemented. (a) Macro UE wideband SINR (b) Home UE wideband SINR.

Figure 13.12 shows UE throughput CDF in the macro–femto deployment with various deployment ratios and $N_f = 6$, when time-domain eICIC is not implemented. Figure 13.12(a) shows macro UE throughput CDF. For non-zero DRs as small as 5%, some macro UEs experience outage, that is, they do not receive any packets (throughput = 0 bits/sec/Hz) for the whole simulation duration. The ratio of UEs experiencing outage increases as DR increases, because the interference from CSG HeNBs increases. Figure 13.12(b) shows home UE throughput

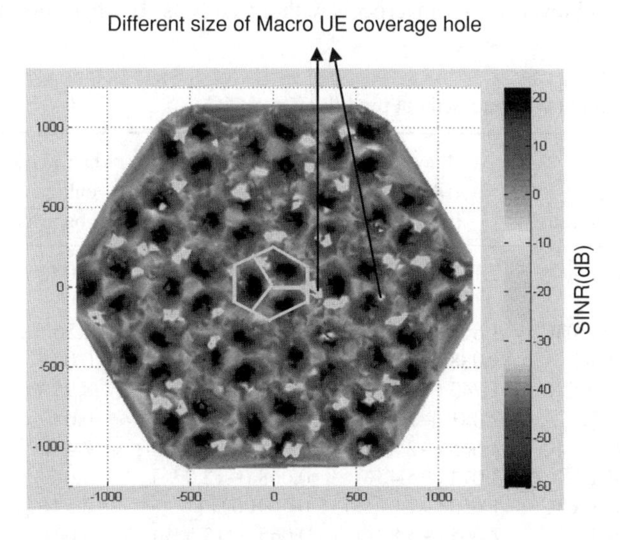

Figure 13.11 Macro UE coverage holes without time-domain eICIC.

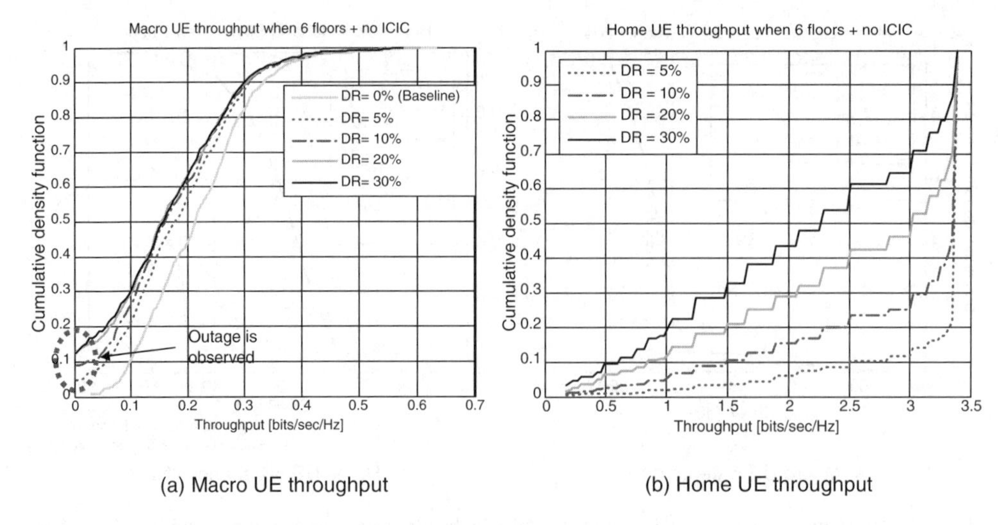

Figure 13.12 Cumulative density function of UE throughput in the macro–femto deployment with various DRs, when time-domain eICIC is not implemented.

CDF. As DR increases, the home UE throughput decreases because the interference from neighbouring HeNBs increases.

Table 13.3 shows throughput results with and without time-domain eICIC in the macro–femto deployment with various deployment ratios and $N_f = 6$. When the time-domain eICIC is implemented, the MeNB's cell average throughput and 5%ile macro UE throughput with DR > 0 (or with some HeNBs are deployed in the macro coverage areas) do not degrade much from those with DR $= 0$, or with macros only. For example, with DR $= 10\%$ the MeNB's cell average throughput and the 5%ile throughput decrease only 5% and 11% respectively from those achieved in the macrocell without HeNBs. Furthermore, the HeNB's cell

Table 13.3 Throughput with and without time-domain eICIC

	ICIC Function	Cell average (MeNB) [bps/Hz]	5 % User (macro UE) [bps/Hz]	Macro UE outage probability	Cell average (HeNB) [bps/Hz]
DR 0 % (only MeNBs)	N/A	2.193	0.077	0.0 %	N/A
DR 5 %	No eICIC	1.863 (−15.1 %)	0.012 (−84.0 %)	4.56 %	3.200
	eICIC	2.088 (−4.8 %)	0.074 (−4.8 %)	0.0 %	3.167 (−1.0 %)
DR 10 %	No eICIC	1.743 (−20.5 %)	0 (−100 %)	8.95 %	2.930
	eICIC	2.081 (−5.1 %)	0.069 (−11.1 %)	0.0 %	2.900 (−1.0 %)
DR 20 %	No eICIC	1.665 (−24.1 %)	0 (−100 %)	12.4 %	2.527
	eICIC	2.074 (−5.4 %)	0.068 (−12.0 %)	0.0 %	2.502 (−1.4 %)
DR 30 %	No eICIC	1.653 (−24.7 %)	0 (−100 %)	12.5 %	1.653
	eICIC	2.065 (−5.8 %)	0.065 (−15.5 %)	0.0 %	2.124 (−0.8 %)

average throughput with time-domain eICIC is maintained to a similar number as without time-domain eICIC. In summary, time-domain eICIC technique greatly improves the macro UE's packet reception in the presence of CSG HeNBs while still maintaining the HeNB's throughput.

13.4 Ongoing Work in Rel-11 LTE-A

The Rel-10 work on eICIC marks an important milestone in the evolution of HetNet technology. For the first time, the operators can effectively overlay and operate their networks with a combination of higher-powered nodes and lower-powered nodes, without suffering significantly from the additional interference that is inherent in this overlay architecture. The two conceptually simple tools developed in Rel-10, the time-domain ICIC and the CA-based eICIC, turned out to be very powerful and robust. In short, the Rel-10 work laid a solid foundation for successful widespread commercialization of HetNet technology across the world.

On the other hand, there were a few areas of further enhancement and optimization that were left out of Rel-10 due to lack of time. These areas of enhancement were identified in the Rel-11 work item proposal on further enhanced ICIC (FeICIC) presented in [12]:

- Finalize the leftover work from Rel-10 on inter-frequency and inter-radio-access-technology (inter-RAT) TDM restricted RRM.
- Based on system performance gains, identify the scenarios for which UE performance requirements in the following two bullets will be specified in terms of, for example, the number of interferers and their relative levels with respect to the serving cell.
- UE performance requirements and possible air-interface changes/eNB signalling to enable significantly improved detection of primary cell ID (PCI) and system information (master information block (MIB), secondary information block (SIB)-1 and paging) in the presence of dominant interferers for FDD and TDD systems, and different network configurations (e.g., subframe offset/no-subframe offset), depending on UE receiver implementations – (RAN1, RAN4, RAN2).
- UE performance requirements and necessary signalling to the UE for significantly improved downlink control and data detection and UE measurement/reporting in the presence of dominant interferers (including colliding and non-colliding reference signals, as well as, multimedia broadcast single frequency network (MBSFN) used as ABS, as well as, ABS configurations) for FDD and TDD systems depending on UE receiver implementations. Improved detection based on air interface enhancements to be considered – (RAN1, RAN4, RAN2)
  - Dominant interference applicable to both macro–pico and CSG scenarios and with or without handover biasing.
- As a second priority, study the following aspects:
  - Data channel ICIC enhancements, such as FDM/TDM coordination and enhanced signalling for resource allocation; or supporting the application of single-carrier time domain ICIC mechanism on the frequency of SCell in carrier aggregation setting.
  - Higher layer enhancements, for example, for idle mode operation, power saving and mobility enhancements (RAN1, RAN2, RAN3, RAN4).

- Uplink enhancements, for example uplink interference mitigation for macro–pico and macro–femto.
- Identify interference scenarios stemming from different UL/DL configurations or muted UL subframe configuration in TDD and corresponding air interface change and CSI reporting requirements.
- Dominant interference impacting on legacy UEs (Rel-8/9/10).

After extensive evaluation and discussion for about one year in the 3GPP RAN1 working group after the WI [12] was first approved, it is becoming increasingly clear that the Rel-11 FeICIC will focus on the following areas of enhancement:

- Support of non-zero power ABS for system capacity enhancement.
- Support of cell range expansion (CRE) with a bias of more than 6 dB
  - Cell acquisition (reception of primary/secondary synchronization signals (PSS/SSS) and primary broadcast channel (PBCH)) in low geometry (< -6dB).
  - Mitigation of dominant interference due to macro-CRS.

The standardization work in 3GPP is still ongoing at the time of writing. We will briefly summarize some of the proposals for FeICIC below.

13.4.1 Support of Non-zero Power ABS

The operation of reduced non-zero power ABS is very simple: instead of not transmitting any PHY-layer control and data in the ABS, the MeNB can choose to transmit at a reduced non-zero power level to UEs that have good geometry. Here the assumption is that the reduced power, while sufficient for the high-geometry UEs connected to the macrocell, will not cause significant interference to pico UEs in a CRE region.

For normal subframes, two parameters were defined and signalled by eNB in Rel-8/9/10 for the purpose of DL power control and allocation among PDSCH channels. The first parameter is a UE-specific P_A value, and it represents the ratio of PDSCH EPRE to cell-specific reference signal (CRS) EPRE in symbols without CRS. The second parameter is a cell-specific P_B value, and it represents the ratio of PDSCH EPRE in symbols with CRS versus symbols without CRS. These two parameters, together with the signalled EPRE of CRS, completely define the power level of PDSCH traffic intended for a given UE. Tables 13.4 and 13.5 show the supported values of P_A and P_B in Rel-8/9/10 LTE. By setting −6dB for P_A and 1/2 for P_B with two or four antenna ports, the eNB can transmit PDSCH EPRE with 25% of CRS ERPE

Table 13.4 The UE-specific P_A

	Value
P_A	3dB, 2dB, 1dB, 0dB, −1.77dB, −3dB, −4.77dB, −6dB

Table 13.5 The cell-specific P_B for one, two or four cell-specific antenna ports

P_B	One antenna port	Two and four antenna ports
0	1	5/4
1	4/5	1
2	3/5	3/4
3	2/5	1/2

in symbols without CRS and 12.5% of CRS in symbols with CRS, which is the minimum PDSCH power level.

In a heterogeneous network scenario [13], it may be beneficial to schedule high-geometry macro UEs in the ABS, since it helps with macrocell throughput while creating negligible interference to UEs in a picocell's CRE region. The way of accomplishing this is for eNB to transmit at a reduced power level in the ABS compared to the normal subframes, and this reduced power level should take into account the location of the victim picocell and its distance to the aggressor macrocell, the load situation in these victim picocells, and so on. From a signalling and protocol design perspective, it is desirable to configure another P_A for ABS, denoted by P_{A_ABS}, that represents the PDSCH EPRE to CRS EPRE for OFDM symbols without CRS in an ABS. The lowest configurable value of P_{A_ABS} needs to be extended to cover sufficiently low target power level in non-zero transmission power ABS.

The signalling and protocol design changes for the cell-specific parameter P_B is slightly more complicated. P_B describes the power ratio between symbols with and without CRS. Regardless of the unicast power allocation for each UE, the CRS EPRE would be the same across the whole downlink bandwidth at all times. In case CRS EPRE is higher than average PDSCH EPRE (or CRS power being 'boosted'), the symbols without CRS can utilize the remaining power not used by the CRS for PDSCH REs. On the other hand, in the symbols with CRS, the PDSCH power will be reduced in order to support the 'boosted' CRS power, and the eNB will use the parameter P_B to tell UEs the extent of this power reduction.

In non-zero transmission power ABS, there could be two options. The first option is similar to the case of P_A, where an additional parameter P_{B_ABS} is configured by the eNB to represent the ratio of PDSCH EPRE between symbols with and without CRS in the ABS. On the other hand, since the target PDSCH power in ABS is likely to be lower than CRS EPRE, the eNB may not need to have the additional reduction of PDSCH transmission power in the symbols with CRS, in order to support the 'boost' of CRS power. Therefore the second option is to not signal an additional P_{B_ABS} for ABS, and for the UE to assume that the ratio of PDSCH EPRE between symbols with and without CRS is one, that is there is no additional PDSCH reduction to support CRS power 'boost', so the UE can always assume $P_{B\_ABS} = 1$ in the ABS. Figure 13.13 shows a conceptual illustration of power allocation in normal subframes and ABS, assuming the additional indication of P_{A_ABS}, but no additional indication of P_{B_ABS} (option 2). With this configuration, in the ABS the UE will experience significantly reduced PDSCH transmission power, although there will be no variation of PDSCH power level in symbols with and without CRS.

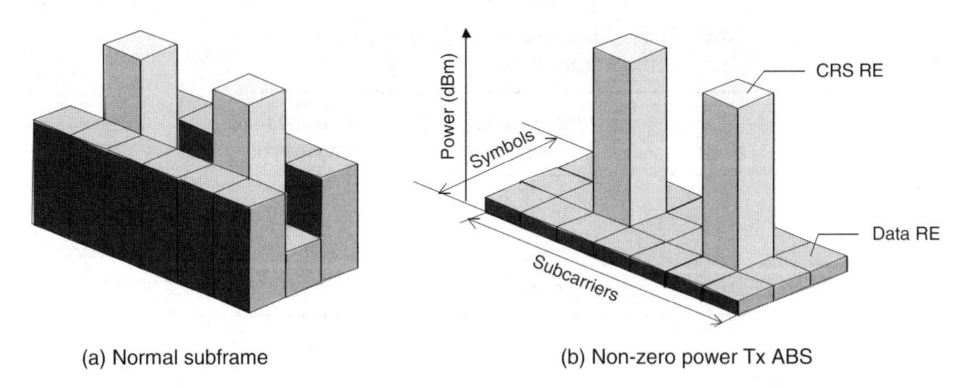

Figure 13.13 Power allocation for normal subframe and ABS.

13.4.2 Network-assisted Cell Acquisition for CRE UE in Low Geometry

In Rel-11, as the pico eNB's CRE region expands to cover more areas, many UEs are expected to be able to detect and attach to these CRE pico base stations in a very low geometry environment such that the geometry is less than −6dB (meaning the interference from macro base station is more than 6dB stronger than the signal the UE is trying to read from the pico base station capable of CRE). If the dominant interferer is a macrocell in such a macro–pico deployment scenario, then the Rel-11 UE needs to have an enhanced interference cancellation (IC) capability, so that it is able to detect and handover to picocells with range expansion. Without the interference cancellation capability, the UE might not detect the picocell with CRE although it is within the extended coverage of the picocell. This is illustrated in Figure 13.14, where only a UE with enhanced interference cancellation capability is able to see weak picocells in its neighbourhood.

While UE interference cancellation is an important technique to resolve the interference problem for HetNets, it is intended to be used for HetNets with time-domain ICIC. It is also important to understand whether there is any undesirable side effect as a result of UEs applying the technique to detect neighbouring cells without CRE or to detect neighbouring cells in a legacy network. To this end, we consider a HetNet scenario with colliding PSS/SSS

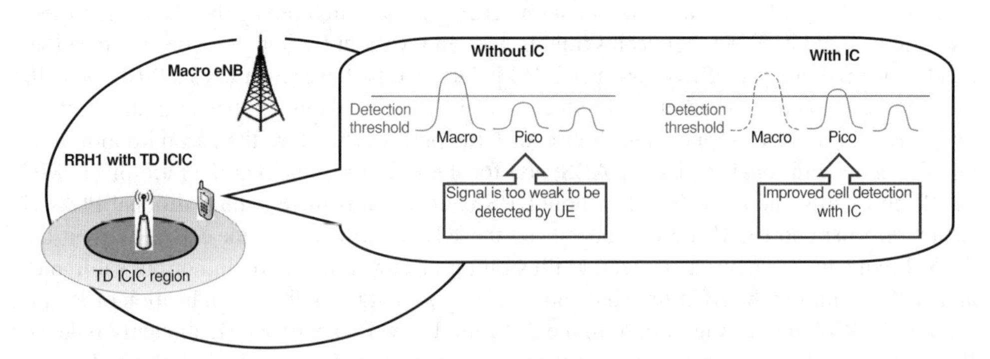

Figure 13.14 Improved neighbouring picocell detection with PSS/SSS interference cancellation.

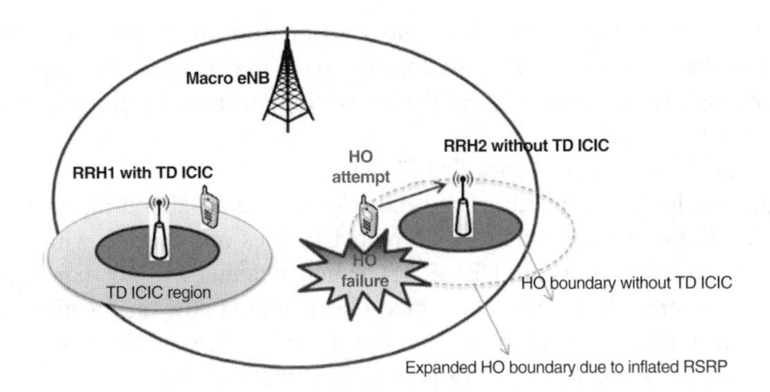

Figure 13.15 Potential HO failure caused by inflated RSRP measurement of neighbouring cell.

signals and non-colliding CRS signals between macro- and picocells, such that the picocell in this particular scenario does not support cell range expansion [14]. As mentioned earlier, applying interference cancellation to PSS/SSS enables detection of cells that would otherwise be missed. Upon detection of unintended cells, that is cells with no CRE, the UE will commence unrestricted measurement on CRS to generate an RSRP result. Since the cell was detected at a distance further away than would normally be the case, the CRS REs of the cell concerned contain potentially much higher interference power (e.g., from the control/data of serving cell). As a result, the RSRP measurement of the cell detected could be inaccurate; for example, it is shown in [12] that the RSRP value may dramatically increase, up to 4dB, due to the strong interference power.

Due to the inflated RSRP value, a reporting event may be triggered by the UE. As a result, the measurement result and the detected cell's PCI may be reported by the UE to the network, which may lead to handover initiation by the network. The handover will likely fail since the target cell's signal is actually too weak. This problem is illustrated in Figure 13.15. To resolve this problem, network assistance could be helpful, as the network can provide a list of cell IDs that contain the information about neighbouring picocells supporting CRE. The UE then checks the cell ID of any detected cell (after interference cancellation) against the list; if the ID is on the list, the UE proceeds to measurement report and prepare for handover; and if the ID is not on the list, the UE simply stops there and moves on to the next candidate cell.

13.4.3 Mitigation of CRS Interference for CRE UE in Low Geometry

A pico UE in low geometry will suffer from strong interference from a macrocell's CRS. The macrocell has to transmit its CRS in all non-MBSFN subframes in order to maintain backward compatibility and support early release UEs. In order to mitigate these dominant CRS interferences, two alternatives have been recently proposed and evaluated in the 3GPP RAN1 working group.

- Receiver-based interference cancelling (Rx-IC)
  The pico UE estimates the CRS from the macrocell and cancels it out prior to PDSCH demodulation. This scheme does not require specification support for mitigation of CRS

interference. However, the scheme is sensitive to phase imperfections due to co-channel interference in the estimated CRS. Also, this approach may cause some channel estimation performance degradation, if the CRS of the interfering macrocell collides/overlaps with the CRS of the victim picocell.

- Transmitter-based interference cancelling (Tx-IC)
  PDSCH rate-matching is applied around the CRS of the neighbouring macrocell(s), and as a result the PDSCH transmitted to the pico UE does not overlap with the CRS from the macrocell. Therefore, the pico UE's PDSCH data channel will not experience any CRS interference. To support this scheme, the pico UEs need to know the location and sequence of the interfering CRS. Some companies have proposed to introduce additional signalling to inform pico UEs of the interfering CRS parameters, which include:
  - the physical cell IDs corresponding to a set of neighbouring cells,
  - the number of CRS ports corresponding to a set of neighbouring cells, and
  - a bitmap conveying the location of the subframes in which the CRS interference from neighbouring cells is cancelled.

Evaluation of these two alternative approaches is currently ongoing, and is expected to be completed during 2012. With the support of either or both of the methods, the CRS interference can be effectively mitigated for Rel-11 and future release LTE-Advanced HetNet systems.

13.5 Conclusion

The Rel-10 work on eICIC marks an important milestone in the evolution of HetNet technology. For the first time the operators can effectively overlay and operate their networks with a combination of higher-powered nodes and lower-powered nodes, without suffering significantly from the additional interference that is inherent in this overlay architecture. The two conceptually simple tools developed in Rel-10, the time-domain ICIC and the CRE for picocells, turned out to be very powerful and robust. The Rel-10 HetNet work laid a solid foundation for successful widespread commercialization of HetNet technology across the world. Meanwhile, ongoing effort of FeICIC in Rel-11, such as non-zero power ABS, low-geometry cell-association with CRE-enabled picos and CRS interference mitigation, will further enhance and optimize the performance of the HetNet system, and make the technology even more attractive for the operators.

References

1. REV-080032, 'LTE-Advanced Technology Candidates', Qualcomm Europe, April 2008.
2. 3GPP TS 36.814, E-UTRA Further advancements for E-UTRA physical layer aspects, v9.0.0, March 2010.
3. RP-100383, 'New Work Item Proposal: Enhanced ICIC for non-CA based deployments of heterogeneous networks for LTE', CMCC, March 2010.
4. 3GPP TS 36.213, E-UTRA Physical layer procedures, v9.3.0, September 2010.
5. 3GPP TS 36.423, E-UTRAN X2 application protocol, 10.4.0, December 2011.
6. 3GPP TS 36.300, E-UTRA and E-UTRAN overall description, v10.6.0, December 2011.
7. 3GPP TS 25.814, Physical layer aspects for E-UTRA, v7.1.0, September 2006.
8. R1-100142, 'System Performance of Heterogeneous Networks with Range Expansion', Samsung, January 2010.

9. R1-103049, 'Performance evaluation of Home eNB ICIC function based on time/frequency domain silencing', Samsung, May 2010.
10. R4–092042, 'Simulation assumptions and parameters for FDD HeNB RF requirements', *Alcatel-Lucent, picoChip Designs*, Vodafone, May 2009.
11. 3GPP TR 36.921, E-UTRA FDD Home eNodeB (HeNB) Radio Frequency (RF) requirements analysis, March 2010.
12. R1-110058, 'Considerations on RSRP/RSRQ Definition', CATT
13. R1-120166, 'Pa and Pb in non-zero Tx power ABS', Samsung
14. R1-120168, 'Discussion on further enhanced non-CA-based ICIC', Samsung

14

Macro–Femto Heterogeneous Network Deployment and Management

Peng Lin, Jin Zhang, Yanjiao Chen, and Qian Zhang
Hong Kong University of Science and Technology, China

Femtocell techniques can address the problem of poor coverage within buildings and increase the network capacity cost-efficiently. Until now, some wireless service providers (WSPs) have launched their femtocell services while there are still plenty of challenges unsettled. In this chapter, we discuss the business mode in macro–femto heterogeneous networks.

We propose three frameworks for heterogeneous network deployment and management according to the deployment types of femtocells: joint deployment, WSP deployment and user deployment. In the joint-deployment framework, the femtocells are provided by the WSP and paid for by users. In the WSP-deployment framework, the WSP does not charge users for the femtocells installed. In the user-deployment framework, users unilaterally purchase femtocells from retail stores and install them. The unique characteristics, corresponding challenges and potential solutions of these frameworks are further investigated to provide a deeper insight systematically.

We also present two schemes for WSP revenue maximization under the WSP-deployment framework. The first scheme jointly handles the interference and user demand satisfaction via cross-tier channel allocation and the second scheme further considers the optimal pricing selection for accessing different networks. Through the two schemes, we give preliminary discussions of more detailed schemes under the WSP-deployment framework.

Heterogeneous Cellular Networks, First Edition. Edited by Rose Qingyang Hu and Yi Qian.

14.1 Introduction

Recent surveys show that the in-building generated phone calls and data traffic are expected to account for 50% and 70% of the total volume respectively in the near future [1]. Mobile operators need to be able to support the huge indoor traffic demand to win the market share. Despite the ever-growing investment of the macrocell base stations (BSs), users still often suffer from low signal strength and poor service quality in indoor environments, especially under 3G cellular networks operating in the high frequency band. The urgent task for mobile operators is to provide a good indoor coverage and high capacity in a cost-effective way. The femtocell technique [2] is believed to be the most promising approach to solving this problem, and it has drawn a good deal of attention from manufacturers, operators and researchers.

Femto-BSs are low-power, small-size and home-placed base stations which provide wireless service to cellular users nearby. They operate in the spectrum licensed to the cellular operator and connect to their local network by wired backhaul. The femto-BSs are able to provide a higher capacity because their users can receive its signals with higher signal-to-noise-ratio due to the short transmitter–receiver distance and low attenuation. Besides, the small-size densely deployed femto-BSs can create a large bundle of spectrum reuse opportunities and thus improve the overall system capacity, which is welcomed by the operators. The femto-BS is similar to a WiFi access point in the sense of providing indoor coverage, but it has advantages over WiFi. Firstly, it does not require a dual-mode mobile device, as these sell at a higher price. Secondly, it can provide guaranteed quality of service (QoS) using a licensed band, while WiFi in the crowded unlicensed band cannot.

Because of these appealing characteristics, femtocell techniques have drawn great attention from many wireless service providers (WSPs). Many well-known WSPs have recently finished small scale trials and formally launched their commercial femtocell services, such as Sprint's 'Airave', Verizon's 'Network Extender' and China Unicom's '3G Inn'. With the initial attempt emerging, there are still lots of challenges to address, from both economic and technical aspects. We believe that the economic challenges, especially the business operating modes including the demands, goals and incentives from the various market participants, are the most fundamental and important. A perfect solution that is then based on an unrealistic scenario will not be acceptable to WSPs and users, and this will contribute little to the development of femtocell industry. Only when these issues are well defined and investigated can we further discuss the corresponding technical solutions reasonably. However, many existing works have simply tried to address the detailed technical problems while giving little attention to the related scenarios. As far as we know, there are few works which systematically analyze the potential business frameworks of the macro–femto hybrid networks.

This chapter will discuss the macro–femto cross-tier heterogeneous networks from a business mode point of view. The unique characteristics and possible solutions are then discussed. We propose three frameworks for the deployment and management of macro–femto hybrid networks: joint-deployment, WSP-deployment and user-deployment frameworks, according to the ownership and operating mode of the femto-BSs. Under each framework, we identify the pros and cons and the challenges from both economic and technical points of view. We also provide possible solutions to solve these challenges. Then, to give concrete examples and illustration, we propose two schemes under a WSP-deployment framework, in which the WSP tries to provide better wireless access service and maximize their revenue by adjusting the spectrum allocation and traffic scheduling. One of the schemes adopts cross-tier channel allocation to jointly manage the interference and demand satisfaction, and the other considers the optimal pricing for accessing different networks.

14.2 Frameworks for Macro–Femto Network Deployment and Management

In this section, we will discuss the frameworks of macro–femto heterogeneous network by clarifying their network scenarios driven by business cases. Under each framework, the state-of-the-art, the economic and business challenges, as well as potential solutions will be further investigated.

14.2.1 Joint-deployment Framework

In this framework, the femto-BSs are provided by the operator and paid for by users for indoor installation, either in a one-off price way or by monthly rental. The femto-BSs just take some minutes to sense the radio environments, localize the positions, register themselves and auto-configure the parameters for the first time, then work stably later. WSP can be aware of their appearance and set/adjust some system parameters and strategies. As the owners, users can freely move and turn on/off the femto-BSs. Apart from the hardware and new services by the femto-BSs, users also have to pay for the landline networks and consumed electrical power. The operator saves lots of money by offloading most of the capital expenditure (CAPEX) and operational expenditure (OPEX) to users. This may explain why this framework is now adopted by most of the mobile operators. It can be further divided into two categories according to whether the macro users are allowed to access the nearby femto-BSs.

Closed access mode

In this mode, a femto-BS just serves its registered users, such as the owners, family members and friends. The unregistered ones, such as U_{m1} and U_{m3} in Figure 14.1, cannot access the femto-BSs even if they are in close proximity. According to the survey [3], home users prefer femtocells with closed access mode. The reason is straightforward in that no femto users would want to share the limited capacity of their femto-BSs and the landline throughput with others, if no incentive is provided. Most of the current business femtocell products support the mode switch such that users can set the mode according to their willingness.

How to optimize the performance of heterogeneous networks is one of the key concerns under this mode. The major technical challenges are the spectrum allocation strategy and interference management, which are partially coupled together. The frequently used two spectrum schemes are orthogonal assignment and co-channel assignment. The orthogonal assignment, which assigns orthogonal spectrum to macro and femto systems, can completely eliminate cross-tier interference. But it limits the available spectrum for both macro and femto systems, thus is not so efficient in spectrum reuse. How to optimally split the spectrum and manage the femto–femto interference has been studied by some works. However, co-channel assignment, which enables macro and femto systems to operate at the same frequency, is more efficient in terms of spectrum use but the interference is more serious. Both cross-tier and co-tier interference exist if using co-channel assignment, which makes it more difficult to analyze.

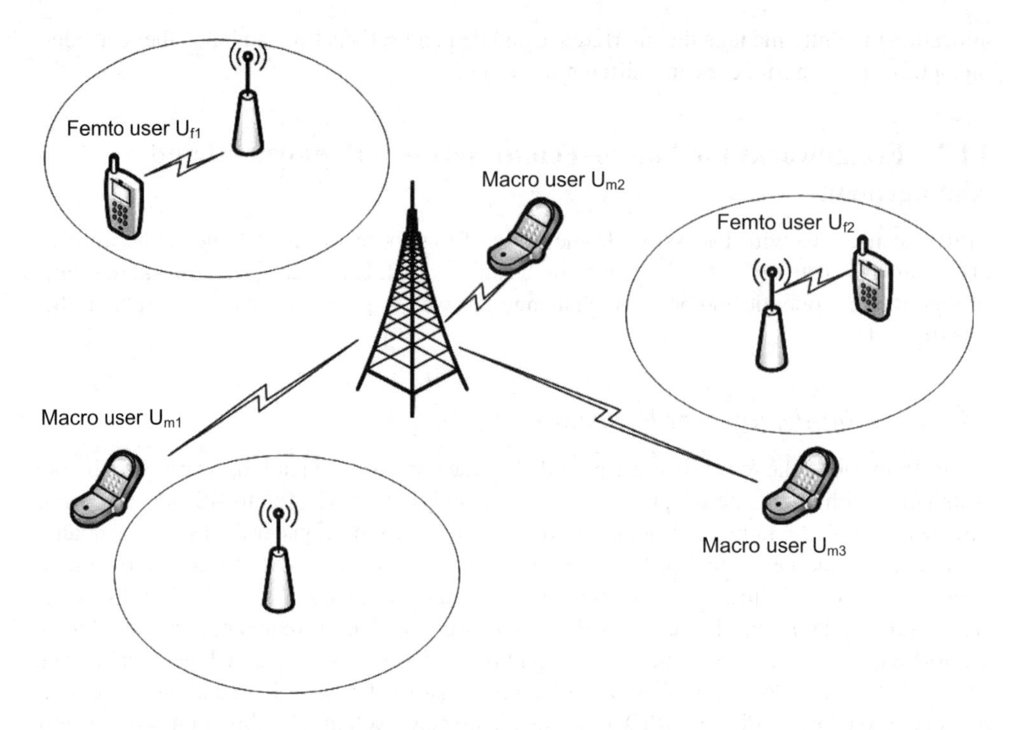

Figure 14.1 Closed access mode.

'Dead zone' is a typical problem in this case, which has been studied by many works. During the downlink transmission stage, macro users located at the edge of the macrocell can receive strong interference from nearby femto-BSs. During the uplink transmission, the situation reverses. The femto users receive excessive interference from the macro users. The users, such as U_{m3} and U_{f2} in Figure 14.1, will repeatedly experience poor signal-to-interference-plus-noise-Ratio (SINR) in the signal dead zones. There have not yet been satisfying conclusions on the choice, such as the optimal performance of two assignments and the influences of different conditions.

Traditional solutions on spectrum assignment and interference management techniques can also work in the case. The distinguishing point for the macro–femto hybrid networks is the two-tier structure and the different physical characteristics, such as the dense deployment, smaller power of femto-BSs, and the strong signal fading environment. These do not exist in traditional cellular networks or wireless local area networks. The WSP has to jointly consider the two-tier performance and interactions, which adds to the difficulty.

Open access mode

As a potential solution to the dead zone problem, open access mode has drawn much attention. In open access mode, the femto-BSs can serve nearby macro users as well as users in white

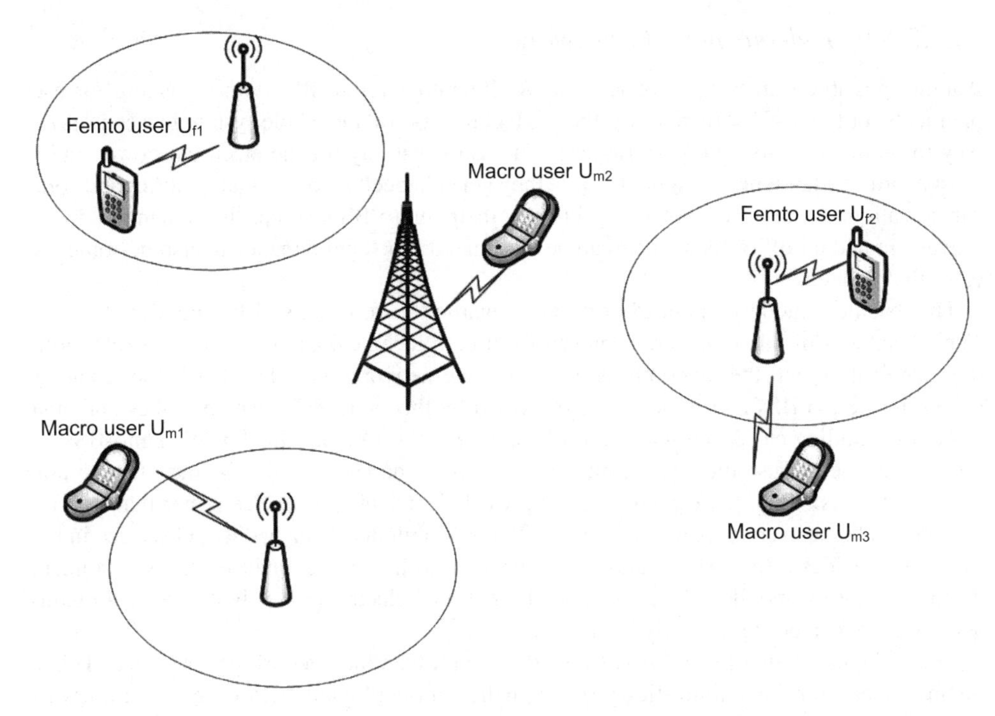

Figure 14.2 Open access mode.

lists. As shown in Figure 14.2, U_{m1} and U_{m3} are allowed to access the nearby femto-BSs in this mode. Research studies reveal that the open access mode is more efficient in improving the system capacity. This is because the short-distance femto transmissions take the place of the long-distance macro ones, which experience serious path losses. It eases the burden of macro-BSs and saves resources for femto-BSs. The dead zone problem can also be avoid.

The first and most important challenge here is how to motivate the holders to open their femto-BSs. It is the indoor users who pay for everything: hardware, landline, monthly plan and electrical power. They are not willing to make free offers to help the operator increase its profit. To motivate the openness of femto-BSs, the WSP should surrender enough profit, and the users should respond based on an overall consideration. In fact, the WSP can provide incentives for the owners to open their femto-BSs by various forms of subsidization. It can provide plenty of choices, such as cashback or a special femto-zone tariff. By weighing the gains from improved capacity and losses from the subsidization, it can achieve the optimality. A femto user's considerations may include its own demand, the attractiveness of the allowance and the demand from nearby macro users. It may have different degrees of concerns about its demand satisfaction and the allowance. It may also have to compete with other femto users for the allowance, as it may not be the only one who can help. In all, both the operator and the users have to trade off the gains and losses. Suitable game and auction models may provide some deep insights into the scenario.

14.2.2 WSP-deployment Framework

Another possible framework is where the WSP deploys femto-BSs without asking for user payments for femto-BSs. In this way, the WSP could control the whole system in a centralized way to optimize the overall performance. The users just pay for the electrical power and at most additional landline throughput, while they get enhanced indoor signal qualities and some subsidization. As users cannot freely change their femto-BS's setup, the dynamics due to movement and on-off of femto-BSs can be eliminated. The deployment can also be somehow carefully planned.

This business mode is quite exciting and regarded as a successful business exploration. The advantages of this mode are achieved from both technical and economic aspects. From the technical aspect, the randomness of the mobile networks is reduced. All the resources (spectrum, femto-BSs and backhauls) are owned by the same WSP, which makes elaborate system scheduling easier. From the economic aspect, the free installations of femto-BSs are quite attractive to customers and competitive to the opponents. And it is important to note that the WSP is taking its responsibility to provide better indoor service rather than passing the buck to home users. Japan's operator Softbank Mobile has launched its 3G service in this mode. It provides a free ADSL plus femto-BS deal to home and small business customers. The users pay for mobile calls as usual and additional electronic fees but the signal quality and coverage can be significantly improved.

The technical challenge in this case is how to optimize the network performance. This is distinguished from the optimization problem in the joint-deployment/closed access framework in the following ways. First, its dynamic factors in the hybrid networks have been reduced to some extent, including the movement and on-off of the femto-BSs. Second, under the open access assumption, users have more choice of subscribing femto-BSs. It is possible to jointly schedule the additional subscription problem with other problems to achieve a better system performance. It is also different from the challenges in joint-deployment/open access framework. Because one has to provide incentive for opening the femto-BSs and the other does not. The overall performance of an optimization is usually considerably better than that of a game where the participants are selfish.

There have been some works focusing on this scenario. Chandrasekhar et al. [4] conducted an uplink capacity analysis and gave an interference avoidance strategy under the stochastic deployment assumption. But their scheme did not guarantee the optimal system performance. Shetty et al. [5] proposed an economic framework to maximize the operator's revenue. However, they considered the simple cases and adopted some unrealistic assumptions such as the linear distributed capacity. We will look into this scenario and propose two schemes on the revenue maximization of the WSP, which will be introduced briefly later.

14.2.3 User-deployment Framework

Under this category, home users deploy femto-BSs out of self-interest. The femto-BSs are in closed access mode, as there is usually no incentive to share the femto-BSs. These femto-BSs are obtained without registration on WSPs. If they work in the licensed band, they are secondary in priority as they have to avoid causing harmful interference to primary macrocell users and vacate channel resources if necessary.

The femto-BSs deployed in a plug-and-play way are quite flexible but in an unplanned manner. The disorder behaves in two aspects. First, the femto-BSs do not have a particular common interest. They compete for available spectrum resource and generate interference to others nearby operating on the same channel. Second, they lack thorough planning and fair coordination from the methodology aspect. These two factors mean that the optimality of the system performance is hardly achievable.

Most current works study the best strategies of the femto-BSs under this scenario. Typical cognitive techniques, such as power control, are adopted to manage the interference and maximize the individual's utility. Jo et al. [6] addressed open and closed loops interference mitigation strategies for femto users. Chandrasekhar et al. [7] modelled the femto users' power control problem as a non-cooperative game and provided a distributed utlity-based SINR adaptation method. The existence and uniqueness of Nash equilibrium was derived in their model. Yun et al. [8] proposed CTRL, a distributed and self-organizing femtocell architecture to manage the femto–macro interference. The CTRL was designed to be compatible with the legacy macrocell system. All their results verified the protection of the macro users' transmissions, but unsurprisingly none could guarantee the performance of the femto users.

However, we argue that this framework may not be as promising as the previous two, although many works have been conducted within it. It may be adopted as an auxiliary way to complement the other modes for unguaranteed services. The major reason is that these femto-BSs do not work in their dedicated spectrum. If they work in the 2G/3G licensed band, even though the cognitive functionalities are executed, the interference to primary systems cannot be completely eliminated. So the licensed system does not allow them to access it. Otherwise, they work on other licensed or unlicensed bands, which needs dual-mode modules in the devices and makes the femtocell not so competitive as the WiFi technique. Furthermore, neither being secondary in licensed band nor sharing the unlicensed band can provide guaranteed QoS.

So far we have proposed and discussed three frameworks of the macro–femto heterogeneous networks. There have been some works discussing the potential problems within the three frameworks, most of which are focusing on the third framework. We will elaborate the second framework and propose two revenue maximization schemes in the next section.

14.3 Revenue Maximization with WSP-deployed Femto-BSs

In this section, we propose two schemes to maximize the WSP's revenue under a WSP-deployment framework. In the first scheme, we adopt the cross-tier channel allocation to jointly manage the interference and provide service and then achieve the optimal revenue. We give a centralized algorithm for a small size system to approach the optimality and distributed algorithm for large size system to obtain a near-optimal result with a constant gap. In the second scheme, optimal pricing and spectrum scheduling are combined to maximize the revenue.

14.3.1 On Cross-tier Channel Allocation

Network scenario and system model

Consider a single macrocellular OFDMA network[1] with a central macro-BS, N_f femto-BSs and N_u subscribing users. Let P_m and P_f be fixed macro and femto access prices on unit throughput. The WSP makes channel allocation of the time-frequency blocks, such that its aggregate revenue from macro-accessing and femto-accessing is maximized.

Assume that the femto-BSs are open accessed. Each user accesses the BS which provides the highest received SINR. Let T_{f_i} be the set of users subscribing to femto-BS B_i and T_m be the set of users subscribing to macro-BS.[2]

All the BSs operate on the WSP's licensed bands such that the external noise can be avoided. Assume that the power levels on channels are fixed for BSs. N_c channels of the macrocell $C = \{C_1, \ldots, C_{N_c}\}$ are divided into orthogonal parts for the BSs. The time-frequency orthogonal allocation is supported by the OFDMA technique.

Problem formulation

We will formulate the scenario by formally describing first the constraints and then the optimization function.

We have two types of constraints: interference constraints to manage the BS-level interference and service provision constraints to schedule channel allocation from the BSs to end users.[3]

1. The interference constraints eliminate the cross-tier interference and alleviate the co-channel femto–femto interference.

   We assume that the macro-BS may interfere with any femto-BS as it transmits at a high power level. So we make the allocation orthogonal between macro and femto tiers. Any pair of femto-BSs is defined to be an interfering pair if their mutual received signal strengths are larger than a predefined threshold β. The β can be chosen according to the system parameters, such as the received SINR of the macro-BS's signal at the edge of the macrocell.

   The channel usages of BSs are expressed as a matrix $\{\alpha_{ij}\}$ where $\alpha_{ij} \in [0, 1]$ represents the time portion of a frame that the C_j is allocated to femto-BS B_i. Let $\alpha_{mj} \in [0, 1]$ be the time portion usage of macro BS on C_j. To manage the interference, the sum of the usages

[1] This scheme intends to optimally redistribute the spectrum that has been allocated to the macro BS in an intra-macrocell way. We take a traditional inter-cellular spectrum plan as the default. This scheme will not cause any serious cross-cellular interference.

[2] In this scheme we define the femto-accessing users as femto users and macro-accessing users as macro users, no matter whether they are located indoors or outdoors.

[3] We consider a cross-tier channel allocation only. It makes the scheme simple and effective at the cost of other performance gains. For example, if applying an advanced diversity transmission scheme, the power cost can be reduced and the data rate will be improved.

of the interfering BSs on the same channel should not exceed the unit one (normalized). The constraints should be satisfied to manage the interference:

$$\alpha_{ij} + \sum_{B_k \in A_i} \alpha_{kj} + \alpha_{mj} \leq 1 \tag{14.1}$$

for all $i = 1, \ldots, N_f$ and $j = 1, \ldots, N_c$. A_i is the set of B_i's 'left of' and interfering femto neighbours. For easy understanding, B_k is in A_i if and only if B_k is on the left of B_i and the two femto-BSs are an interfering pair.

The macro-BS is able to interfere with all femto-BSs, so every femto-BS has to involve the $\{\alpha_{mj}\}$ in its constraints. Thus we have in total $N_f N_c$ constraints in this form as the interference avoidance constraints.

This formulation of interference avoidance can guarantee an interference-free channel allocation. The argument is very simple that we just start allocation from the macro BS and then the leftmost femto-BS. For every other BS, we examine all its 'left of' interfering neighbours and assign channel usages from where the neighbours finish. Its definition naturally guarantees the feasibility of the allocation. However, this 'left of' formulation may not fully reuse the spectrum spatially.

To calculate the potential interference from the non-interfering femto-BSs, we consider the worst case by defining that the possible upper-bound of the aggregate interference I_i at femto-BS B_i should be its received signal strength from all non-interfering femto-BSs. We make an approximation that users of B_i received the same aggregate interference I_i as B_i for the reason that femto-accessing users are usually closed to the femto-BSs and far away from other femto-BSs.

2. The service provision constraints make sure that the total spectrum resource consumed by users subscribing to a BS does not exceed the amount allocated to the BS. Denote $\nu_{ij} \in [0, 1]$ as the time portion that the user U_i is served by its associating BS on channel C_j. The following service provision constraints hold for all i, j,

$$\sum_{U_k \in T_{f_i}} \nu_{kj} \leq \alpha_{ij}, \tag{14.2}$$

$$\sum_{U_k \in T_m} \nu_{kj} \leq \alpha_{mj}, \tag{14.3}$$

for all $i = 1, \ldots, N_f$ and $j = 1, \ldots, N_c$.

User U_i asks to satisfy its demand d_i at the beginning of the frame. It would not pay for additional throughput that exceeds d_i. The WSP can charge it $\min\{t_i, d_i\}P$, where $t_i \geq 0$ is the provided throughput and P is P_f or P_m, depending on its subscribing BS. The throughput of user U_i is the aggregation of Shannon channel capacities through time $\sum_{j=1}^{N_c} \nu_{ij} W_j \log(1 + \gamma_{ij})$, where the W_j is the bandwidth of channel C_j and γ_{ij} is the SINR of U_i on channel C_j. γ_{ij} can be approximated by the $\frac{p_f H_{ij}}{I_k + N_0 W_j}$ for a femto user accessing to BS B_k, and $\frac{p_m H_{ij}}{N_0 W_j}$

for a macro user, where the p_f and p_m are femto- and macro-BSs' transmission power levels, H_{ij} is the attenuation factor and N_0 is the thermal noise power density.

$$\begin{aligned} \text{maximize } & P_m \sum_{macro\ \ user\ \ U_i} \min\left\{\sum_{j=1}^{N_c} \nu_{ij} W_j \log(1+\gamma_{ij}), d_i\right\} \\ & + P_f \sum_{femto\ \ user\ \ U_i} \min\left\{\sum_{j=1}^{N_c} \nu_{ij} W_j \log(1+\gamma_{ij}), d_i\right\} \\ \text{subject to } & \alpha_{ij} + \sum_{B_k \in A_i} \alpha_{kj} + \alpha_{mj} \leq 1 \text{ for all } i, j. \\ & \sum_{U_k \in T_{f_i}} \nu_{kj} \leq \alpha_{ij}, \sum_{U_k \in T_m} \nu_{kj} \leq \alpha_{mj} \text{ for all } i, j. \\ \text{variables } & \{\alpha_{ij}\}, \{\alpha_{mj}\}, \{\nu_{ij}\} \in [0, 1]. \end{aligned} \tag{14.4}$$

The overall revenue consists of the parts from macro and femto service provision and is summarized as Problem (14.4). Later, we will discuss the algorithm for solving this problem.

Centralized and distributed algorithms

Problem (14.4) is convex and thus can be efficiently solved by a mature technique [9], referred to as centralized method (CM). The computational complexity is polynomial of the problem scale. The centralized protocol is designed as follows. The BSs pass their optimization-related parameters to a central node and the computation is executed on the single node. The parameters are the demand vector of associating users and their SINR levels. The central node will return the results in the form of channel usages $\{\alpha_{ij}\}$, $\{\alpha_{mj}\}$, $\{\nu_{ij}\}$ to corresponding BSs. The total number of values exchanged in this case is limited by $\delta_{CM} = (N_f + 3N_{f_u})N_c$, where N_{f_u} is the number of femto users. The parameters and results can be passed through the wired backhaul or the wireless interface. The use of a wired interface will increase the packets delay and uncertainty due to internet traffic conditions. The information exchange can also be implemented on the wireless link. Either frequency division or time division can be employed. Various technique such as cooperative relay and interference cancellation can be used to improve the communication efficiency. The CM algorithm can efficiently find the optimal results at a low communication and computational cost given that the system scale is not very large.

Where there are a great many users in the macrocell, the CM algorithm can be impractical as the communication and computational cost increases rapidly. We will propose a distributed algorithm LD based on the Lagrangian decomposition to serve for this case.

The basic idea of the Lagrangian is to take the constraints into account by augmenting the objective function with a weighted sum of the constraint functions [9]. For this problem, we only need to put the interference avoidance constraints into the Lagrangian [10], because we will decompose the problem into sub-problems at each BSs later to reduce the computational cost on single node. Only the coupled (interference avoidance) constraints where variables

$\{\alpha_{ij}\}$ and $\{\alpha_{mj}\}$ of different BSs are coupled together, need to be moved to the objective function. The service provision constraints involve local variables for each BS, and thus can be preserved in the sub-problems.

Now we decouple the original problem. Define the Lagrangian associated with the problem (14.4) as:

$$\begin{aligned} \text{maximize } & L(\{\alpha_{ij}\}, \{\alpha_{mj}\}, \{\nu_{ij}\}, \{\lambda_{ij}\}) \\ & = P_m \sum_{macro\ U_i} \min\left\{\sum_{j=1}^{N_c} \nu_{ij} W_j \log(1+\gamma_{ij}), d_i\right\} \\ & + P_f \sum_{femto\ U_i} \min\left\{\sum_{j=1}^{N_c} \nu_{ij} W_j \log(1+\gamma_{ij}), d_i\right\} \\ & - \sum_{j=1}^{N_c}\sum_{i=1}^{N_f} \lambda_{ij}(\alpha_{ij} + \sum_{B_k\in A_i} \alpha_{kj} + \alpha_{mj} - 1) \\ \text{subject to } & \sum_{U_k\in T_{f_i}} \nu_{kj} \leq \alpha_{ij}, \sum_{U_k\in T_m} \nu_{kj} \leq \alpha_{mj} \text{ for all } i,\ j. \end{aligned} \tag{14.5}$$

where $\lambda_{ij} \geq 0$ is the Lagrange multiplier with the constraint $\alpha_{ij} + \sum_{B_k\in A_i} \alpha_{kj} + \alpha_{mj} \leq 1$.

[10] developed a general decomposition method for utility maximization problems where the objective function should be strictly concave. As the objective function (14.5) does not satisfy this condition, we add the auxiliary terms to make it strictly concave such that we can apply the decomposition technique. The terms themselves should be strictly concave, for example $-\epsilon \sum_{i=1}^{N_f}\sum_{j=1}^{N_c} \alpha_{ij}^2$ where $\epsilon > 0$. All the constraints are kept unchanged. The Lagrangian in (14.5) becomes:

$$L(\{\alpha_{ij}\}, \{\alpha_{mj}\}, \{\nu_{ij}\}, \{\lambda_{ij}\}) - \epsilon \sum_{i=1}^{N_f}\sum_{j=1}^{N_c} \alpha_{ij}^2 \tag{14.6}$$

The dual function of (14.6) is defined as the minimum value of its Lagrangian over $\{\alpha_{ij}\}$:

$$g(\{\lambda_{ij}\}) = \sup_{\{\alpha_{ij}\}} \left(L(\{\alpha_{ij}\}, \{\alpha_{mj}\}, \{\nu_{ij}\}, \{\lambda_{ij}\}) - \epsilon \sum_{i=1}^{N_f}\sum_{j=1}^{N_c} \alpha_{ij}^2 \right) \tag{14.7}$$

The strong duality of (14.6) can be guaranteed by the Slater's constraint qualification. In fact, the strong duality always holds here as (14.6) is feasible. So the solution to the dual problem:

$$\begin{aligned} & \text{minimize } g(\{\lambda_{ij}\}) \\ & \text{subject to } \{\lambda_{ij}\} \geq 0 \end{aligned} \tag{14.8}$$

is exactly the same as the one to the primal problem (14.6) with the unchanged constraints.

Now the decomposition method can be applied to (14.8) to conduct two levels of optimization.

At the lower level, we arrange the terms in (14.7) with local variables in the same BS together to form the sub-problems. The local variables for femto-BS B_i are $\{\alpha_{ij}\}$ and $\{\nu_{kj}|U_k \in T_{f_i}\}$ for all j. Similarly, the local variables for the macro-BS are $\{\alpha_{mj}\}$ and $\{\nu_{kj}|U_k \in T_m\}$ for all j. Each femto-BS B_i solves the subproblem:

$$\begin{aligned} \text{maximize } & P_f \sum_{U_i \in T_{f_i}} \min \left\{ \sum_{j=1}^{N_c} \nu_{ij} W_j \log(1+\gamma_{ij}), d_i \right\} \\ & - \sum_{j=1}^{N_c} (\lambda_{ij}(\alpha_{ij} - 1) + \sum_{B_i \in A_k} \lambda_{kj}\alpha_{ij}) - \epsilon \sum_{j=1}^{N_c} \alpha_{ij}^2 \\ \text{subject to } & \sum_{U_k \in T_{f_i}} \nu_{kj} \leq \alpha_{ij} \text{ for all } j. \\ & \text{variables } \{\alpha_{ij}\}, \{\nu_{kj}|U_k \in T_{f_i}\} \text{ for all } j. \end{aligned} \tag{14.9}$$

to get the optimal value $g_i(\{\lambda_{ij}\})$. Similarly, the macro-BS solves the problem to get the optimal value $g_m(\{\lambda_{ij}\})$

$$\begin{aligned} \text{maximize } & P_m \sum_{U_i \in T_m} \min \left\{ \sum_{j=1}^{N_c} \nu_{ij} W_j \log(1+\gamma_{ij}), d_i \right\} \\ & - \sum_{j=1}^{N_c} \sum_{i=1}^{N_f} \lambda_{ij}\alpha_{mj} - \epsilon \sum_{j=1}^{N_c} \alpha_{mj}^2 \\ \text{subject to } & \sum_{U_k \in T_m} \nu_{kj} \leq \alpha_{mj} \text{ for all } j. \\ & \text{variables } \{\alpha_{mj}\}, \{\nu_{kj}|U_k \in T_m\} \text{ for all } j. \end{aligned} \tag{14.10}$$

The BSs can optimize (14.9) and (14.10) in a parallel manner as the related variables for their sub-problems are all local information.

At the higher level, we need to update the Lagrange multipliers by solving the problem:

$$\begin{aligned} \text{minimize } & \sum_{i=1}^{N_f} g_i(\{\lambda_{ij}\}) + g_m(\{\lambda_{ij}\}) + \sum_{i=1}^{N_f} \sum_{j=1}^{N_c} \lambda_{ij} \\ \text{subject to } & \{\lambda_{ij}\} \geq 0. \end{aligned} \tag{14.11}$$

The purpose of this step is to ensure the consistency of the Lagrange multipliers on the neighbouring BSs. As the dual function $g\{\lambda_{ij}\}$ is non-differentiable, we can take the sub-gradient method to decouple (14.11). In particular, a simple way to achieve it is:

$$\lambda_{ij}(t+1) = [\lambda_{ij}(t) + s(t)(\alpha_{ij}(t) + \sum_{B_k \in A_i} \alpha_{kj}(t) + \alpha_{mj}(t) - 1)]^+ \tag{14.12}$$

where $s(t) > 0$ is the step size in the t-th iteration. Generally speaking, the value of $\{s(t)\}$ can affect the convergence, converging speed and the gap to the optimal solution, thus should be carefully selected. Theoretically $s(t)$ is suggested to be constant (small enough) or diminishing $\frac{1+m}{t+m}$, where m is a fixed non-negative number [10].

Now we give the distributed algorithm LD. It takes several iterations of two-level optimization to give the near-optimal solution. We denote the variables and parameters in iteration t with additional suffix (t).

Algorithm 1 Distributed algorithm based on Lagrangian decomposition

Require: non-negative initial value of partial Lagrange multiplier $\{\lambda_{ij}(0)\}$ and $\epsilon(0)$

1: $t \leftarrow 0$
2: **repeat**
3: Given the $\{\lambda_{ij}(t)\}$, each BS solves its sub-problems.
4: Every femto B_i sends $\{\alpha_{ij}(t)\}$ to all BSs in A_i. They receive the neighbouring femto-BSs' messages. They also send $\{\alpha_{ij}(t)\}$ and $g_i(\{\lambda_{ij}\})$ to the macro-BS.
5: The macro-BS gathers the information to computes the $\epsilon(t+1)$. It sends $\epsilon(t+1)$ and $\{\alpha_{mj}(t)\}$ to all femto-BSs.
6: Femto B_i updates its Lagrange multipliers by (14.12).
7: $t \leftarrow t+1$
8: **until** the $g_i(\{\lambda_{ij}\})$ and $g_m(\{\lambda_{ij}\})$ are changed by no more than a predefined threshold.
9: The BSs unify the $\{\alpha_{ij}\}$ to strictly obey the interference constraints. The correction of the optimal value by auxiliary terms is conducted.

The two levels of sub-problem optimizations are iteratively conducted and by the end we will achieve the optimal solution to (14.6), but our problem of interest is (14.5). We can make compensation by subtracting the auxiliary terms, but the bias of the solution to the original problem (14.5) cannot be completely corrected. Thus the ϵ should not be too large to induce much deviation. In the simulation, we also find that if the ϵ is too small, the convergence is very slow. We suggest making ϵ dynamic:

$$\epsilon(t+1) = \frac{\eta M(t)}{\sum_{i=1}^{N_f} \sum_{j=1}^{N_c} \alpha_{ij}(t)^2}, \tag{14.13}$$

where $M(t)$ is the optimal system revenue in the current iteration and η is a small and positive constant. The idea is to make the auxiliary terms proportional to the current system revenue $M(t)$.

For the selection of parameters such as $s(t)$ and $\epsilon(t)$, we have to trade off the gap to the optimal value and the convergence speed. We can binary search the domains of the parameters to find the proper values such that the gap is tolerable and the convergence is fast enough. The results of searching can be used for a long time as our algorithm is not sensitive to these parameters. The initial values of $\{\lambda_{ij}\}$ can be zero.

The sub-problems are also solved by the same optimization technique. The LD reduces the computational overhead as the scale of a sub-problem is much smaller than the original one. We can analyze the communication cost as follows. At step 4, each B_i sends out $N_c(N_{b_i}+1)+1$

real numbers, where N_{b_i} is the cardinality of A_i. At step 5, the macro BS sends out $N_f(N_c + 1)$ real numbers. Considering that LD stops after limited iterations in practice, let the number be N_i. Thus the total real numbers exchanged in the process is about $\delta_{LD} = N_i(N_c \sum_{i=1}^{N_f} N_{b_i} + 2N_cN_f + 2N_f)$,[4] which is irrelevant to the number of femto-accessing users N_{f_u}. Comparing the δ_{CM} and δ_{LD}, we find that LD is preferable when there are numerous femto-accessing users as the communication cost may not increase too much and the computational cost on a single BS is significantly reduced.

Evaluation

We conduct simulations to the advantages of hybrid macro–femto networks.

Consider the macrocell with radius 500 metres and the macro BS deployed in the centre. The $N_f = 20$ femtocells and $N_u = 100$ users are located randomly. We assume that each femto-BS has at least one indoor user located at home and the remaining $N_u - N_f = 80$ users are located anywhere in the cellular. A line of two femto-BSs means a potential interfering relationship between them. The other default values of the parameters are: $N_c = 20$, $W_i = 0.2$ MHz for $i = 1, \ldots, 10$ and 0.4 MHz for $i = 11, \ldots, 20$, $P_m = 1$ per Mb, $P_f = 0.3$ per Mb, $p_m = 0.2$ watt on each channel, $p_f = 0.1$ watt on each channel, $N_0 = -174$ dBm/Hz, $\{\lambda_{ij}(0)\} = 0$, $\epsilon(0) = 0$, $\eta = 0.02$, $s(t) = \frac{1}{t}$. The femto user U_i's demand d_{f_i} is set to be random at [0, 5] Mb and the macro user U_i's demand d_{m_i} be random at [0, 1] Mb. The performance results are all averaged over 100 evaluations to smooth the random factors such as the shadowing effect, demand diversity and location variation.

Assume that the outdoor and indoor path-losses are $28 + 35\log_{10}(r)$ dB and $38.5 + 20\log_{10}(r)$ dB respectively, where r is the transmitter–receiver distance in metres. The wall loss $L_w = 10$ dB will be counted if the BS and the user are separated by walls. The shadow fading S is log-normal distributed with 8 dB standard deviation for outdoor and 4 dB for indoor.

We compare our scheme with some baseline schemes, which include fixed allocation and no femto-BSs scheduling. In the fixed allocation, the macro-accessing and femto-accessing are pre-allocated ω and $1 - \omega$ portion of spectrum respectively. A femto-BS shares the $1 - \omega$ portion with its interfering neighbours equally. In the traditional scenario without femto-BSs, macro-BS uses all the spectrum to serve users. Figures 14.3 and 14.4 are the results under $d_{f_i} \in [0, 5]$, $d_{m_i} \in [0, 1]$ and $d_{f_i} \in [0, 20]$, $d_{m_i} \in [0, 4]$ respectively. These two figures show that our scheme always achieves the highest system revenue. If there are a lot of femto-BSs and indoor users, the allocations involving femto BSs are definitely better than with macro-allocation only. They also show that the performance gap between fixed allocation schemes and our scheme increases with the number of femto-BSs in the large demand case.

14.3.2 On Optimal Pricing and Spectrum Partition

Network scenario and system model

Consider the same two-tier networks. The WSP offers macrocell and femtocell services with prices P_m and P_f per unit capacity respectively. The WSP has set its ultimate goal as maximizing its profit M via pricing selection and spectrum allocation.

---

[4]If utilizing the broadcasting nature of wireless networks, the necessary transmission times will be less.

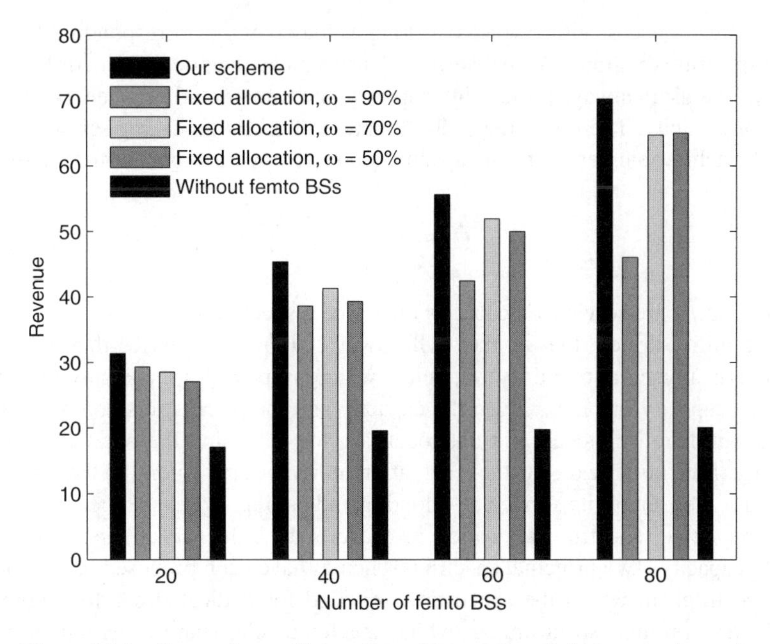

Figure 14.3 Comparison with baseline schemes when default users demand.

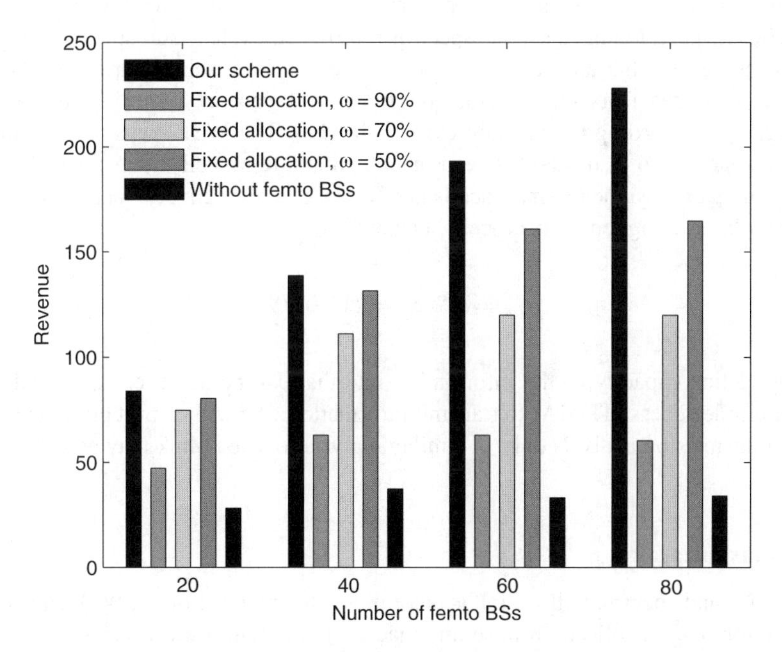

Figure 14.4 Comparison with baseline schemes when larger users demand.

Users are free to access either service as long as they pay the corresponding fee. Once the prices and spectrum distribution are determined, users make their decision in order to optimize their utility. The users always prefer high capacity and low price. Let γ represents the user's valuation for capacity. Therefore, the utility function of each user taking service $j = m, f$ is defined to be achievable benefit from capacity subtracted by the payment for the service:

$$U_j = \gamma c_j - p_j. \tag{14.14}$$

in which c_j and p_j are service j's capacity and price respectively.

Users are utility driven, that is, they will always choose the service that gives the most utility. There is a reservation utility U_0, below which users will not take any type of service whatever the capacity or price. Given the capacity and the price, users compared the utility they can obtain from femtocell and macrocell service, choosing the one that generates the higher utility if the utility exceeds the reservation utility. Users respond to the price and QoS, like voting for their favourite service by adopting and paying for that service.

Apart from price, spectrum allocation also exerts an influence on users' determination through the capacity. Two principal spectrum schemes have been proposed in previous works: (1) Split spectrum, in which the spectrum is divided for dedicated use for femtocells and macrocell; (2) Common spectrum, in which the femtocells share a certain proportion of spectrum with the macrocell [11]. We presume that the WSP acquires a split spectrum scheme, where macro-BS and femto-BSs operate on different frequencies and do not interfere. All the femto-BSs share the same frequency, utilizing integrated channel allocation strategy to avoid femto–femto interference. We assume that the femtocell acquires fixed channel allocation strategy (FCA), partitioning the total spectrum into k equivalent sub-spectra for exclusive usage in each cell so that any femto-BSs that reuse the same sub-spectrum are far enough away to be prevented from femto–femto interference. k can only take integer values 1, 3, 4, 7, 9, 12, ... according to a certain expression [12]. The WSP owns a fixed amount of spectrum resource S to distribute between macrocell and femtocell. Let S_m and S_f represent the spectrum that is assigned to macrocells and femtocells respectively; let α denote the ratio of spectrum that is assigned to macrocells. Define:

$$S_m = \alpha S, S_f = (1 - \alpha)S. \tag{14.15}$$

Here we define capacity as the information rate that can be achieved. Users adopt time-division multiple access (TDMA) to transmit information. We will restrict our attention to the downlink transmission analysis only. A similar analysis for the uplink may be performed.

Revenue maximization

After the price and spectrum allocation is determined, the revenue of the WSP can be derived as the sum of money it collects from selling macrocell and femtocell services.

$$M = p_m E(C_m) + p_f E(C_f), \tag{14.16}$$

where C_m and C_f are the total capacity for macrocell and femtocell service respectively. Let c_m and c_f denote the capacity that each user can achieve from macrocell and femtocell service. Further, we assume that c_m and c_f independently follow an exponential distribution with parameter λ_m and λ_f.The distribution is on the interval $[0, +\infty)$ and the probability density function (pdf) is as follows:

$$f(c_j;\lambda_j) = \begin{cases} \lambda_j e^{-\lambda_j c_j} & c_j \geq 0 \\ 0 & c_j < 0 \end{cases} \tag{14.17}$$

Figure 14.5 shows the pdf of exponential distribution with different λ.

The capacity varies with the user's location, the quality of the channel and the transmission power. We can obtain the user's expected capacity $E(c_j)$ from adoption of the service j by taking the expectation over all possible capacity trajectories:

$$E(c_j) = \frac{1}{\lambda_j}. \tag{14.18}$$

For the macrocell, we make the simplifying assumption that $E(c_m)$ is a function of spectrum S_m only. For the femtocell, we assume that $E(c_f)$ is proportional to the employed spectrum and inversely proportional to the user density. For the femtocell, we assume that the capacity

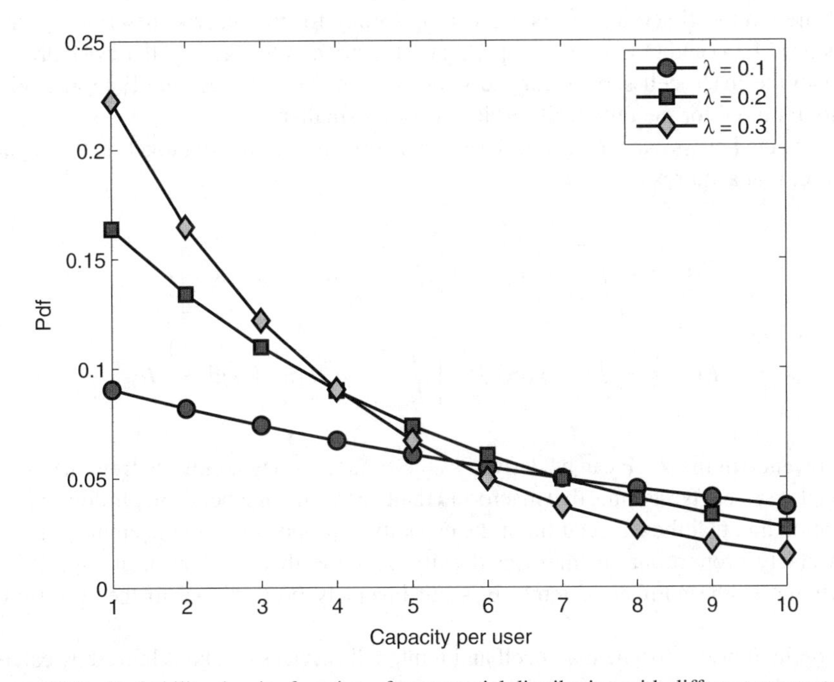

Figure 14.5 Probability density function of exponential distribution with different parameters.

is also proportional to the number of femto-BSs but inversely proportional to the spectrum reuse factor. As a result, the parameter λ_m, λ_f is determined by:

$$
\begin{aligned}
E(c_m) &= \frac{1}{\lambda_m} = \frac{\beta \alpha S}{\rho}, \\
E(c_f) &= \frac{1}{\lambda_f} = \frac{\beta n(1-\alpha)S}{k\rho}.
\end{aligned}
\tag{14.19}
$$

The parameter ρ denotes the user density. The parameter β represents the spectral efficiency, which can be viewed as bits per second per hertz supported by the system [13]. It can be estimated by summing over all users in the system, divided by the channel bandwidth. Spectral efficiency is affected not only by the single user transmission technique, but also by the multiple access schemes and radio resource management techniques utilized. We make the simplifying assumption that β is a predetermined constant.

Once the prices are fixed, in order to attract users to take wireless service of any type, the utility gained must be beyond the reservation utility. For this reason, the minimum capacity of macrocell and femtocell service, denoted by c_{m0} and c_{f0} respectively, is determined by forcing (14.14) to be the reservation utility:

$$
c_{j0} = \frac{1}{\gamma}(U_0 + p_j), \; j = m, f. \tag{14.20}
$$

With the capacity satisfying (14.20), if the utility provided by the macrocell service surpasses that of the femtocell service, users are willing to pay for the macrocell service, vice versa. Equations (14.14) and (14.20) also imply that if the price of the femtocell service drops, users will swap the macrocell service for the femtocell service. At the same time, the minimum capacity required for the femtocell service becomes smaller.

The expected gross capacity for each service is obtained by adding together the capacity of every user that acquires that service:

$$
\begin{aligned}
E(C_m) &= \int_{c_{f0}}^{+\infty} f(c_f;\lambda_f)\left[\int_{U_m > U_f}^{+\infty} c_m f(c_m;\lambda_m)c_m\right] dc_f, \\
E(C_m) &= \int_{c_{m0}}^{+\infty} f(c_m;\lambda_m)\left[\int_{U_f > U_m}^{+\infty} c_f f(c_f;\lambda_f)dc_f\right] dc_m.
\end{aligned}
\tag{14.21}
$$

The revenue of the WSP can be derived as the sum of money it collects from macrocell and femtocell services. We assume that macro and femto capacity independently follow exponential distribution and that the expectation of the capacity is proportional to the employed spectrum and inversely proportional to the user density. Assume that the femtocell capacity is also proportional to the number of femto-BSs but inversely proportional to the spectrum reuse factor.

The optimal prices for the macrocell and femtocell services can be achieved by calculating the first-order partial derivatives of M with respect to p_j, $j = m, f$ and assigning them to be 0. Then, by running through all the possible values of spectrum allocation ratio from 0.01 to 0.99, the optimal one that delivers the maximum revenue can be revealed.

Evaluation

We compare our joint price and spectrum allocation (JPSA) scheme with the fixed spectrum allocation (FSA) scheme. Figure 14.6 shows that when the number of femto-BSs is around 33, JPSA outperforms FSA by approximately $1 - 6.8$ times under different fixed spectrum distribution ratios. The revenue of FSA approaches that of JPSA at the beginning because the fixed spectrum allocation ratio advances to the optimal one. However, the revenue degrades rapidly as the fixed spectrum allocation ratio drifts away from the optimal one. If the spectrum allocation is fixed, the maximum revenue for the WSP only increases when there are a few femto-BSs but drops dramatically when there are large numbers of femto-BSs. Nevertheless, with JPSA the maximum revenue keeps augmenting with the increase of femto-BSs. One reason is that the potential of spectrum spatial reuse is well exploited by the femtocell. In consequence, it provides strong economic incentive for the WSP to promote the femtocell service.

Figure 14.7 displays the optimal prices for macrocell and femtocell services respectively when the spectrum allocation ratio α is set to be the optimal value. On the one hand, the price of macrocell service keeps going up. The reason is that the WSP assigns more spectrum to macrocell when the number of femto-BSs increases since the femtocell can reuse spectrum more efficiently. As a result, the capacity of the macrocell becomes higher, pushing up the

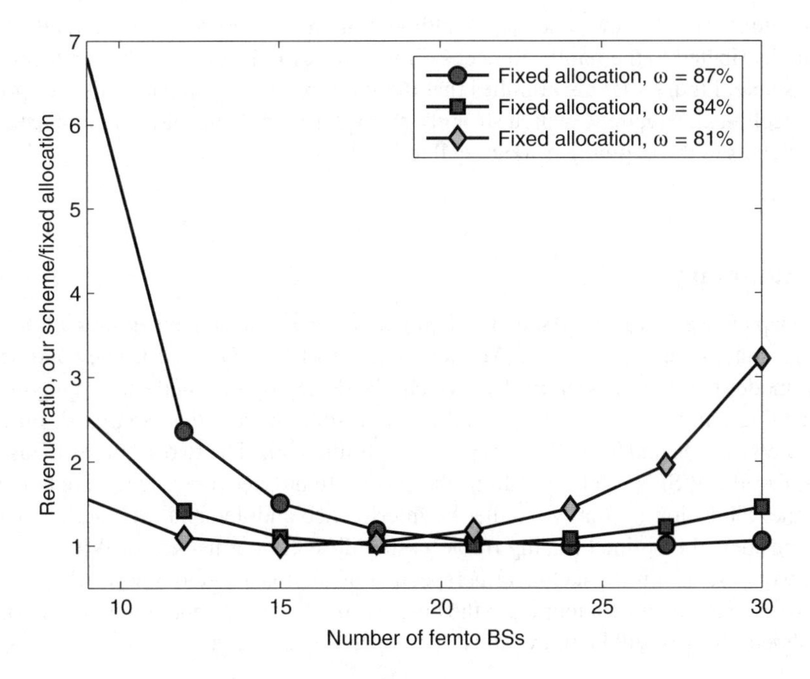

Figure 14.6 WSP's revenue vs. number of femto-BSs.

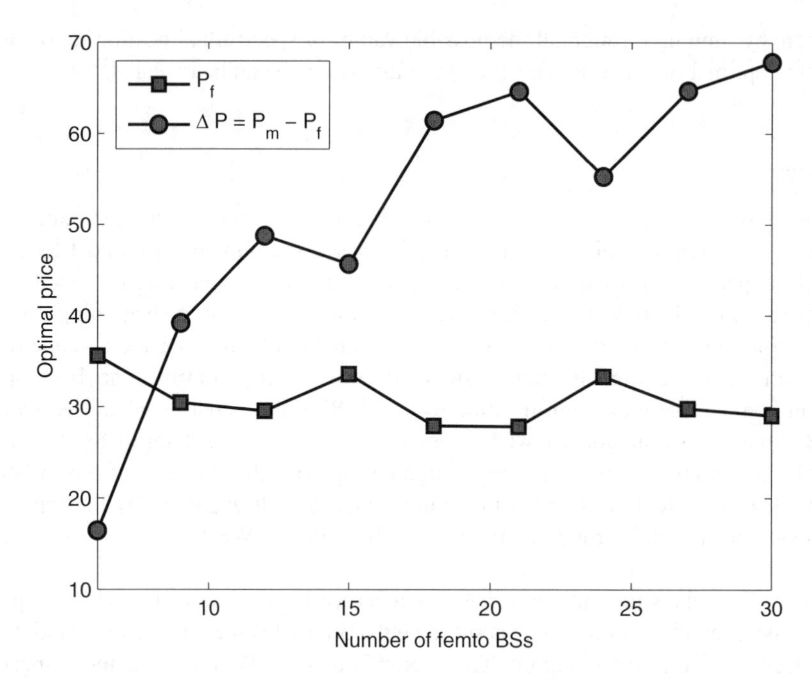

Figure 14.7 Optimal price vs. number of femto-BSs.

corresponding price. On the other hand, although there are more and more femto-BSs, the spectrum distributed to the femtocell keeps diminishing. So the price of the femtocell service is rather stable. Figure 14.7 also implies that the femtocell tends to charge a lower price than the macrocell in all situations, which strongly motivates users to acquire the femtocell service and contributes to the booming of femtocell technology.

14.4 Summary

In this chapter, we have discussed the deployment and management issues of femtocells in two-tier macro–femto networks. We have proposed three potential frameworks from the business modes point of view: joint-deployment, WSP-deployment and user-deployment. The corresponding uniqueness, challenges and possible solutions are then analyzed. Further, we present two schemes under the WSP-deployment framework. The two schemes focus on how to maximize the WSP's total revenue in the macro–femto networks. One adopts the cross-tier channel allocation to jointly handle the interference and demand satisfaction. The other further considers the optimal pricing for accessing different tier networks. While most of the existing works focus on the technical perspective of the macro–femto networks, we believe that the works paying more attention to the business model and combining both economic and technical perspectives will be more valuable to operators and users.

References

1. G. Mansfield, 'Femtocells in the US Market – Business Drivers and Femtocells in the US Market – Business Drivers and Consumer Propositions', in *FemtoCells Europe 2008*, June 2008.
2. 3GPP, '3GPP Technical Report 25.820 V8.2.0', Sept 2008.
3. M. Latham, 'Consumer attitudes to femtocell enabled in-home services C insights from a European survey', in *FemtoCells Europe 2008*, June 2008.
4. V. Chandrasekhar and J. Andrews, 'Uplink capacity and interference avoidance for two-tier cellular networks', in *IEEE GLOBECOM*, 2007, pp. 3322–3326.
5. N. Shetty, S. Parekh and J. Walrand, 'Economics of femtocells', in *Proceedings of IEEE Globecom*, 2009, pp. 6616–6621.
6. H. Jo, C. Mun, J. Moon and J. Yook, 'Interference mitigation using uplink power control for two-tier femtocell networks', *IEEE Transactions on Wireless Communications*, vol. 8, no. 10, p. 4907, 2009.
7. V. Chandrasekhar, J. Andrews, T. Muharemovic, Z. Shen and A. Gatherer, 'Power control in two-tier femtocell networks', *IEEE Transactions on Wireless Communications*, vol. 8, no. 8, pp. 4316–4328, 2009.
8. J. Yun and K. Shin, 'CTRL: A Self-Organizing Femtocell Management Architecture for Co-Channel Deployment', in *ACM Mobicom*, 2010.
9. S. Boyd and L. Vandenberghe, *Convex optimization*. Cambridge University Press, 2004.
10. D. Palomar and M. Chiang, 'A tutorial on decomposition methods for network utility maximization', *IEEE Journal on Selected Areas in Communications*, vol. 24, no. 8, pp. 1439–1451, 2006.
11. J. Hobby and H. Claussen, 'Deployment options for femtocells and their impact on existing macrocellular networks', *Bell Labs Technical Journal*, vol. 13, no. 4, pp. 145–160, 2009.
12. L. Ortigoza-Guerrero and A. Aghvami, *Resource allocation in hierarchical cellular systems*. Artech House, 2000.
13. S. Verdú and S. Shamai, 'Spectral efficiency of CDMA with random spreading', *IEEE Transactions on Information Theory*, vol. 45, no. 2, pp. 622–640, 1999.

15

Field Trial of LTE Technology

Bo Hagerman,[1] Karl Werner,[1] and Jin Yang[2]
[1]*Ericsson Research, Sweden*
[2]*Verizon Communications Inc., USA*

15.1 Introduction

Heterogeneous networks will expand the capacity of current deployed commercial wireless networks by fully exploring layered deployment structures. Macro-type nodes form the backbone of today's deployed networks, and this will remain so even when layered architectures are introduced. Hence, in order to understand the potential benefits of heterogeneous networks, it is essential to understand the basic characteristics of macro-type network deployments.

Advanced wireless networks are multiple input/multiple output (MIMO) enabled. Multiple antennas for reception and transmission at the radio base station (eNB) and the user equipment (UE) is a key enabler for the high performance offered by the 3rd Generation Partnership Project (3GPP) long term evolution (LTE) standard [1–8]. Understanding the performance characteristics of MIMO technology in practical environments is therefore critical to understanding the performance of the network as a whole. Field test results in a pre-commercial macro network deployment have demonstrated network performance gain by introducing MIMO. The trial results are also useful for the understanding of MIMO performance in a micro layer of a heterogonous network. The results of the trial are summarized in [9].

The LTE standard [3] supports multiple antenna technologies that improve both link- and system-level performance in a wide range of scenarios. Multiple antenna transceivers require certain conditions on the radio channel to be fulfilled in order to work well. One of these conditions is that the radio channels experienced by the different receive antennas must show sufficiently low correlation. The longer the wavelength at the carrier frequency compared to the size of the terminal, the more challenging it is to fulfil this condition from a UE antenna design perspective. [10–12] illustrate basic results from field trials on 2.6 GHz carrier frequency evaluating antenna configurations for HSPA 2×2 and LTE with up to 4×4 MIMO schemes.

Heterogeneous Cellular Networks, First Edition. Edited by Rose Qingyang Hu and Yi Qian.

A trial was designed to investigate how well downlink MIMO that is using rank-two transmission, works in a cellular system on 750 MHz carrier frequencies. Various realistic UE prototype form factors are tested with 2×2 MIMO, that is two receive branches at the UE and two transmit branches at the eNB. This is the typical setup for today's deployed LTE networks. The deployed eNB antenna systems are dual-polarized antennas. The trial was performed in a radio environment as similar as possible to what UEs will experience in a live LTE network. In order to achieve this, different load conditions were created, hand dummies were used, and areas with different town architectures (urban and suburban) were used for the measurements. Hence, results regarding the interaction of hands holding the device and vehicle components surrounding it were also obtained. The campaign is intended to investigate how the MIMO performance depends on the radio environment and the antenna design at the used carrier frequency. It is not intended to evaluate any particular vendor of UEs nor the absolute network performance.

In this chapter, we will address realistic 4G LTE network radio characteristics and performance under both urban and suburban radio environments. The relationship between LTE network throughputs and signal-to-noise ratio (SNR) are presented under both loaded and unloaded network traffics. The impacts of different UE form factors are highlighted. Significant benefits of MIMO at 750 MHz are demonstrated. Section 15.2 will provide an overview of the field trial, and in Section 15.3 the measurement results are presented. A summary and comparison of the results are given in Section 15.4 with conclusion provided in Section 15.5.

15.2 Field Trial Overview

The LTE trial network is compliant with 3GPP Rel-8 standards [3, 4]. The trial network architecture is illustrated in Figure 15.1. UEs are connected to network over-the-air through eNB. The control plane signals are connected to mobility management entity (MME), while the data plane user traffics are processed by serving gateway (SGW), then routed through packet data network gateway (PDN-GW) to various data applications. Authentication and policy are enforced by home subscriber server (HSS), policy and charging rules function (PCRF). The trial network consists of 10 cell sites with 26 sectors. Transmission power from

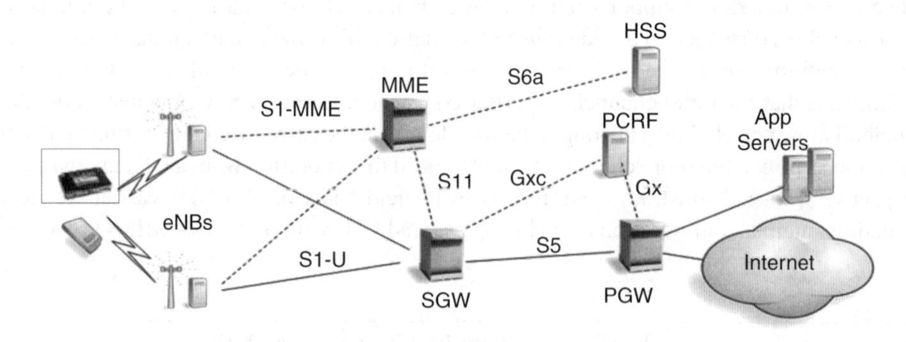

Figure 15.1 LTE network architecture for the trial.

the two antennas serving each sector is 20 watt each. The coverage area includes both typical North American urban and suburban topologies.

15.2.1 UE Antennas

At the tested frequency band (746–756 MHz) the wavelength is 0.4 m. A typical smartphone is smaller than 0.2 m along its dominating dimension.

Three different dual antenna UE mockups were tested in the trial: Smartphone 1 (SP1) and Smartphone 2 (SP2) were designed with the size of a smartphone, 73×150 mm (~3×6 in.). Featurephone 1 (FP1) was designed with a smaller form factor, 65×115 mm (~2.5×4.5 in.). The antenna realization – such as radiating element type, orientation and positioning – are designed on purpose for high and low antenna branch correlation in order to evaluate its impact on performance in the network. In particular, SP2 has a very high antenna branch correlation in order to serve as a 'no MIMO' (that is no multi-layer transmission) reference case. In addition, a reference antenna combination comprising two pole antennas was also used. The reference was designed to have very good correlation properties. Compared to the mockups, the reference was designed with much looser limitations on the form factor.

Figure 15.2 shows pictures of the mockups, the reference and their mounting in, and on top of, the test vehicle. Three use modes are included in the trial: hands-on-roof, hands-in-van, and free-space-on-roof. The hands-on-roof and hands-in-van use modes include hand dummies with the same radio characteristics as a hand. The mockups were placed in a two-hand grip: 'browse mode'. In the hands-in-van use mode, the mockups and hands were placed to the right of the driver's seat; in the hands-on-roof use mode they were placed on the roof of the car. Note that the body style of the vehicle is an important aspect when characterizing the in-vehicle radio channel, hence results obtained may only be valid for the type used in the trials: a panel truck with no windows behind the mockup. In the free-space-on-roof use mode, the mockups were placed on a block of foam that has similar radio characteristics to air. The blocks were then placed on the roof of the car in the same way as in the hands-on-roof use mode. Pictures exemplifying the use modes and UE mockups are shown in Figure 15.2. The cables connecting the antenna mockups to the receiver circuitry were the same independent of use mode and type of antenna used.

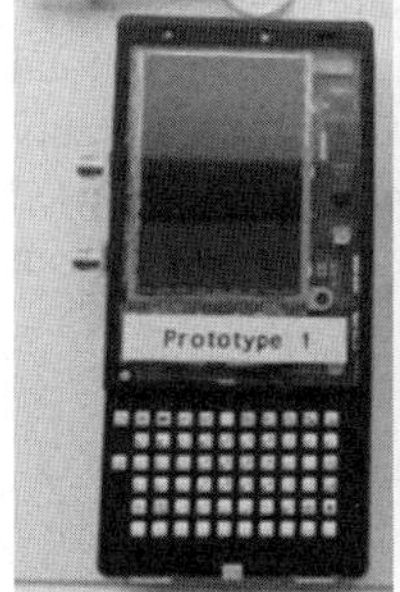

Figure 15.2 Field trial setup. *Left:* SP1 UE antenna mockup (SP2 has identical external appearance). *Centre:* FP1 type of UE antenna mockups in the hands-in-van use mode. *Right:* roof of measurement van with reference antennas mounted on the rear, and mockups mounted on the front.

15.2.2 Network Configuration and Field Trial Setup

Measurement drives were conducted with antennas mounted according to use mode, see previous section. In each measurement drive, multiple modem boards were used. One of them was connected to the reference antenna (always mounted on the van roof with no hand interaction) and the others were connected to UE mockups. Either the two smartphones (SP1 and SP2) or the featurephone (FP1) was tested in each test drive. The network implemented round robin scheduling and data transmission was made using UDP with a bitrate selected to ensure full buffers for all UEs at all times. Furthermore, the drive routes were selected and the network was configured to ensure that there were no handovers during the test drives. Hence all UEs were in the same cell at all times.

The network implemented the possibility to transmit user plane data in selected cells by scheduling downlink transmission to non-existent UEs. This feature was used in the trial to create interference from neighbouring cells in a reproducible way in order to investigate the impact of network load on performance. Two load settings were used for the neighbouring cells: full user plane load and no user plane load. It should be noted that some interference was present also in the latter case as reference signals and control information was continuously transmitted in all cells at all times.

Drives were performed on two measurement routes: urban and suburban. The area used for the suburban measurement route is characterized by single- or two-storied houses (residential and business). It has a typical suburban character. The terrain is generally quite flat but the measurement route has a leg back and forth up a rather steep hillside towards strong interferers. The duration of the suburban measurement route was about 16 minutes and the drive speed mostly kept below 20 mph ($\sim$35 km/h).

The urban measurement route is of a typical North American urban character: The route is in an area dominated by high-rise buildings in a regular grid of streets. In general, the interference level is significantly higher than in the suburban drive route. The urban measurement route took around 9 minutes. Traffic had a much larger impact of the drives on this route due to the much higher occurrence of traffic lights and congestion. Drive speed was slightly higher but mostly under 30 mph ($\sim$45 km/h). GPS information was logged, and post-processing was performed with the purpose of removing some of the traffic effects that disturb comparability. With such normalization, repeatability was in general good. On each measurement route and with each use mode, two test drives were performed, with and without neighbouring cell load. The results presented in this chapter are based on data from [3 use modes] $\times$ [2 measurement routes] $\times$ [2 network load settings] $\times$ [2 mockup arrangements (smartphone or featurephone)], in total 24 test drives. Measurement results were logged by the UE modems and they are inevitably depending on vendor-specific algorithms. However, the same type of modem board were used in all tests. Hence absolute numbers are of limited interest, and focus is on comparison between the use modes and mockups.

15.3 Measurement Results

Figure 15.3(a) shows measured average received signal strength indicator (RSSI) values for mockup FP1 in all measurement drives. Note that these values include attenuation not only due to propagation losses and antenna efficiency but also due to cables and connectors. The attenuation caused by the hands holding the device in the hands-on-roof use mode as

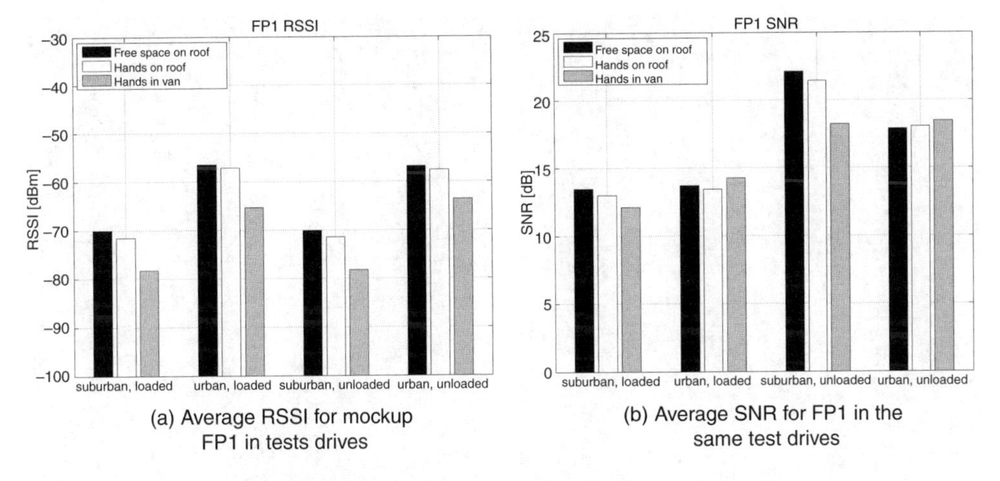

Figure 15.3 Test LTE network radio characteristics.

well as the attenuation caused by surrounding vehicle components in the hands-in-van use mode are evident in these bars. The signal strength loss due to use mode is consistent over the measurement drives. Figure 15.3(b) shows estimated SNR levels over the measurement drives. It is clear that the distinct drops in signal strength due to hand and van attenuation are not visible in the SNR, with the only exception being drives on the suburban measurement route in the unloaded case. Clearly, signal strength only determines SNR in scenarios with low interference. Hence, in loaded interference limited scenarios, antenna efficiency is less important.

Figure 15.4 shows the CDFs for SNR over two test drives on the suburban measurement route with the hands-in-van use mode. The SNR varies in an interval between 2 dB and 20 dB in the loaded scenario, and between 5 dB and 30 dB in the unloaded scenario. Comparing the right

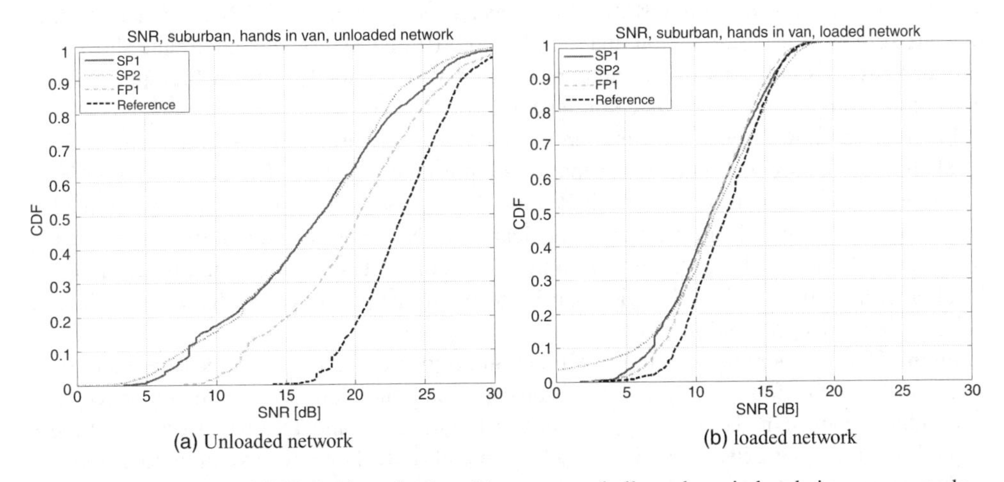

Figure 15.4 CDFs of SNR for the suburban drive route and all mockups in hands-in-van use mode.

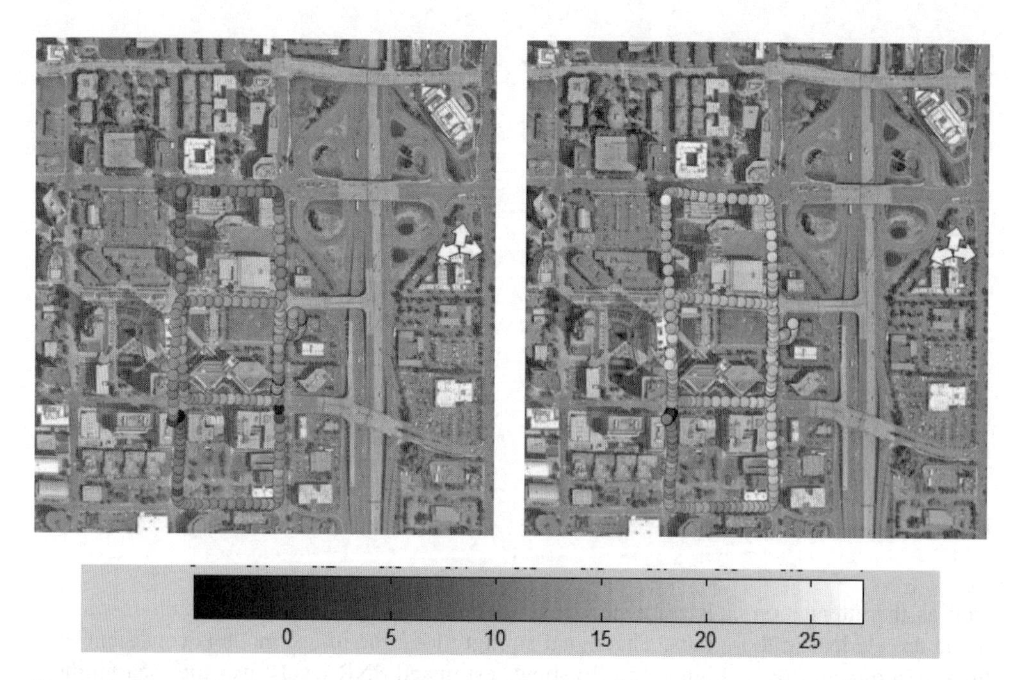

Figure 15.5 Detailed map over the urban measurement route (reference antenna). *Left*: loaded network. *Right*: unloaded network. Base station position and direction of the sector antenna main-beams are shown by arrows in upper right corner. *Source*: Original map: © 2005 Google Maps.

and left plots show how SNR is pushed down by the interference. SNR is interference limited in the loaded scenario; there are only small differences between the in-car mounted mockup-antennas and the reference antenna mounted on the roof. Conversely, in the unloaded scenario, the reference antenna clearly has the highest SNR. This is expected given its placement on the roof of the van. Similarly, in the unloaded case, the FP1 mockup gets higher SNR than the two smartphone mockups (SP1 and SP2). This is because the antennas of FP1 are placed so that they are subject to less attenuation due to the hand. SP1 and SP2 on the other hand show very similar SNR; this indicates similar antenna efficiency. In the loaded case, where interference is limiting, all mockups experience similar SNR.

Figure 15.5 shows SNR as a function of position in the urban measurement route. SNR levels for both loaded and unloaded scenarios are presented. As was clear also from the CDFs described above, SNR levels are pushed down by the increased interference in the loaded scenario. Furthermore, from the maps it is clear that SNR levels cannot be determined using the distance to serving base station and to interferers only; it is a complicated function of the local environment. It is, for example, clearly visible in the plot how the shadowing effect of tall buildings changes SNR drastically.

Figure 15.6 shows rank selection as a function of position on the same measurement route as Figure 15.5. Obviously rank = 2 selections are much more common in the unloaded case due to the higher SNR. However, by comparing Figures 15.5 and 15.6 we also see that there is no direct mapping between rank = 2 selection and SNR. Note, for example, that SNR on the two upper east–west-bound streets is higher than SNR on the two lower east–west-bound

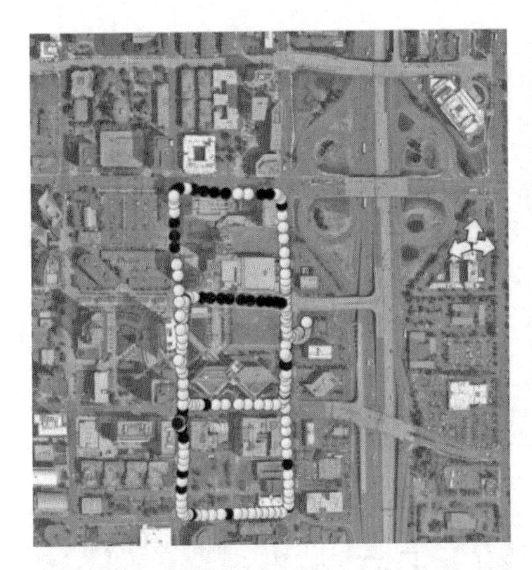

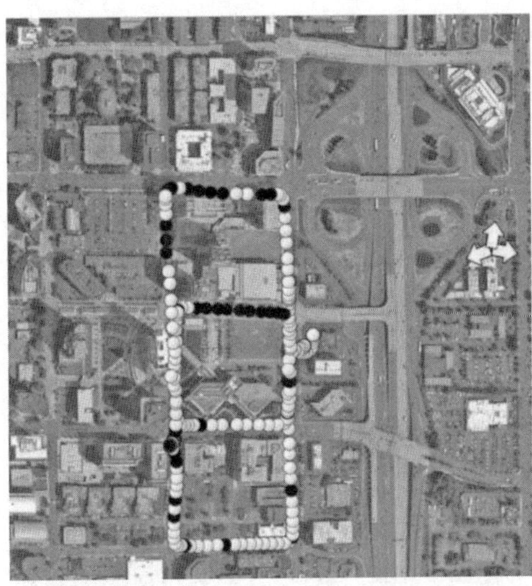

Figure 15.6 Detailed map over rank selection over the urban measurement route (reference antenna). Black dots indicate no MIMO (rank = 1), white dots indicate MIMO (rank = 2). *Left*: loaded case. *Right*: unloaded case. Base station position and direction of the sector antenna main-beams are indicated by arrows (upper right corner). *Source*: Original map: © 2005 Google Maps.

streets. By comparing the rank selection ratio for the same two pairs of streets we see that rank-two selections are more common on the lower pair of streets. Prevalence of good channel conditions for MIMO is not directly correlated to SNR conditions (although high SNR is necessary for MIMO, that is using rank-two transmission).

Figures 15.7 show SNR as a function of position on the suburban measurement route, both for loaded and unloaded cases. Given the town architecture with much lower buildings, it is not surprising to see fewer shadowing effects here compared to the corresponding maps for the urban route (Figure 15.5). Figure 15.8 shows rank selection statistics. In the unloaded case, rank = 2 is selected more or less throughout the route. In the unloaded case, rank-two selection is still dominating. In this environment, SNR seems to be relatively well correlated to rank-two selection.

Figure 15.9 shows CDFs of normalized throughput for the same measurement drives as in Figure 15.4. In the unloaded scenario, the reference outperforms the mockups. In the loaded scenario, mockups FP1 and SP1 perform similar to the reference: SNR is more interference-limited and vehicle attenuation is less important. The higher received power in the unloaded scenario translates into an SNR difference and hence the throughput is better. While FP1 shows much higher SNR in the unloaded case than SP1 (Figure 15.4) it shows similar throughput (Figure 15.9). On the other hand, SP1 has similar SNR to SP2 but much better throughput. Differences are smaller in the loaded case but SP2 stands out in this case as well with much lower throughput than the others. With the possibility of rank-two transmission (MIMO), UEs experiencing the highest SNR do not necessarily enjoy the highest throughput; it is not possible to map SNR directly to throughput, because antenna correlation and channel properties are equally important.

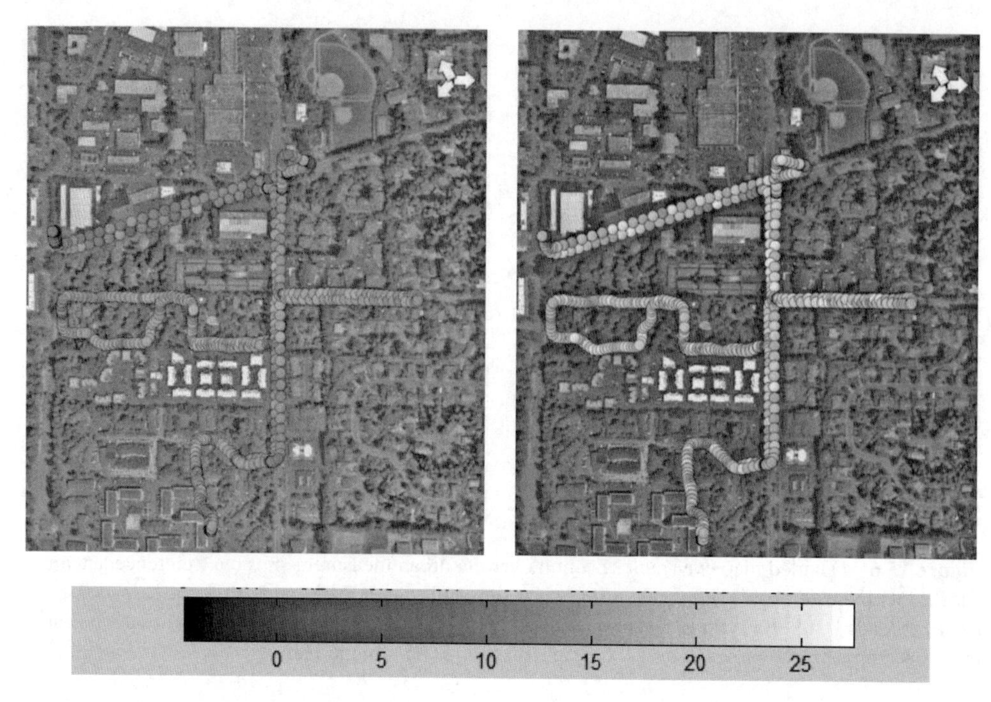

Figure 15.7 Detailed map over the suburban measurement route. *Left*: loaded network. *Right*: unloaded network. Base station position and direction of the sector antenna main-beams are indicated by arrows (upper right corner). *Source*: Original map: © 2005 Google Maps.

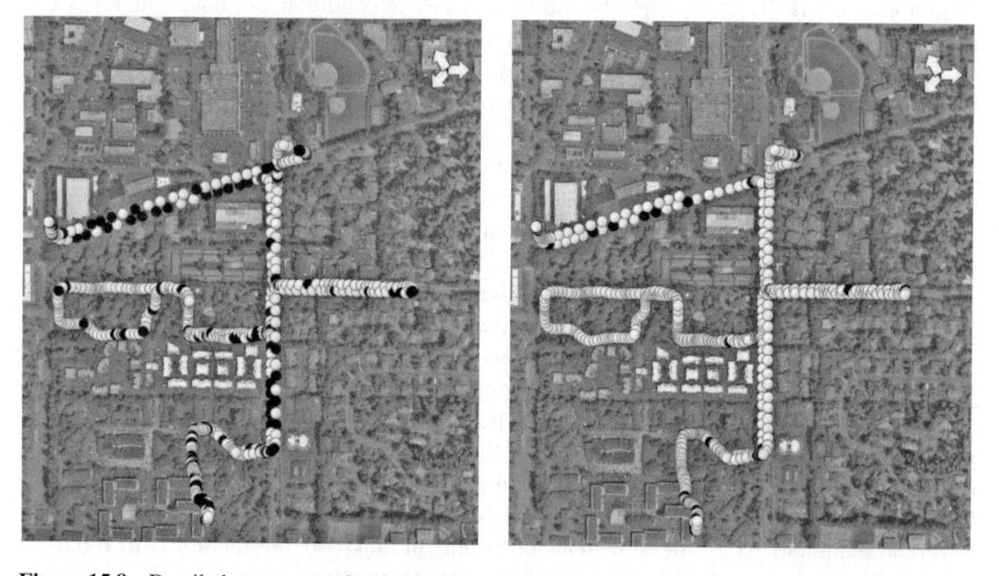

Figure 15.8 Detailed map over rank selection over the suburban measurement route. Black dots indicate no MIMO (rank = 1), white dots indicate MIMO (rank = 2). *Left*: loaded case. *Right*: unloaded case. Base station position and direction of the sector antenna main-beams are indicated by arrows (upper right corner). *Source*: Original map: © 2005 Google Maps.

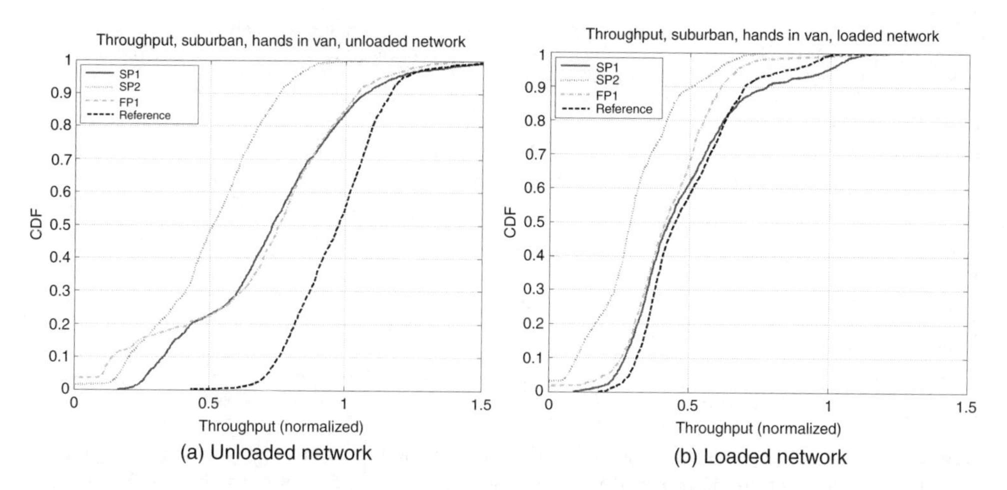

(a) Unloaded network (b) Loaded network

Figure 15.9 CDFs of throughput for the suburban drive route and all mockups with the hands-in-van use mode.

Figure 15.10 shows throughput plotted against SNR for an aggregation of all test-drives with the hands-in-van use mode. The left plot shows throughput plotted against own SNR. The right plot shows throughput plotted against reference antenna SNR. The former way of plotting hides the effect of SNR differences between UEs while the latter includes it. It is clear from the right plot that the SNR drop associated with placement inside the van causes performance to deteriorate for both FP1 and SP1, and more so for SP1. However, the left plot does not show this drop. This indicates that correlation properties are not affected as much as received power (and hence SNR in power limited scenarios) by placement inside the van. This is particularly true for SP1, which on the other hand has lower efficiency than FP1 when held by hands. SP2 is clearly outperformed: the difference between SP2 on the one hand and

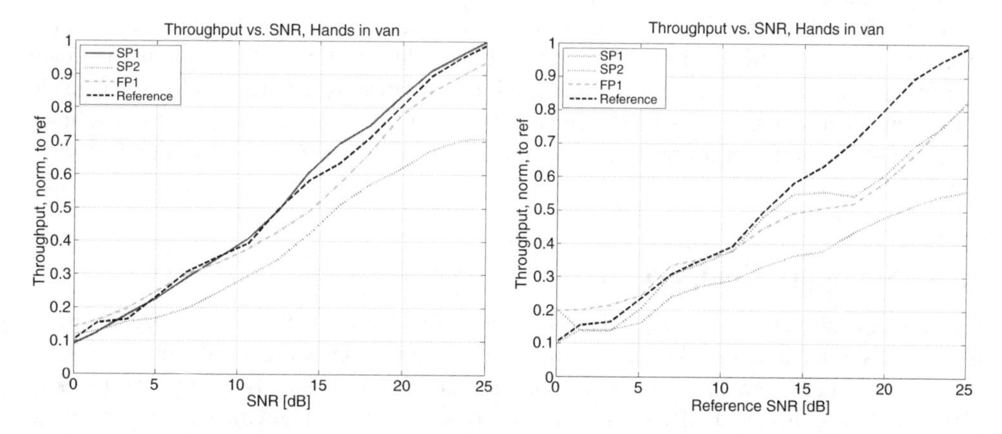

Figure 15.10 Throughput as a function of SNR for the hands-in-van use mode for all tested mockups. *Left*: measured by same antennas. *Right*: measured by the roof mounted reference antenna.

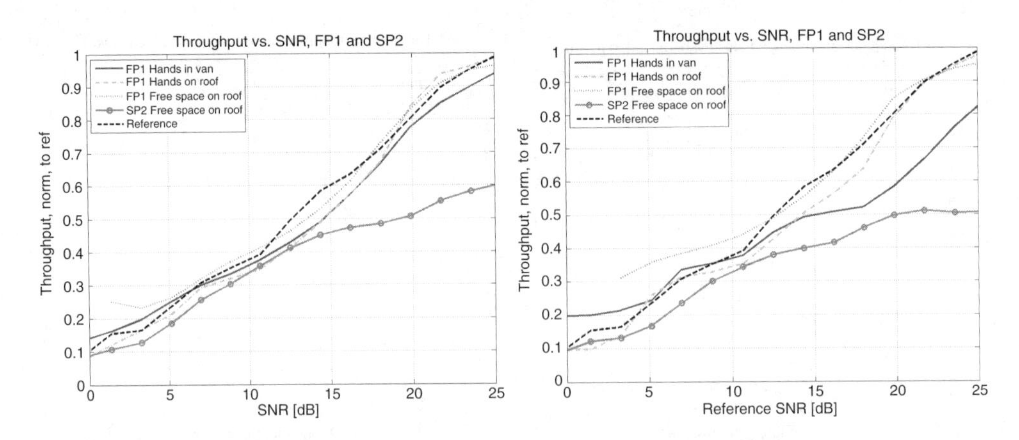

Figure 15.11 Throughput as a function of SNR for all use modes with the FP1 mockup and (for extra reference) SP2 mockup in the free-space-on-roof use mode. Left and right plots show SNR reference similar to 15.10.

FP1 and SP1 on the other hand is larger than the difference between the mockups and the reference.

Figure 15.11 shows the mockup FP1 in all use modes. When plotted against reference antenna SNR (right plot) there is as expected a performance drop associated with hands holding the device (compare free-space-on-roof and hands-on-roof use modes) and with placement inside the van (compare hands-on-roof and hands-in-van use modes). However, when plotted against own SNR (left plot), difference is slim. This indicates that correlation properties do not deteriorate as much as SNR when adding hands or placing the mockup inside the van. Figure 15.11 also shows SP2 in the free-space-on-roof use mode. Obviously, SP2 is outperformed (even by FP1 when placed inside van), in particular at high SNR, and this is mostly due to poor correlation properties and to a smaller extent due to lower antenna efficiency yielding lower SNR.

Figure 15.12 shows the ratio of rank-two selection as a function of SNR (measured by each mockup). SP2 is practically never selecting rank-two while SP1 and FP1 select multi-stream transmission almost as often as the reference. It is also interesting to note that placement inside van and with hands does not change this picture very much. Note that the way of plotting hides the effect of SNR differences between mockups as well as between mockups and reference.

15.4 Summary Comparison

Figure 15.14 shows mean normalized throughput (normalization with reference in each drive) for all measurement drives. As a complement, Figure 15.13 shows rank-two selection ratios for the same measurement drives. Mockups SP1 and FP1 consistently outperform mockup SP2. The UEs connected to SP1 and FP1 are able to select multi-stream transmission to a large extent while the UE connected to SP2 in practice never selects multi-stream transmission. In measurement drives with large load, resulting in high interference, SP1, FP1 and the reference

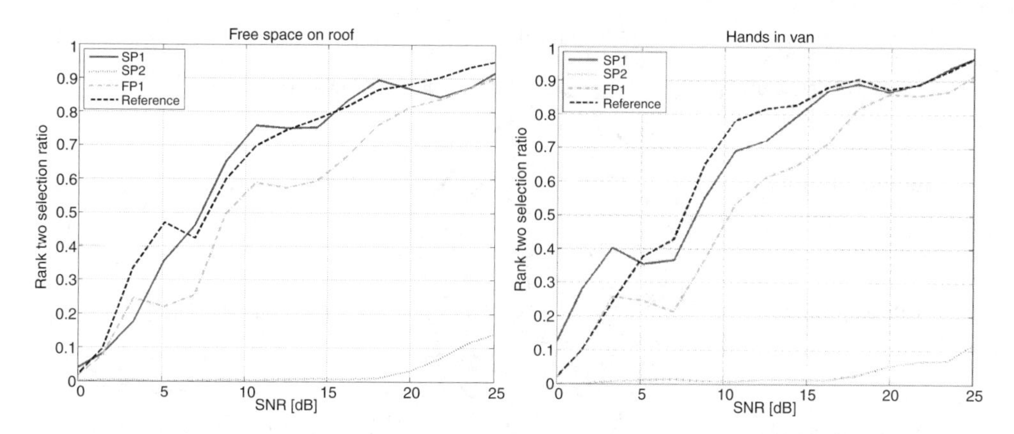

Figure 15.12 Rank-two selection ratio as a function of SNR. *Left*: free-space on roof use mode. *Right*: hands-in-van use mode.

perform similarly. The reference is significantly better only when compared to hands-in-van use mode in the least interference limited scenario (suburban, no load).

SP1 is generally able to select two stream transmission more often than FP1. This is consistent with the larger physical size of SP1. However, this difference is not as evident in throughput: there is a range of channel conditions where rank-two and rank-one transmission can be expected to perform similarly.

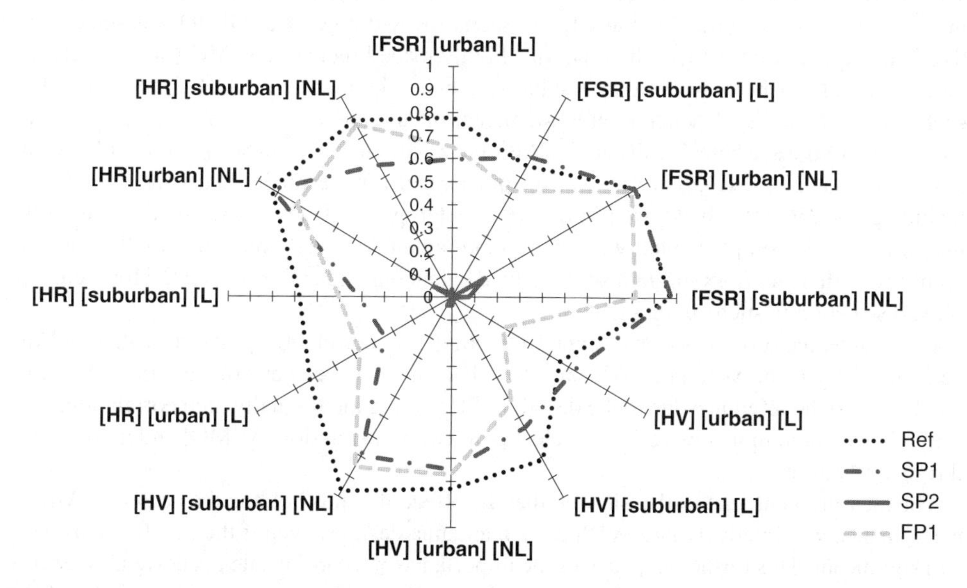

Figure 15.13 Average rank two (MIMO) selection ratio for free space on roof (FSR), hands in van (HV), hands on roof (HR) use modes in combination with loaded (L) and unloaded (NL) network settings. Measurements using reference and all mockups are included.

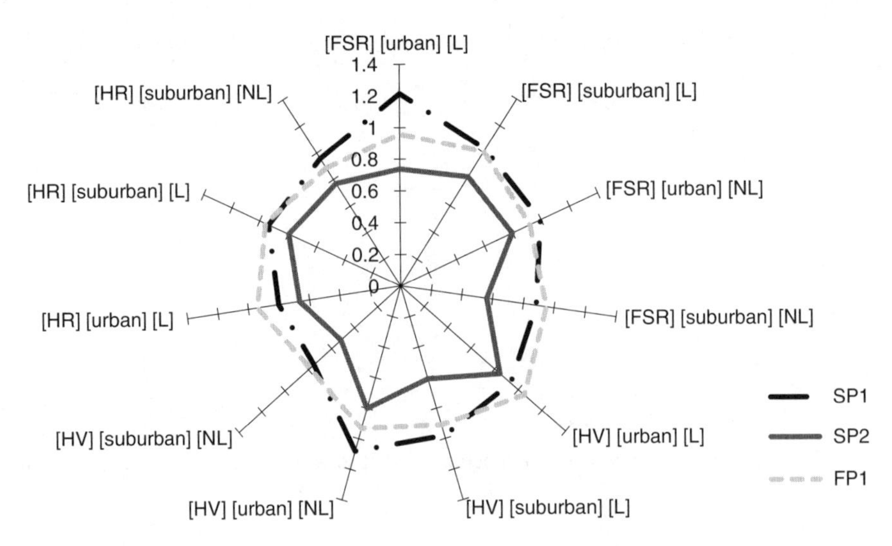

Figure 15.14 Throughput normalized to reference antenna throughput (on same drive). Measurements for free-space-on-roof (FSR), hands-in-van (HV), hands-on-roof (HR) use modes in combination with loaded (L) and unloaded (NL) network settings are included. Measurements using reference and all mockups are included. The measurement data for [HR][urban][NL] was corrupt due to backhaul issues.

15.5 Conclusion

The trial demonstrated significant benefits of MIMO in 4G LTE network operating at 750 MHz, in terms of measured throughput at UEs. In scenarios with low load, MIMO was selected by the terminal more than 80% of the time. In a highly loaded scenario, MIMO was still selected more than half of the time. Similar results have been obtained at 2.6 GHz, see [11]. While such results are highly dependent on implementation and on network deployment – there are channel conditions where MIMO and SIMO have similar performance – they still give an indication that 2×2 MIMO works well in practice in a wide range of carrier frequencies, including the 750 MHz band used in the trial. Heterogeneous networks – that is, where the macro nodes are complemented with many more antenna nodes – may potentially establish scenarios with better SNR through network load balancing, indicating a potential for increased MIMO selection in such deployments.

If the antenna system is not designed targeting low correlation properties, then MIMO performance will be very poor. Mockup SP2 had much lower performance than the other mockups and the reference due to the design of SP2 giving intentionally high correlation. This is manifested both in its low ability to select rank-two transmission (MIMO) and in the actual throughput figures.

The measurement results also showed that the effect of hands holding the device on MIMO performance was relatively large with certain antenna designs, even if the MIMO gains were still significant. This further emphasizes the importance of a careful antenna design, especially at lower carrier frequencies. Mockup FP1 had a smaller performance loss than mockup SP1: it showed less attenuation due to hands, which yields better SNR at low network load levels. With MIMO-enabled systems, antenna design becomes even more critical to performance.

This is equally true for both the eNB and the UE side. The trial also indicates the importance of interference management when deploying dense network. As seen in the trial, in many situations (but far from all), SINR is determined by interference levels, not absolute signal strength. Furthermore, in a MIMO-enabled system, performance is not only determined by the SNR experienced by the receiver. It also depends on the correlation properties of the radio channel; these are highly dependent on the local environment. This fact is clearly visible in the data obtained in the trial. While these aspects are of great importance in today's macro-type wireless networks, they will continue to play an important role when planning deployment of heterogeneous networks where the macro nodes are complemented with many more antenna nodes. eNB nodes need to be carefully designed, placed and coordinated to ensure proper management of interference. Control channel and user channel resource management among cells are essential to the overall user and network performance. Heterogeneous network trials and analyses are ongoing to validate various means to improve network performance.

References

1. E. Dahlman, S. Parkvall, J. Sköld and P Beming, *3G Evolution: HSPA and LTE for mobile broadband*, second edition, Academic Press.
2. M. Riback, J. Karlsson, Initial Performance Measurements of LTE, Ericsson Review, March 2008.
3. 3GPP TS 36.211 3rd Generation Partnership Project, Technical Specification Group Radio Access Network; Evolved Universal Terrestrial Radio Access (E-UTRA); Physical Channels and Modulation (Release 8).
4. 3GPP TS 36.213 3rd Generation Partnership Project, Technical Specification Group Radio Access Network; Evolved Universal Terrestrial Radio Access (E-UTRA); Physical layer procedures (Release 8).
5. W.C. Jakes, *Microwave Mobile Communications*, New York, Wiley, 1984.
6. Helmut Bölcskei, Moritz Borgmann, and Arogyaswami J. Paulraj, 'Impact of the Propagation Environment on the Performance of Space-Frequency Coded MIMO-OFDM', *IEEE J. Select Areas Commun.*, Vol. 21, no. 3, pp. 427–439, April 2003.
7. 3GPP TS 36.300, 3rd Generation Partnership Project, Technical Specification Group Radio Access Network; Evolved Universal Terrestrial Radio Access (E-UTRA) and Evolved Universal Terrestrial Radio Access Network (E-UTRAN) Overall description.
8. 3GPP TS 23.401, 3rd Generation Partnership Project, Technical Specification Group Services and System Aspects; GPRS enhancements for E-UTRAN access'.
9. B. Hagerman, K. Werner and J. Yang, MIMO Performance at 700MHz: Field Trials of LTE with Handheld UE, Proceedings of VTC Fall 2011, San Francisco 2011.
10. M. Riback et al., 'MIMO-HSPA Tested Performance Measurements' in Proceedings of PIMRC 2007, Athens, 2007.
11. K. Werner, J. Furuskog, M. Riback and B. Hagerman, Antenna configurations for 4×4 MIMO in LTE – Field Measurements, Proceedings of VTC Fall 2010, Taipei 2010.
12. J. Furuskog, K. Werner, M Riback and B. Hagerman, Field trials of LTE with 4×4 MIMO, Ericsson Review, issue 2, 2010.

Index

Heterogeneous Cellular Networks, First Edition. Edited by Rose Qingyang Hu and Yi Qian.